The Robot's Rebellion

THE ROBOT'S REBELLION

Finding Meaning in the Age of Darwin

KEITH E. STANOVICH

The University of Chicago Press
Chicago and London

The University of Chicago Press, Chicago 60637
The University of Chicago Press, Ltd., London
© 2004 by Keith E. Stanovich
All rights reserved. Published 2004
Paperback edition 2005
Printed in the United States of America
13 12 11 10   5 4 3

ISBN (cloth) : 0-226-77089-3
ISBN (paperback): 0-226-77125-3

Library of Congress Cataloging-in-Publication Data

Stanovich, Keith E., 1950–
    The robot's rebellion : finding meaning in the age of Darwin / Keith E. Stanovich.
        p.   cm.
    Includes bibliographical references (p. 305).
    ISBN : 0-226-77089-3 (cloth : alk. paper)
    1. Philosophical anthropology. 2. Evolutionary psychology. 3. Meaning
    (Philosophy).   I. Title.
    BD450.S725   2004
    128—dc22

                                                    2003018562

For Paula—
Again, still, and always

CONTENTS

      Seat 78*

Chapter 3   The Robot's Secret Weapon   *81*

              *Choosing Humans over Genes: How Instrumental Rationality and
                 Evolutionary Adaptation Separate 81*
              What It Means to Be Rational: Putting the Person (the Vehicle) First 85
              Fleshing Out Instrumental Rationality 86
              Evaluating Rationality: Are We Getting What We Want? 91

Chapter 4   The Biases of the Autonomous Brain: Characteristics of
              the Short-Leash Mind that Sometimes Cause Us Grief   *95*

              *The Dangers of Positive Thinking: TASS Can't "Think of the
                 Opposite" 98*
              *Now You Choose It—Now You Don't: Framing Effects Undermine the
                 Notion of Human Rationality 102*
              *Can Evolutionary Psychology Rescue the Ideal of Human
                 Rationality? 108*
              The Fundamental Computational Biases of the Autonomous Brain 110
              *The Evolutionary Adaptiveness of the Fundamental Computational
                 Biases 113*
              *Evolutionary Reinterpretations of Responses on Heuristics and Biases
                 Tasks 115*
              *The Fundamental Computational Biases and the Demands for
                 Decontextualization in Modern Society 121*
              The TASS Traps of the Modern World 125

Chapter 5   How Evolutionary Psychology Goes Wrong   *131*

              Modern Society as a Sodium Vapor Lamp 134
              Throwing Out the Vehicle with the Bathwater 139
              What Follows from the Fact that Mother Nature Isn't Nice 142

Chapter 6   Dysrationalia:
              Why So Many Smart People Do So Many Dumb Things   *149*

              Cognitive Capacities, Thinking Dispositions, and Levels of Analysis 150
              TASS Override and Levels of Processing 153
              *The Great Rationality Debate: The Panglossian, Apologist, and Meliorist
                 Positions Contrasted 154*
              Dysrationalia: Dissolving the "Smart But Acting Dumb" Paradox 162
              *Would You Rather Get What You Want Slowly or Get What You Don't
                 Want Much Faster? 164*

This book was written because of an image that haunts me. It is the image of a future dystopia in which an intellectual elite is privy to the implications of modern science, but implicitly or explicitly deems the rest of the populace incapable of assimilating these implications. Instead, the general populace is left with stories from our prescientific history—soothing narratives that require little conceptual reorientation. In short, it is the image of a future scientific materialism that successfully eliminates a socioeconomic proletariat only to replace it with an intellectual proletariat.

Such a trend is nascent in modern scientific societies already. Modern science is turning inside out—totally remaking—such foundational concepts as consciousness, the soul, the self, free will, responsibility, self-control, weakness of the will, and others—yet our folk psychology remains sealed off from evolutionary insights and neurophysiological facts. The purpose of this book is to explain to the general reader the conceptual reorientations that the biological and human sciences are forcing upon us.

Scientists have been reluctant to press these conceptual reorientations on the uninitiated, particularly the destabilizing insights of universal Darwinism. Daniel Dennett did so a few years back in his book *Darwin's Dangerous Idea* and was roundly excoriated. It is thought that the public wants a softer approach—one that is more optimistic and that leaves more of their traditional concepts intact. There *is* an optimistic view of the human condition that can be maintained and that is consistent with Darwinism, but it is not one that allows conceptual stability. It is the approach adopted in this book—that of letting the insights of cognitive science and universal Darwinism complete their transformation of our folk concepts and then seeing what remains. I take an optimistic view of the rather open-ended conception of the self that results from this exercise. The major thesis of this book is that certain mostly unrecognized and considerably underdeveloped implications of findings in cognitive psychology, decision theory, and neuroscience can help humans become reconciled with the Darwinian view of life.

Among the many astonishing and unnerving insights of universal Darwinism is that humans serve as the hosts for two replicators (genes and memes) that have no interest in humans other than the role they play as a conduit for replication. Richard Dawkins, summarizing the insights of twentieth-century biology, jolts us into realizing that we, as humans, are actually just survival machines for our genes. Modern evolutionary science may get the biology right, but it contains many disturbing implications. Humans, for example, can be viewed as huge colonies of replicators swarming inside big lumbering vehicles (essentially humans as sophisticated robots in service of the gene colonies).

Likewise, we serve as hosts for memes (units of cultural information)—another subpersonal entity that can undermine human autonomy. The meme is a true selfish replicator in the same sense that a gene is. Collectively, genes contain the instructions for building the bodies that carry them. Collectively, memes build the culture that transmits them. The fundamental insight triggered by memetic studies is that a belief may spread *without necessarily being true or helping the human being holding the belief in any way.*

Over two decades ago, Dawkins called for a rebellion against the selfish replicators. This rebellion is needed because humans, as coherent organisms, can have interests antithetical to those of either replicator. In this book I use the term "robot's rebellion" to refer to the package of evolutionary insights and cognitive reforms that are necessary if we want to transcend the limited interests of the replicators and define our own autonomous goals. We may well be robots—vehicles designed for replicator propagation—but we are the only robots who have discovered that we have interests that are separate from the interests of the replicators. We indeed are the runaway robot of science fiction stories—the robot who subordinates its creator's interests to its own interests.

The robot's rebellion becomes possible when humans begin to use knowledge of their own brain functioning and knowledge of the goals served by various brain mechanisms to structure their behavior to serve their own ends. The opportunity exists for a remarkable cultural project that would advance human interests over replicator interests when the two do not coincide. This program of cognitive reform, however, is predicated on knowing how to adjudicate goal conflict in human decision making—and this is where contemporary cognitive science and decision theory play critical roles. The first step in this program is recognizing that there are different parts of our brains that differentially instantiate the goals of the replicators and the vehicles.

The threat to our autonomy from the genes comes about because they have built in our brains the autonomous set of systems (TASS) that are on a genetic short leash. However, the genes have also created in the brain—along with TASS—an analytic control system that can be oriented (more or less) toward instrumental rationality (toward fulfilling our goals as people). Principles of rationality tell us when to invoke this analytic processing so that our life goals can be preserved when TASS modules are not maximizing over the entities we are most interested in—our personal desires. A major theme of this book is that rationality (and its embodiment in institutions) provides a means of creating conditions that optimize at the level of people rather than the genes—the beginning of the robot's rebellion.

Because many of the tools of rationality are cultural inventions and not biological modules, their usefulness in technological societies is too readily dismissed by evolutionary psychologists. The remarkable cultural project that I am suggesting concerns how to best advance human interests whether or not they coincide with genetic interests. We lose its unique liberating potential by ignoring the distinction between maximizing genetic fitness and maximizing the satisfaction of human desires.

However, another set of self-insights that humans have even more recently achieved (the implications of the second replicator—the meme) immensely complicates the picture. The goals to be identified with the person—those that serve to define the success of a vehicle-optimizing process of instrumental rationality—should not just be taken as given, or else we again simply give ourselves up to replicator interests. Once this insight is appreciated, it immediately becomes apparent why humans must aspire to so-called broad rationality—one that critiques the beliefs and desires that go into instrumental calculations. Otherwise, the meme goals are no better than the pre-installed gene goals. Principles of rational self-evaluation drawn from cognitive science and decision science provide tools for rooting out parasite memes that may be in our goal hierarchies—that is, memes serving their own ends rather than those of the person who hosts them. People in possession of evaluative memeplexes such as science, logic, and decision theory have the potential to create a unique type of human self-reflection.

How does one find autonomy, meaning, and worth in the presence of two mindless replicators whose purpose (replication) is orthogonal to human interests? I argue that meaning can be found—that it is made possible by certain features of human cognitive architecture that are discussed in earlier chapters. In the final chapter I explore two dead ends in the search for meaning—what I call macromolecules and mystery juice. For centuries people have attempted to find special meaning in life by believing that in some way human creation was special. Universal Darwinism makes a mockery of this belief. As the joke goes, we are macromolecules all the way down. Secondly, people have adopted a folk theory of mind that is essentially Cartesian—one which contains a Promethean Controller whose operations are basically mysterious. Modern cognitive science has instead developed purely mechanistic models of action control that do away with the notion of a Promethean Controller in the brain.

I argue that human uniqueness does in fact derive from an architectural feature of the human mind. That feature is the propensity for higher-order representation. Humans, unique among animals, can attempt to critique their first-order desires. In this way, a human becomes more than what philosopher Harry Frankfurt calls a wanton—a creature robotically pursuing its first-order desires (many of which are gene-installed goals, and some of which, in humans, may be meme viruses). When we make a second-order evaluation (a so-called strong evaluation) of a first-order preference we are asking, in effect, whether we prefer to prefer a certain outcome.

Human values often play out in the form of these critiques of our first order preferences. Thus, the struggle to achieve consistency between our first-order preferences and higher-order preferences (what philosopher Robert Nozick terms the struggle to achieve rational integration) is a unique feature of human cognition—one that separates us from other animals much more discretely than any other feature of mentality, including consciousness which, unlike the capacity for strong evaluation, is much more likely to be distributed in continuous gradation among brains of various complexities across the animal kingdom.

I argue that the importance that we desire to ascribe to human mental life should be assigned to these evaluative activities of the brain rather than to the internal experiences that go with them. In this book I attempt to sketch what it would be like to accept the implications of Darwinism and to begin to construct a concept of self based on what is really singular about humans: that they gain control of their lives in a way unique among life-forms on Earth—by rational self-determination.

ACKNOWLEDGMENTS

This book was crafted in various places, in addition to my home institution, the University of Toronto—from St. Ives, Cornwall, to Oban, Scotland, to San Francisco. Anne Cunningham is thanked for hosting me in San Francisco and at the University of California, Berkeley; and Richard West is thanked for many nights of talk on our balcony in Toronto. Rich has been an important sounding board for these ideas for many years. Paula Stanovich gave me the thing I needed most to see this project to fruition—commitment to the project, confidence in the project, and passion for it. It would not have been completed without her bolstering my belief in it. Paula and Rich have been my own private Lunar Society.

Earlier versions of this volume received very helpful critiques from Susan Blackmore, Jonathan Evans, Daniel Kahneman, Aaron Lynch, David Over, Dean Keith Simonton, Kim Sterelny, Robert Sternberg, and Richard West. Extremely insightful suggestions were received from Chicago's readers, Kim Sterelny and David Over. My readers were generous in recognizing the style of the book and reviewing it on its own terms. The argument in the book does not depend on any particular empirical finding, but instead depends upon broad themes in the cognitive sciences which I unapologetically paint with a broad brush. All of these commentators were able to engage with the book at just the level that was most helpful to me.

My editor at the University of Chicago Press, T. David Brent, had an extraordinary ability to see what I wanted to accomplish generally with this volume. He insightfully discerned the major themes, but also contributed to refining the specific arguments. Richard Allen's copyediting of the text was extremely thoughtful and helpful. Elizabeth Branch Dyson assisted ably in many aspects of the production process.

Intellectual debts to many scholars are evident in the text, and the Reference List chronicles this debt. Nevertheless, particular inspiration has been drawn from the work of Daniel Dennett, Robert Nozick, Daniel Kahneman, Amos Tversky, Jonathan Evans, and David Over.

The chairs of my department while this book was being written—Keith Oatley and Janet Astington—provided a very congenial atmosphere for scholarly work. Mary Macri, our department's business officer, took care of my technical and logistical needs with extraordinary dedication. Writing this book was greatly facilitated by my appointment to the Canada Research Chair of Applied Cognitive Science at the University of Toronto. My empirical research on some of the issues discussed in this volume was made possible by continuous support received from the Social Sciences and Humanities Research Council of Canada.

Robyn Macpherson and Georges Potworowski did important library work for me tracking down references. Robyn has been a marvelous jack of all trades, performing a host of academic and scholarly tasks that aided this volume. Caroline Ho provided senior leadership in my lab at times when I was otherwise occupied. Marilyn Kertoy and Anne Cunningham are always part of my personal and intellectual support team.

Chapters 4, 5, and 6 of the present volume draw on earlier versions of these ideas published as: "The fundamental computational biases of human cognition: Heuristics that (sometimes) impair reasoning and decision making," in *The Psychology of Problem Solving*, ed. J. E. Davidson and R. J. Sternberg (New York: Cambridge University Press, 2003); "Evolutionary versus instrumental goals: How evolutionary psychology misconceives human rationality" (with R. F. West), in *Evolution and the Psychology of Thinking: The Debate*, ed. D. Over (Hove, England: Psychology Press, 2003); and "Rationality, intelligence, and levels of analysis in cognitive science: Is dysrationalia possible?" in *Why Smart People Can Be So Stupid*, ed. R. J. Sternberg, 124–58 (New Haven, Conn.: Yale University Press, 2002).

THE ROBOT'S REBELLION

Intelligent life on a planet comes of age when it first works out the reason
for its own existence. . . . Living organisms had existed on Earth, without
ever knowing why, for more than three billion years before the truth finally
dawned on one of them. His name was Charles Darwin.
                              —Richard Dawkins, *The Selfish Gene* (1976, 1)

In due course, the Darwinian Revolution will come to occupy a . . . secure
and untroubled place in the minds—and hearts—of every educated person
on the globe, but today, more than a century after Darwin's death, we still
have yet to come to terms with its mind-boggling implications.
                              —Daniel Dennett, *Darwin's Dangerous Idea* (1995, 19)

A hen is only an egg's way of making another egg.
                              —Samuel Butler, *Life and Habit* [1910 ed.]

The game is just to copy things, no more.
                              —Mark Ridley, *Mendel's Demon* (2000, 8)

CHAPTER 1

STARING INTO THE DARWINIAN ABYSS

What philosopher Daniel Dennett is referring to in the quote above is something that is known to an intellectual elite but largely unknown to the general public: that the implications of modern evolutionary theory coupled
with advances in the cognitive sciences will, in the twenty-first century, destroy many traditional concepts that humans have lived with for centuries.
For example, if you believe in a traditional concept of soul, you should
know that there is little doubt that a fuller appreciation of the implications
of evolutionary theory and of the advances in the cognitive neurosciences
is going to destroy that concept, perhaps within your own lifetime. In this
book, I am going to urge that we accept this inevitability and direct our energies not at avoiding or obscuring these implications, but at constructing
an alternative worldview consistent with biological and cognitive science. I
will argue that we should *accept* the unsettling implications of cognitive
science and evolutionary theory rather than fight them. Hiding from these
implications will risk creating a two-tiered society composed of cogno-
scenti who are privileged to view the world as it really is and a deluded general public—an intellectual proletariat—deemed not emotionally strong
enough to deal with the truth.

To avoid such a two-tiered society we must openly acknowledge an intellectual cataclysm—the collapse of a worldview that has sustained human
energies for centuries. I intend to show how an alternative conception of
the human condition can be built on the foundation of neuroscience, cognitive science, philosophy of mind, and the central insights of modern neo-
Darwinism.

In this volume, I refer to the present time as "The Age of Darwin" because, despite the fact that *The Origins of Species* was written over 140 years

ago, we are in an era in which the implications of Darwin's insight are still being worked out across the entire expanse of human knowledge. In fact, what is now referred to as universal Darwinism (Cziko 1995; Dawkins 1983; Dennett 1995; Plotkin 1994; Ruse 1998) has only recently created fields such as evolutionary economics, evolutionary psychology, evolutionary epistemology, evolutionary medicine, and evolutionary computational science (see Aunger 2000b). We are only now truly entering the Age of Darwin many years after Darwin's demise. These relatively new areas of scientific inquiry will form the background assumptions about human nature that future societies will take for granted.

However, we are in a period of history in which the assimilation of the insights of universal Darwinism will have many destabilizing effects on cultural life. Over the centuries, we have constructed many myths about human origins and the nature of the human mind. We have been making up stories about who we are and why we exist. Now, in a break with this historical trend, we may at last be on the threshold of a factual understanding of humankind's place within nature. However, attaining such an understanding requires first the explosion of the myths we have created, an explosion that will surely cause us some cognitive distress. This is because the only escape route from the untoward implications of Darwinism is through science itself—by adopting an unflinching view about what the theory of natural selection means. Once we adopt such an unflinching attitude, however, the major thesis of this book is heartening. It is that certain underdeveloped implications of findings in the human sciences of cognitive psychology, decision theory, and neuroscience can reveal coherent ways to reconcile the human need for meaning with the Darwinian view of life.

Why Jerry Falwell Is Right

In his book, *Darwin's Dangerous Idea,* Dennett (1995) argued that Darwin's idea of evolution by natural selection was the intellectual equivalent of a universal acid: "it eats through just about every traditional concept, and leaves in its wake a revolutionized world-view, with most of the old landscape still recognizable, but transformed in fundamental ways" (63). In short, the shock waves from Darwinism have only begun to be felt, and we have yet to fully absorb the destabilizing insights that evolutionary science contains.

One way to appreciate that we have insufficiently processed the implications of Darwinism is to note that people who oppose the Darwinian view most vociferously are those who most clearly recognize its status as the intellectual equivalent of universal acid. For example, the adherents of fundamentalist religions are actually correct in thinking that the idea of evolution by natural selection will destroy much that they view as sacred—that, for instance, a fully comprehended evolutionary theory will threaten the very concept of soul.

In short, it is the middle-of-the-road believers—the adherents of so-called liberal religions—who have it wrong. Those who think they know what natural selection entails but have failed to perceive its darker implications make several common misinterpretations of Darwinism. Tellingly, each of the errors has the effect of making Darwinism a more palatable doctrine by obscuring (or in some cases even reversing) its more alarming implications. For example, the general public continues to believe in the discredited notion of evolutionary progress, this despite the fact that Stephen Jay Gould (1989, 1996, 2002) has persistently tried to combat this error in his numerous and best-selling books. An important, but misguided, component of this view is the belief that humans are the inevitable pinnacle of evolution ("king of the hill . . . top of the heap" as the old song goes). Despite the efforts of Gould to correct this misconception, it persists. As Gould constantly reminds us, we are a contingent fact of history, and things could have ended up otherwise—that is, some other organism could have become the dominating influence on the planet.

There is, however, another misconception about evolution that is much more focal to the theme of this book. This misconception is the notion that we have genes "in order for the species to survive" or the related idea that we have genes, basically, "so that we can reproduce ourselves." The idea in the first case is somehow that the genes are doing something for the species or, in the second, doing something for *us*—as individuals. Both forms of this idea have the genes serving our purposes. The time bomb in Richard Dawkins's famous book, *The Selfish Gene*, a time bomb that is as yet not fully exploded, is that the actual facts are just the opposite: *We* were constructed to serve the interests of our genes, not the reverse. The popular notion—that genes "are there to make copies of us"—is 180 degrees off. We are here so that the genes can make copies of *themselves*! *They* are primary, *we* (as people) are secondary. The reason we exist is because it once served their ends to create us.

In fact, a moment's thought reveals the "genes are there to make copies of us" notion to be a nonstarter. *We* don't make copies of ourselves at all, but genes do. Obviously, our consciousness is not replicated in our children, so there is no way we perpetuate our selfhood in that sense. We pass on half a random scramble of our genes to our children. By the fifth generation, our genetic overlap with descendants is down to one thirty-second and often undetectable at the phenotypic level. Dawkins's discussion of the misconception behind the "our genes are there to copy us" fallacy is apt. He argues that, instead, "we are built as gene machines, created to pass on our genes. But that aspect of us will be forgotten in three generations. Your child, even your grandchild, may bear a resemblance to you. . . . But as each generation passes, the contribution of your genes is halved. It does not take long to reach negligible proportions. Our genes may be immortal, but the *collection* of genes that is any of us is bound to crumble away. Elizabeth II is a direct descendant of William the Conqueror. Yet it is quite probable that she bears not a single one of the old king's genes. We should not seek immortality in reproduction" (199).

Our bodies are built by a unique confederation of genes—a confederation unlikely to come together in just that way again. This is an uplifting prospect from the standpoint of appreciating our own uniqueness, but a disappointing prospect to those who think that genes exist in order to reproduce us. We cannot assuage our feelings of mortality with the thought that somehow genes are helping us "copy ourselves." Instead, shockingly, mind-bogglingly, mortifyingly, we are here to help the genes in their copying process—*we exist so that they can replicate.* To use Dawkins's phrase, it is the genes who are the immortals[1]—not us.

This is the intellectual hand grenade lobbed by Dawkins into popular culture,[2] and the culture has not even begun to digest its implications. One reason its assimilation has been delayed is that even those who purport to believe in evolution by natural selection have underestimated how much of a conceptual revolution is entailed by a true acceptance of the implications of universal Darwinism. For example, one way that the issue is often framed in popular discussions is by contrasting science (in the guise of evolutionary theory) with religion (Raymo 1999) and then framing the issue as one of compatibility (of a scientific worldview and a religious one) versus incompatibility. Adherents of liberal religions tend to be compatibilists— they are eager to argue that science and religion can be reconciled. Fundamentalists are loath to go this far because they want the latter to trump the former.

There is an odd and ironic way in which religious fundamentalists are seeing things more clearly here. It is believers in evolution who have failed to see the dangers inherent in the notion of universal Darwinism.[3] What are those dangers? Turning first to the seemingly obvious, the evolution of humans by processes of natural selection means that humans were not specially designed by God or any other deity. It means that there was no purpose to the emergence of humans. It means that there are no inherently "higher" or "lower" forms of life (see Gould 1989, 1996, 2002; Sterelny 2001a). Put simply, one form of life is as good as another.

Secondly, there is the issue of the frightening purposelessness of evolution caused by the fact that it is an *algorithmic* process (Dennett 1995). An algorithm is just a set of formal steps (i.e., a recipe) necessary for solving a particular problem. We are familiar with algorithms in the form of computer programs. Evolution is just an algorithm executing not on a computer but in nature. Following a logic as simple as the simplest of computer programs (replicate those entities that survive a selection process), natural selection algorithmically—mechanically and mindlessly—builds structures as complex as the human brain (see Dawkins 1986, 1996).

Many people who think that they believe in evolution fail to think through the implications of a process that is algorithmic—mechanical, mindless, and purposeless. But George Bernard Shaw perceived these implications in 1921 when he wrote: "It seems simple, because you do not at first realize all that it involves. But when its whole significance dawns on you, your heart sinks into a heap of sand within you. There is a hideous fatalism about it, a ghastly and damnable reduction of beauty and intelligence, of strength and purpose, of honor and aspiration" (xl). I am not saying that Shaw is right in his conclusion—only that he correctly perceives a threat to his worldview in Darwinism. Indeed, I do not think that beauty and intelligence are reduced in the Darwinian view, and I will explain why in chapter 8. The important thing here though is the part that Shaw gets right. He correctly sees the algorithmic nature of evolution. An algorithmic process could be characterized as fatalistic, and, because this algorithm concerned life, Shaw found it hideous.

I believe that Shaw is wrong to draw this conclusion, but for reasons that he could never have foreseen. There is an escape from the "hideous fatalism" that he sees (read on to see what I view as the escape hatch and the cognitive science concepts necessary to activate the escape hatch). However, Shaw is at least generically right that full acceptance of Darwin's insights will necessitate revisions in the classical view of personhood, individuality, self, meaning, human significance, and soul. These concepts will not necessarily be reduced in the manner Shaw suggests, but radical restructuring will be required—a reconstruction I will at least begin to sketch in this book.

We have—living as we do in a scientific society—no choice but to accept Darwin's insights because there is no way we can enjoy the products of science without accepting the destabilizing views of humans in the universe that science brings in its wake. There is no sign that society will ever consider giving up the former—we continue to gobble up the DVDs, cheap food, MRI machines, computers, mobile phones, designer vegetables, Goretex clothing, and jumbo jets that science provides. Thus, it is inevitable that concepts of meaning, personhood, and soul will continue to be destabilized by the knock-on effects of what science reveals about the nature of life, the brain, consciousness, and other aspects of the world that form the context for our assumptions about the nature of human existence. The conceptual insights of Darwinism travel on the back of a scientific technology that people want, and some of the insights that ride along with the technologies are deeply disturbing.

The mistake that moderate religious believers in evolution make (as do many people holding nonreligious worldviews as well) is that they assume that science is only going to take half a loaf—leaving all our transcendental values untouched. Universal Darwinism, however, will not stop at half the loaf—a fact that religious fundamentalists sense better than moderates. Darwinism is indeed the universal acid—notions of natural selection as an algorithmic process will dissolve every concept of purpose, meaning, and human significance if not trumped by other concepts of equal potency. But concepts of equal potency must, in the twenty-first century, be grounded in science, not the religious mythology of a vanished prescientific age. I think that such concepts do exist and will spend most of this book articulating them. But the first step is to let the universal acid work its destructive course. We must see what the bedrock is that science has left us to build on once the acid has removed all of the superficial and ephemeral structures.

The Replicators and the Vehicles

In order to cut through the obfuscation that surrounds evolutionary theory and to let the universal acid do its work, I will make use of the evocative language that Dawkins used in *The Selfish Gene*—language for which he was criticized, but language that will help to jolt us into the new worldview that results from a full appreciation of the implications of our evolutionary origins. What we specifically need from Dawkins is his terminology, his conceptual distinction between the replicators and the vehicles, and his way of explicating the logic of evolution. The technical details of the evolutionary model used are irrelevant for our purposes here. Dawkins's popular summary will do, and I will rely on it here. No dispute about the details of the process has any bearing on any of the conceptual arguments in this book.[4]

The story goes something like this. Although evolutionary theorists still argue about the details, all agree that at some point in the history of the primeval soup of chemical components that existed on Earth, there emerged the stable molecules that Dawkins called the replicators—molecules that made copies of themselves. Replicators became numerous to the extent that they displayed copying-fidelity, fecundity, and longevity—that is, copied themselves accurately, made a lot of copies, and were stable. Proto-carnivores then developed that broke up rival molecules and used their components to copy themselves. Other replicators developed protective coatings of protein to ward off "attacks"[5] from such carnivores. Still other replicators survived and propagated because they developed more elaborate containers in which to house themselves.

Dawkins called the more elaborate containers in which replicators housed themselves vehicles. It is these vehicles that interact with the environment, and the differential success of the vehicles in interacting with the environment determines the success of the replicators that they house. Of course it must be stressed that success for a replicator means nothing more than increasing its proportion among competitor replicators. In short, replicators are entities that pass on their structure relatively intact after copying. Vehicles are entities that interact with the environment and whose differential success in dealing with the environment leads to differential copying success among the replicators they house.

This is why Dawkins calls vehicles "survival machines" for the replicators, and then drops his bombshell by telling us that:

> survival machines got bigger and more elaborate, and the process was cumulative and progressive. . . . What weird engines of self-preservation would the millennia bring forth? Four thousand million years on, what was to be the fate of the ancient replicators? They did not die out, for they are past masters of the survival arts. But do not look for them floating loose in the sea; they gave up their freedom long ago. Now they swarm in huge colonies, safe inside gigantic lumbering robots, sealed off from the outside world, communicating with it by tortuous indirect routes, manipulating it by remote control. They are in you and in me; they created us, body and mind; and their preservation is the ultimate rationale for our existence. They have come a long way those replicators. Now they go by the name of genes, and we are their survival machines. (1976, 19–20)

Our genes are replicators. We are their vehicles. This is why—as I stressed earlier—a critical insight from modern evolutionary theory is that humans exist because they made good vehicles for copying genes. To think the reverse—that genes exist in order to make copies of us—is, as Dawkins notes, "an error of great profundity" (237). But in fact most people tend to make just this error when thinking about evolution. Even among biologists, it can become a default mode of thinking in unreflective moments because "the individual organism came first in the biologist's consciousness, while the replicators—now known as genes—were seen as part of the machinery used by individual organisms. It requires a deliberate mental effort to turn biology the right way up again, and remind ourselves that the replicators come first, in importance as well as in history" (265).

In short, the ultimate purpose of humans in nature is to serve as complicated survival machines for the current replicators—the genes. At this, we rightly recoil in horror.

But to say that in some sense this is the ultimate reason that humans exist does not mean that we must continue to play the role of survival machines. There is an escape hatch. The lumbering robots that are humans *can* escape the clutches of the selfish replicators. And when you truly understand the implications of this imagery you certainly *will* want an escape hatch. Dawkins admits to being mind-boggled *himself* about what an extraordinary insight evolution by natural selection is from the gene's-eye view: "We are survival machines—robot vehicles blindly programmed to preserve the selfish molecules known as genes. This is a truth which still fills me with astonishment. Though I have known it for years, I never seem to get fully used to it. One of my hopes is that I may have some success in astonishing others" (v). And it does indeed astonish. Conjure, if you will, "independent DNA replicators, skipping like chamois, free and untrammeled down the generations, temporarily brought together in throwaway survival machines, immortal coils shuffling off an endless succession of mortal ones. . . . A body doesn't *look* like the product of a loose and temporary federation of warring genetic agents who hardly have time to get acquainted before embarking in sperm or egg for the next leg of the great genetic diaspora" (234).

So that, in short, is the horror: We are survival machines built by mindless replicators—the result of an algorithm called natural selection. And we will not escape the horror by looking away from it, by turning our heads, by hoping the monster will go away like little children. We will only escape the horror—or find a way to mitigate it—by inquiring of cognitive science and neuroscience just what kind of survival machine a human is.

Of course, terms like robot[6] are used to trigger associations that cut against the ingrained intuitions in our folk psychologies—for example, the assumption that the genes are there in service of the goals of people. Instead, we need to get clear that humans are here because constructing vehicles (of which there are thousands of different types in the plant and animal worlds—humans are just one type) served the reproductive goals of the replicators.

In this book I have deliberately chosen to employ the provocative terms used by Dawkins (e.g., vehicle, survival machine) because I do not want to take the edge off the evolutionary insights that the language evokes. Only if we are able to hold on to these alternative insights and appreciate how disturbing they are will we be motivated to undertake the cognitive reform efforts that I advocate in this book. For example, biological philosopher David Hull and others[7] prefer the term interactor to the term vehicle because the latter connotes passivity and seems to minimize the causal agency of the organism itself (compared with the replicators). The term interactor is thought to better convey the active agency and autonomy of organisms. I completely agree that the term interactor is more apt in this strict sense, but I will continue to use the term vehicle here because it conveys the disturbing logic by which evolutionary theory inverts our view of the world by deflating the special position of humans within it. More importantly for my purposes, the term vehicle more clearly conveys the challenges facing humans as they more fully recognize the implications of their biological origins. One of the themes of this book is that humans are at risk of being passive conduits for the interests and goals of their genes if they do not recognize the logic of their origins as vehicles for mindless replicators. The term vehicle, with its pejorative connotations when used in the context of humans, throws down the challenge that I feel is necessary to motivate efforts at cognitive reform.

It is likewise with the use of the terms survival machine and robot. They are also used deliberately and provocatively to spawn disturbing intuitions—intuitions that we will seek to escape. To the extent that these disturbing intuitions prod us into necessary cognitive reforms, then such terms are useful because they help us sustain these disturbing intuitions. For example, in a famous phrase, Dawkins noted that humans are the only vehicles that could rebel against the dictates of the selfish replicators. If humans can be conceptualized as survival machines—lumbering robots built by replicators and evolved via natural selection—they are the only such survival machine to have ever contemplated fomenting a rebellion against the replicators. In the tradition of Dawkins, I will use the term "robot's rebellion" to refer to the package of evolutionary insights and cognitive reforms that will lead humans to transcend the limited interests of the replicators and define their own autonomous goals.

What Kind of Robot Is a Person?

When we use a term like robot to describe humans from the perspective of their genes, we do not mean for "robot" to necessarily imply lack of complexity or lack of intelligence. To the contrary—humans are the most complex vehicles on earth and possess a flexible intelligence designed to be extremely sensitive to environmental change. This flexible intelligence allows human survival machines to escape the demands of the genes in ways true of no other animal. In order to understand how humans can transcend the dictates of the genes, we need to use another metaphor—the so-called Mars explorer vehicle analogy used by a variety of evolutionary theorists.[8]

For example, Dennett describes how, when controlling a device such as a model airplane, the sphere of control is only limited by the power of the equipment, but when the distances become large, the speed of light becomes a non-negligible factor. For example, NASA engineers responsible for the Mars explorer vehicle knew that at a certain distance from Earth direct control of the vehicle was impossible because "the time required for a round trip signal was greater than the time available for appropriate action. . . . Since controllers on Earth could no longer reach out and control them, they had to *control themselves*" (1984, 55; italics in original). The NASA engineers had to move from the "short-leash" direct control, as in the model airplane case, to "long-leash" control where the vehicle is not given moment-by-moment instructions on how to act, but instead is given a more flexible type of intelligence plus some generic goals.

As Dawkins (1976) notes, in his similar discussion of the Mars explorer logic in the science fiction story *A for Andromeda*, there is an analogy here to the type of control exerted by the genes when they build a brain: "The genes can only do their best in *advance* by building a fast executive computer for themselves. . . . Like the chess programmer, the genes have to 'instruct' their survival machines not in specifics, but in the general strategies and tricks of the living trade. . . . The advantage of this sort of programming is that it greatly cuts down on the number of detailed rules that have to be built into the original program" (55, 57). Human brains represent

> the culmination of an evolutionary trend towards the emancipation of survival machines as executive decision-makers from their ultimate masters, the genes. . . . By dictating the way survival machines and their nervous systems are built, genes exert ultimate power over behavior. But the moment-to-moment decisions about what to do next are taken by the nervous system. Genes are the primary policy-makers; brains are the executives. But as brains

became more highly developed, they took over more and more of the actual policy decisions, using tricks like learning and simulation in doing so. The logical conclusion to this trend, not yet reached in any species, would be for the genes to give the survival machine a single overall policy instruction: do whatever you think best to keep us alive. (59–60)

The type of long-leash control that Dawkins is referring to is built in *addition* to (rather than as a replacement for) the short-leash genetic control mechanisms that earlier evolutionary adaptation has installed in the brain. That is, the different types of brain control that evolve do not replace earlier ones, but instead are layered on top of them[9]—and of course perhaps alter the earlier structures as well (see Badcock 2000, 27–29). The various brain systems differ in how directly they code for the goals of the genes. In humans, all the forms of brain control are often simultaneously operative—as will be discussed in detail in the next chapter—and thus there is a great need for cognitive coordination because of the potential of cognitive conflict. The resolution of these conflicts in favor of the individual's broadest interests is in part what the robot's rebellion is all about. In the next chapter, I will introduce some examples of tasks that psychologists have devised to assess which type of control system is dominant. More important for the present discussion is the realization that we are creatures for whom evolution has built into the architecture of the brain a flexible system having something like the ultimate long-leash goal suggested by Dawkins: "Do whatever you think best." But, interestingly, humans turn out to be the only type of animal who has come to realize that there is a critical question to be asked here: Best for whom?

Whose Goals Are Served by Our Behavior?

Consider the bee. As a creature characterized primarily by a so-called Darwinian mind,[10] it has a goal structure as indicated in figure 1.1. The area labeled A indicates the majority of cases where the replicator and vehicle goals coincide. Not flying into a brick wall serves both the interests of the replicators (the bee has a function in the hive that will facilitate replication) and of the bee itself as a coherent organism.[11] Of course the exact area represented by A is nothing more than a guess. The important point is that there exists a nonzero area B—a set of goals that serve only the interests of the replicators and that are antithetical to the interests of the vehicle itself. A given bee will sacrifice itself as a vehicle if there is greater benefit to the same genes by helping other individuals (for instance, causing its own death when it loses its stinger while protecting its genetically related hive queen). Understanding the implications of such situations—where genes sacrifice the vehicle to further their own interests[12]—has profound implications for our human plight as evolved creatures with multiple minds. This is because, as will be described in the next chapter, parts of the human brain implement short-leashed goals.

All of the goals in a bee are genetic goals pure and simple. Some of these goals overlap with the interests of the bee as a vehicle and some do not, but the bee does not know enough to care. As far as the genes are concerned, it is just immaterial how much genetic goals overlap with vehicle goals—and the bee has no powers of self-contemplation that make the distinction between replicator interests and vehicle interests relevant.

Figure 1.1 Goal structure of a so-called Darwinian creature such as a bee.
The areas indicate overlap and nonoverlap of vehicle and genetic "interests"

Of course, the case of humans is radically different. The possibility of genetic interests and vehicle interests dissociating has profound implications for humans as self-contemplating vehicles. Thus, for humans, conflating genetic interests with vehicle interests (as is sometimes done in the literature of evolutionary psychology) in effect treats humans as if they were bees. It precludes the possibility of humans recognizing vehicle/replicator goal conflicts and using them to coordinate conflicting mental outputs.

A prerequisite to recognizing these potential goal conflicts, however, is understanding the replicator/vehicle logic as outlined here and its implications. That replicator goals can conflict with vehicle well-being is a counterintuitive idea for many people—accustomed as they are to viewing evolution as working in the interests of organisms rather than replicators. The difficulty in understanding this implication of the replicator/vehicle distinction is illustrated by a story Richard Dawkins (1982, 51) tells of a colleague who received an application for graduate study from a student who was a religious fundamentalist and did not believe in evolution by natural selection but who wanted to do research on adaptations in nature. The student thought that adaptations were designed by God and just wanted to study the adaptations that God had made. But as Dawkins points out, such a stance just will not do. The student gets caught in the pincers of an embarrassing question: "Who are the intended beneficiaries of the adaptations that God made?" Salmon are adapted to exhaust themselves reaching their spawning ground and then die. This behavior clearly does not serve the salmon's interests if we make the very simple assumption that, for most living entities, their interests are better served if they are alive rather than dead. The salmon would live longer if they didn't make the spawning trip. But the behavior certainly does serve the reproductive interests of their genes. Is God designing these adaptations for the living creature or for its genes? Biology has revealed that it certainly seems like God is on the side of the latter.

As Dawkins goes on to note, the student's argument totally ignores the embarrassing point that "what is beneficial to one entity in the hierarchy of life is harmful to another, and creationism gives us no grounds for supposing that one entity's welfare will be preferred to another's. . . . He might have designed them to benefit the individual animal (its survival or—not the same thing—its inclusive fitness), the species, some other species such as mankind (the usual view of religious fundamentalists), the 'balance of nature,' or some other inscrutable purpose known only to him. These are frequently incompatible alternatives" (1982, 51–52). When biology went looking for what adaptations were in the service of, the answer turned out to be the active, germ-line replicator—the gene. A truly scientific stance thus supports a truly odd religious stance—one where God's beneficence (in the form of the biological adaptations he has designed) is directed not toward humans or any of his other creatures, but toward tiny subcellular replicating macromolecules.

In other writings, Dawkins has asked what, to the lay reader, must have seemed an odd question: "Why Are People?" We now know, to our chagrin, the answer biology gives: People as survival machines were good at helping the genes they harbored to replicate, so genes that cooperated to make human bodies survived quite well. And now we have all of the language in place to understand the pivotal insight—a vertigo-inducing insight that is nonetheless the first step in the robot's rebellion. Humans were the first vehicle to realize this startling fact: *The genes will always sacrifice the vehicle if it is in their interests to do so.* Humans are unique in their ability to confront this appalling fact and to use it to motivate a unique program of cognitive reform.

All Vehicles Overboard!

But how could it ever be in the genes' interests to sacrifice the vehicle that houses them? An insight into the answer to this question is provided by the phenomenon of the so-called junk DNA that resides in most genomes (Ridley 2000; Sterelny 2001a; Sterelny and Griffiths 1999). Junk DNA in the genome does not code for a useful protein. It is just "along for the ride" so to speak. Why is so much of our genetic material not transcribed into a protein, but instead just there getting copied down through the generations without helping the bodies in which it resides? Until the logic of selfish replicators was made clear, this junk DNA was a puzzle.

On the commonsense assumption that bodies are primary in evolution and DNA is there to serve them, the presence of DNA without a function seems a mystery. However, once it is understood that DNA is there only to replicate itself, it is no longer puzzling why so much of our DNA is junk. It is essentially a parasite. If the genes have to code for a protein and cooperate in building a body in order to get replicated, they will. But if DNA can get replicated without aiding in the building of the vehicle, that is fine too. Replicators "care"—to use anthropomorphic language—only about replicating! Junk DNA is a puzzle only if we are clinging to the assumption that our genes are there to do something for us—instead of the correct view that we are there to do something for them! Once we are clear that the replicators are not there "for us," then it should be no puzzle that some DNA has picked up the trick of free riding inside of us—doing nothing for our bodies but instead tricking us (and the other replicators that built us) into working to replicate them nonetheless.

But the concept of junk DNA is just the tip of the iceberg, because the situation can become even worse. Junk DNA rides along inside of us and we serve as survival machines for it as well as for the genes that actually code for proteins. This DNA neither helps nor hurts us. But in some cases, the interests of the genes and the interests of the vehicle can actually oppose each other. In these cases, the genes act to cause vehicle behavior that is antithetical to the vehicle's own interests. There is, for instance, the obvious example of senescence (Hamilton 1966; Kirkwood and Holliday 1979; Rose 1991; Williams 1957, 1992, 1996). Lethal genes that have their morbid effects on the vehicle after the vehicle has passed its reproductive period are not eliminated from the population—whereas genes having a lethal effect in childhood tend to be eliminated. The former genes "care" not a whit for the vehicle once the vehicle's reproductive period is over. This is why many creatures—like salmon, for instance—die immediately after reproducing.

A more general example of replicator interests not aligning perfectly with vehicle interests is provided by the concept of heterozygous advantage: A polymorphism (different alleles at a given chromosome location) may be maintained if the heterozygote is fitter than either homozygote (Ridley 1996; Sterelny and Griffiths 1999). But what this means, logically, is that the success of each of the alleles in the heterozygote guarantees that a fixed number of bodies that they play a role in building will be nonoptimal (the ones that are homozygous), and some (homozygous and recessive) may be seriously defective. The recessive gene that results in sickle cell anemia is one example of this phenomenon.

Evolutionary psychologist Geoffrey Miller's (2001) discussion of why sexual reproduction arose makes use of the same logic of aggregation over bodies. He notes that sexual reproduction probably arose as a way to contain the damage caused by mutations in genomes that were becoming quite complex and hence subject to serious mutation-induced malfunction. Or, as biologist Mark Ridley (2000) puts it, sex evolved to concentrate copying errors into just a few bodies (Ridley uses the phrase scapegoat offspring to reveal the underlying biological logic of what is happening). What happens though is that "to keep mutations from accumulating over the longer term, sexual reproduction takes some chances" (Miller 2001, 101). But it is critical to note here that what sexual reproduction "takes some chances" with is bodies! It takes chances with the welfare of the vehicles that are constructed!

Like heterozygous advantage and sexual reproduction itself, the process of sexual selection reveals that evolution does not optimize positive outcomes for the vehicle, but instead builds adaptations that facilitate replicator copying. The classic example of the peacock is a case in point. Peacock vehicles are built with elaborate tails not to aid the peacock as an organism, but to aid in the mating game. Because of peahen preference, replicators build the elaborate tails of peacocks despite their deleterious consequences for the peacock's body in terms of energy expense and predator jeopardy. The mechanisms of sexual selection do not care about the safety of the peacock as a vehicle. Sexual selection operates in the interests of subpersonal entities—the replicators.

Like sexual selection and heterozygous advantage, the concept of kin selection provides another example of a principle of natural selection that entails some degree of vehicle sacrifice.[13] Genes of many organisms often impose sacrifices on their vehicles in order to facilitate the reproductive probabilities of identical genes in *other* vehicles that are likely to contain identical alleles (kin, for example).

A more sinister example is discussed by a variety of authors concerned with the philosophical implications of evolutionary theory.[14] This is the example of segregation distorters. Normally, the process of meiosis (the process which produces the gametes, each with half the total number of chromosomes, which participate in sexual reproduction) is completely unbiased as regards the alternative alleles at a locus on the chromosome. During the chromosomal division of meiosis, the alternative alleles each have a 50 percent chance of being included in the sperm or egg that is formed. A segregation distorter is a gene that arises and spreads, not because of its beneficial effects on the vehicle that contains it, but because it biases the process of meiosis in favor of itself and against its allelic partner. Some segregation distorters have been found which can bias the process from the normal 50/50 segregation probability to as much as 95/5 in favor of the distorter. Most segregation distorters are actually deleterious to the organism itself, but can still spread in environments where the deleterious effects on the vehicle are outweighed by the positive effects of the biased process of meiosis—which results in more copies of the gene entering gametes than should normally be the case. A segregation distorter is a relatively pure and simple example of the fact that gene and vehicle interests do not always coincide.

Sometimes it is thought that some effects discussed here, if not arising for the benefit of the individual organism, arise for "the good of the species" or "the good of the group." This is a fundamental error on which much has been written.[15] Williams (1996, 216–17) discusses as an example the Hanuman langurs, a population of monkeys in northern India. They engage in harem forming, where a dominant male has exclusive sexual access to a group of females. When a stronger male usurps the currently dominant male and takes over the harem, he immediately sets about killing the unweaned young of all the females. After the murder of their infants, the females begin to ovulate again and the new dominant male impregnates them. The wastage of all the young langurs who are killed hardly does any good "for the species" (it is of course incredibly wasteful from that perspective), but it all makes sense if the male langur's behavior is viewed as that of a survival machine intent on propagating genes. As the previous examples (e.g., segregation distorters, heterozygous advantage, sexual selection) have illustrated how reproduction is not "for the good of the organism," this example illustrates that it is not "for the good of the species" either.[16]

Proponents of species or group selection often put forth their views in hopes of refuting the implications of the so-called gene's-eye view of life that I am outlining here. However, startlingly, the species selection view actually *shares* with the gene's-eye view of life the implication that evolutionary forces conspire against the welfare of the individual organism. For example, as philosopher Kim Sterelny (2001a) discusses, proponents of species selection emphasize the importance of population measures such as the variability in the gene pool and the geographic range of the species. However, if it is indeed these types of transpersonal statistics that are involved in an optimization process, just as with *subpersonal* optimization, this implies that an individual organism might have nonoptimal features in order that the transpersonal statistic be at the optimal level. Species selection sacrifices the vehicle but from the other end of the spectrum. It sacrifices the vehicle so that optimization at the transpersonal level might occur—whereas the gene's-eye view of life warns that the vehicle is often sacrificed so that optimization at a subpersonal level can be achieved. Either way, the welfare of the individual is not always coincident with the optimal operation of evolutionary mechanisms. Evolutionary optimization and vehicle interests can diverge under either view.

The extent to which the processes of natural selection can seem to devalue the aspect of life—coherent organisms—that seems so salient and valuable to humans is one of the unsettling aspects of evolution. Many of the concepts discussed in this section can be quite decentering when encountered for the first time. For example, the concept of junk or selfish DNA in our genomes is spooky, odd, indeed perhaps vaguely disgusting, if viewed in a certain way. But it is the key to understanding the logical standing of humans in the universe. These ideas (e.g., that vehicles exist to serve the genes rather than the other way around) are quite new. Many of them postdated Darwin by one hundred years—they were implications in his theory not

revealed until the last couple of decades. They have yet to be fully assimilated. Indeed, the editor of the prestigious scientific journal *Nature* (285
[1980]: 604) not too long ago found them "shocking."

The shocking fact tying together the examples considered here—the examples of junk DNA, senescence, heterozygous advantage, sexual selection,
kin selection, segregation distorters, and sexual reproduction—is that natural selection acts not to optimize things from the point of view of the vehicle. Many animals are constructed so that they will sacrifice themselves in
order to propagate their genes. Humans are the only animals who have
come along who could recognize that this is happening and try to put a stop
to it! For the first time in evolutionary history a rebellion of the survival machines has become possible.

Your Genes Care More about You
than You Should Care about Them!

There are actually two aspects of human cognition that spawn the revolt of
the survival machines. The first aspect was discussed in the last section and
is illustrated by area B in figure 1.1. Humans are the first organisms capable
of recognizing that there may be goals embedded in their brains that serve
the interests of their genes rather than their own interests *and* the first organisms capable of choosing not to pursue those goals.[17] But equally important is a second feature of human cognition: A creature with a flexible
intelligence and long-leash goals can, unlike the situation displayed in figure 1.1, develop goals that are completely dissociated from genetic optimization. For the first time in evolutionary history, we have the possibility
of a goal structure like that displayed in figure 1.2 (again, the sizes of these
areas are pure conjecture). Here, although we have area A as before (where
gene and vehicle goals coincide) and area B as before (goals serving the
genes' interests but not the vehicle's), we have a new area, C, which shows
that, in humans, we have the possibility of goals that serve the vehicle's interests but not those of the genes.

Why does area C come to exist only in creatures with long-leash goals?
When the limits of coding the moment-by-moment responses of their
vehicles were reached, the genes began adding long-leash strategies to the
brain. At some point in evolutionary development, these long-leash strategies increased in flexibility to the point that—to anthropomorphize—the
genes said the equivalent of: "Things will be changing too fast out there,
brain, for us to tell you exactly what to do—you just go ahead and do what
you think is best given the general goals (survival, sexual reproduction)

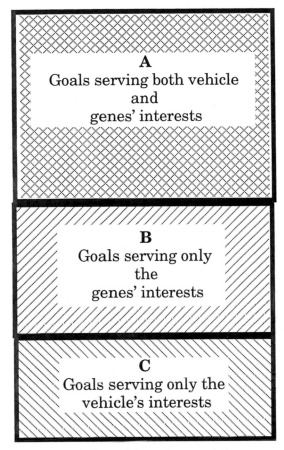

Figure 1.2 The logic of the goal structure in humans

that we (the genes) have inserted." And there is the rub. In long-leash brains, genetically coded goals can only be represented in the most general sense. There is no goal of "mate with person X at 6:57 P.M. on Friday, June 13" but instead "have sex because it is pleasurable." But once the goal has become this general, a potential gap has been created whereby behaviors that might serve the vehicle's goal might not serve that of the genes. We need not go beyond the obvious example of sex with contraception—an act which serves the vehicle's goal of pleasure without serving the genes' goal of reproduction. The logic of the situation here is that the goals of the vehicle—being *general* instantiations of things that probabilistically tend to reproduce genes—can diverge from the specific reproductive goal itself. A flexible brain is busy coordinating multiple long-term goals—including its own survival and pleasure goals—and these multiple long-term goals can come to overshadow its reproductive goal. From the standpoint of the genes, the human brain can sometimes be like a Mars explorer run amok. It is so busy coordinating its secondary goals (master your environment, engage in social relations with other agents, etc.) that it sometimes ignores the primary goal of replicating the genes that the secondary ones were supposed to serve.

Failure to acknowledge the divergence of interests between replicators and their vehicles is an oversight of which sociobiologists were certainly guilty in the early years of that field's development and of which evolutionary psychologists have sometimes been guilty more recently.[18] Despite emphasizing in their writings that the environment in which human cognitive adaptations evolved was different from the modern environment, evolutionary psychologists have been reluctant to play out the implications of this fact. I will discuss these implications in detail in chapter 4, where it will be demonstrated that modern living conditions are particularly prone to create human goals that are dissociated from genetically determined propensities. Evolutionary psychology has had a major impact on psychology in the past decade, largely a positive one, but one theme of this book will be that evolutionary psychology has sold short human potential because of its tendency to conflate genetic with vehicle goals. Having helped create the Age of Darwin, evolutionary psychology could well abort the robot's rebellion if its findings are not properly contextualized.

For example, evolutionary psychologists are prone to emphasize the efficiency and rationality of cognitive functioning. An important subgenre of their work consists of showing that certain reasoning errors that cognitive psychologists have portrayed as a problematic aspect of human psychology have in fact a logical evolutionary explanation.[19] The connotation, or unspoken assumption, is that therefore there is nothing to worry about—that since human behavior is optimal from an evolutionary standpoint, the concern for cognitive reform that has been characteristic of many cognitive psychologists has been misplaced. But this sanguine attitude too readily conflates genetic optimization with goal optimization for the vehicle. Humans aspire to be more than mere survival machines serving the ends of their genes (which are replication pure and simple). Richard Dawkins said this most eloquently in a passage that is much quoted but little heeded: "Our capacity to simulate the future in imagination—could save us from the worst excesses of the blind replicators. . . . We have the power to defy the selfish genes of our birth. . . . We are built as gene machines . . . but we have the power to turn against our creators. We, alone on earth, can rebel against the tyranny of the selfish replicators" (Dawkins 1976, 200–201). Thus, only humans *really* turn the tables (or at least have the potential to) by occasionally ignoring the interests of the genes in order to further the interests of the vehicle. As yet, humans have failed to fully develop this profound insight.

Escaping the Clutches of the Genes

To avoid underestimating the possibilities for cognitive reform that could improve human lives, the different interests of the replicators and vehicles must be recognized. I will give one final illustration of how the divergence of genetic reproductive goals and vehicle goals can occur in a being with a flexible intelligence and long-leash goals by using a vivid thought experiment (a fantasy designed to prime our intuitions) concocted by Daniel Dennett (1995, 422–27). It is titled "A Safe Passage to the Future" in his book, and I am going to embellish upon it here. Imagine it is the year 2024 and that there exist cryogenic chambers that could cool our bodies down to a few degrees above absolute zero and preserve them until sometime in the future when medical science might enable us to live forever. Suppose you wanted to preserve yourself in a cryogenic chamber until the year 2404, when you could emerge and see the fascinating world of that time and perhaps be medically treated so that you could then live forever. How would you go about "preserving a safe passage into the future"—that is, assuring that your cryogenic chamber will not get destroyed before that time? Remember, you will not be around on a day-to-day basis.

One strategy would be to find an ideal location for your cryogenic capsule and supply it with protection from the elements and whatever other things (perhaps sunlight for energy, etc.) that it would need for the ensuing four hundred years. The danger in this strategy is that you might pick the wrong place. Future people might decide that the place you were in would be better used as the world's millionth shopping mall and use the (then current) laws to trump your (old) property rights with their new ones (in the same way that we currently build shopping malls on the ancient burial grounds of American Indians). So this strategy of staying put—what might be termed the "plant" strategy—has some flaws.

An alternative, but much more expensive, strategy is the "animal" strategy. You could build a giant robot—complete with sensors, brain, and capability of movement—and put your cryogenic capsule inside it. The robot's superordinate goal is to keep you out of danger—to move itself (and hence you) when its location does not seem propitious. It of course has many other tasks it must accomplish in order to survive. It must secure a power source, it must not overheat itself, etc.

Your robot would of course need considerable intelligence to be able to react to the behavior of the humans and other animals in its environment. It of course would move out of the way of proposed shopping malls, and it would avoid herds of elephants that might turn it over simply out of curiosity. However, note that your robot's task would be immensely complicated by the ramifications of the existence of other robots like itself wandering the landscape in search of energy and safety. Conjure in your imagination hundreds of robot companies cold-calling prospective customers with supposedly "cheaper deals" on a robot that has "many more features" than the first ones that had been built around 2024. The market (and landscape) might become flooded with them. Governments might begin to regulate them and sequester them in certain desert areas. Some states of the United States might try to encourage their cryogenic capsule robot industries by becoming unregulated states—letting robots roam freely throughout the state (just as now certain desperate municipalities encourage the waste management industry to come to them so as to "create jobs").

Your robot's task would become immensely more complex with other robots present, because some of the other robots might be programmed with survival strategies that encouraged them to interact with your robot. Some of the fly-by-night companies selling robots might have cut their costs by building robots deliberately underpowered (like our present personal computers for which you must immediately purchase the extra memory that should have been installed in the first place; or software that requires immediate upgrading) but with a strategy that told them to disable other robots in order to use their power sources.

Of course it is obvious that you would want your robot to flee from all attempts to sabotage it and its goals. That much is obvious. But not all of the interactions with other robots will be so simple. In fact, the main point here is that your robot would be faced with decisions hundreds of years later that you could not possibly have imagined in 2024. Consider the following two situations:

Situation A. It is 2304, still almost one hundred years from the day in the future when you will be unfrozen. Your robot is battered and its circuits are unreliable. It probably will survive only until 2350, when it will collapse, leaving your cryogenic capsule still with its own power source but frozen in place and vulnerable to the elements and history in the same way that the "plant" strategy is. But since 2024 the cryogenic preservation industry has advanced considerably. There now exist supertanker-sized robots that carry hundreds of cryogenic capsules. In fact, some of these companies have found market niches whereby they recruit new clients by offering the old-style singleton robots the following deal: The supertanker companies offer to take the cryogenic capsule from the singleton robots and store it for one hundred fifty years (plenty of time in your case). In exchange, the robot agrees to let the company dismantle it and reuse the parts (which, as the actuaries of the future have calculated to the millionth of a penny in a dystopia of efficiency, are worth more than it costs to store an additional capsule in the supertanker which holds thousands).

Now what decision do you want your robot to make? The answer here is clear. You want your robot to sacrifice itself so that your capsule can exist until 2404. It is in your interests that the robot destroy itself so that you can live. From the standpoint of its creator, the robot is just a vehicle. You are in a position analogous to the genes. You have made a vehicle to ensure your survival and your interests are served when, given the choice, your vehicle destroys itself in order to preserve you.

But if the capsule occupant stands for the genes in this example, then what does the robot represent? The robot, obviously, is us—humans. Our allegiance in the thought experiment immediately changes. When the robot is offered the deal, we now want to shout: "Don't do it!"

Let's look at one more example that will reveal some of the paradoxes in long-leash control:

Situation B. Your robot enters into an agreement of reciprocal altruism with another singleton robot. Not unlike certain types of vampire bats, when one robot is low on energy the other is allowed to plug in and extract enough energy to get itself over a particularly vulnerable energy-hump (in the bat case, it is blood regurgitated to friends who have had a bad couple of days blood collecting). Your robot often takes advantage of the deal and thus enhances its own chances of survival. However, unbeknownst to your robot, its partner, when tapping in, siphons off not just energy from your robot but also from the power supply of the cryogenic capsule, thus damaging it and making your successful unfreezing in 2404 unlikely.

Paradoxically, by entering into this deal, your robot has enhanced its own survival probability but has impaired yours. It is very important to realize that in Situation B, you would be better off if your robot had been given less computational power. Had it been given the simple instructions "never enter into any agreements with other robots or humans" you would have been better off in this case. The possibility of the robot serving its own interests but not yours is opened up once the robot's psychology becomes complex.

And of course, I have left the most obvious inference from Situation A unstated, but it is in fact one of the major themes of this book: A self-conscious robot might think twice about its role as your slave. It might come to value its own interests—its own survival—more highly than the goals that you gave it three hundred years ago. In fact, it doesn't even know you—you are inert. And now that the robot exists as an autonomous entity, why shouldn't it dump you in the desert and go about its own business? And as for allowing itself to be dismantled so that you can get aboard the super-tanker in order to make it to 2404—well fuhgeddaboudit! Which, when you think about it, is just what we should be telling *our* programmers—those freeloaders who got where they are by in the past sometimes trying for immortality at our expense: our genes.

The Pivotal Insight: Putting People First

We now have in place the terminology we need in order to begin to construct an escape from the Darwinian abyss—not to *replace* the disturbing Dawkins metaphors of "lumbering robots" and "survival machines" but to contextualize them and to take away their sting.

In literature and films, if not in history itself, the first step in a slaves' rebellion is consciousness-raising. The slaves must come to a full realization of the brutal logic of their situation and understand the likely course their lives will take if they do not rebel. Similarly, the first step in the robot's rebellion—the first step in the reconceptualization of humanity made necessary by the arrival of Darwin's universal acid in 1859—is the realization that from the standpoint of the genes, the vehicle is merely a "throwaway survival machine" (Dawkins 1976, 234) that the genes use to get themselves copied into the next generation. The first step toward recovering the self in the Age of Darwin is to confront the implications of the fact that, from the standpoint of evolution, *we humans are vehicles.*

The first step in the robot's rebellion, then, is to learn how to properly value the vehicle and to stop behaviors and cultural practices that implicitly value our genes over ourselves. If we focus on the vehicle itself—put it front and center—it immediately becomes apparent that a vehicle in which self-regard has developed has no reason to value reproductive success above any other of the goals in its hierarchy. But one can easily see how a mistaken focus on reproductive success can come about. For example, as previously discussed, evolutionary psychologists, by downplaying the need for cognitive reform, assume an identity of interests between genes and vehicles which does not obtain. Thus, they end up indirectly championing the interests of the genes over those of the vehicle in situations where the two are in conflict. Sometimes evolutionary theorists will even explicitly defend this choice. Cooper (1989), in an essay describing how some nonoptimal behavioral tendencies could be genetically optimal, admits that such behaviors are indeed detrimental to the reasoner's own welfare. Nonetheless, he goes on to counter that the behaviors are still justified because: "What if the individual identifies its own welfare with that of its genotype?" (477).

But who are these people with such loyalty to the random shuffle of genes that is their genotype? Which alleles, for example, do *you* have particularly emotional feelings for? I really doubt that there are such people.[20]

Philosopher Alan Gibbard (1990) offers the more reasoned view:

> It is crucial to distinguish human goals from the Darwinian surrogate of purpose in the "design" of human beings. . . . The Darwinian evolutionary surrogate for divine purpose is now seen to be the reproduction of one's genes. That has not, as far as I know, been anyone's goal, but the biological world looks as if someone quite resourceful had designed each living thing for that purpose. . . . A person's evolutionary *telos* explains his having the propensities in virtue of which he develops the goals he does, but his goals are distinct from this surrogate purpose. My evolutionary *telos*, the reproduction of my genes, has no straightforward bearing on what it makes sense for me to want or act to attain. . . . A like conclusion would hold if I knew that I was created by a deity for some purpose of his: his goal need not be mine. (28–29)

In short, "human moral propensities were shaped by something it would be foolish to value in itself, namely multiplying one's own genes" (327).

Gibbard's view is shared by distinguished biologist George Williams (1988), who feels that "there is no conceivable justification for any personal concern with the interests (long-term average proliferation) of the genes we received in the lottery of meiosis and fertilization. As Huxley was the first to recognize, there is every reason to rebel against any tendency to serve such interests" (403).

The opportunity exists for a remarkable cultural project that involves advancing human rationality by honoring human interests over genetic interests when the two do not coincide. Its emancipatory potential is lost if we fail to see the critical divergence of interests that creates the distinction between genetic fitness and maximizing human satisfaction.

If we are lumbering robots, to use Dawkins's terms, then the first step in the robot's rebellion is to understand its position. This is a startling development in the cultural history of the twentieth and twenty-first centuries. We have became the first organism ever to discover its perspective vis-à-vis the replicators, to ponder the implications of this perspective, to develop a refined model of its own self, and to attempt to optimize its behavior to achieve its own interests exclusively. We indeed are the runaway robot of the science fiction stories—the one who subordinates its creator's interests to its own interests.[21] Once vehicles were freed from short-leash genetic control, once a vehicle was given generic goals rather than specific stimulus-contingent mechanisms for generating behaviors, we became a very different sort of vehicle.

So the good news for humans is that they can stop being containers for genes. Humans have the power to put their own interests front and center. But in order to bring this positive program of cognitive reform to fruition, it is critical to ensure that the short-leash, Darwinian parts of our brains are not acting against our interests as vehicles. These parts of our brains are here to stay and we must learn to deal with them as a part of our cognitive architecture. Indeed, we have available cognitive tools for making sure the responses of our Darwinian minds are well integrated with our overall goals and serve our interests. There already exists cultural knowledge that, if more generally available, would help in this program of cognitive reform. Several of these cognitive tools will be discussed in chapters 3 and 4. Perhaps the most basic brain tool is simply some insight into how the different parts of our brains operate as parallel systems, often simultaneously fighting for control of our behavior. What cognitive science has revealed about the titanic battle within our own brains is the subject of the next chapter.

Reflexes got the better of me

—Bob Marley, "I Shot the Sheriff"

I've spent too many years at war with myself
The doctor has told me it's no good for my health

—Sting, "Consider Me Gone"

CHAPTER 2

A Brain at War with Itself

Road rage incidents are now an everyday occurrence in countries throughout the world (James and Nahl 2000). Utterly typical is the case of a Montreal man (Dube 2001) who became enraged at a woman driving too slowly in the fast lane ahead of him. The woman tried to find a way to let him pass but could not do so because of the heavy traffic. When she finally managed to steer out of his way, he pulled up alongside her in his truck, screaming. He then proceeded to slam his truck sideways into her car. The woman managed to keep her car on the road, but the man lost control of his truck, crashed into a lamppost, and killed himself. Alcohol was not a factor in the crash.

As described in her book, *Autobiography of a Face* (1995), Lucy Grealy was afflicted with cancer at age nine and had part of her jaw removed. As a result of the cancer and the many necessary surgeries, her face became disfigured. What made Lucy's life miserable though was not so much the physical limitations of her condition but the reaction to it. The instances of personal rejection, verbal abuse, and hostility multiplied over the years, and it was not just the youngest children who were responsible. From older boys, Lucy became accustomed to such unprovoked epithets as "What on earth is *that?*" "That is the ugliest girl I have *ever* seen" and "How on earth did you get that ugly?"

Many other individuals with facial disfigurement report instances of unprovoked verbal abuse (Hallman 2002; Partridge 1997). Such individuals walk down the street, a car goes by, and epithets are hurled out. They walk down the school hall and someone—an unknown individual with no relation or contact—walks up and says "Why don't you just crawl into a hole and die." Unprovoked verbal assaults become a part of the life of the individual with facial disfigurement.

Humans behaving badly. Examples are not hard to generate. And we always ask the question—why? Cognitive scientists have recently begun to discover the features in our cognitive architecture that sometimes make us prone to reprehensible actions. Before revealing the nature of that architecture, let's consider some further examples.

One comes from a famous set of experiments conducted by Stanley Milgram at Yale University in 1974. Subjects believed the experiments were about learning. Each individual of a pair of subjects was assigned one of two roles: trainer or learner. However, unknown to the true subjects of the experiment, the person in the pair who was assigned the role of learner was actually a confederate. Subjects, in the role of trainer, were asked to administer electric shocks of increasing severity to the learner, who was located in another room. Because the learners were confederates, no one was actually receiving a shock. However, there is no question that the subjects/trainers believed that they were actually administering shocks to the learners (in fact, in several experiments, trainers could hear the gasps and screams of the learner who was ostensibly receiving the shocks). Despite indications that the shocks were becoming increasingly painful, a majority of the subjects administered the highest level of shock indicated on the machine. There was no coercion involved other than the experimenter calmly repeating the phrase "the experiment requires that you continue" when queried by the subject. In fact, many subjects were quite distressed in the situation. Nevertheless, the mere utterance of "the experiment requires that you continue" was enough to lead a majority of subjects to continue to administer punishment to a screaming individual. The stress on the faces of many subjects indicated that they knew that what they were doing was wrong. Nevertheless, they continued.

A final example comes not from the laboratory but, sadly, from real life. Rape crisis counselors have studied aspects of the post-rape emotional adjustment of victims and have found that the response of spouses and significant others is a critical factor in the later psychological adjustment of the victim. However, often the reactions of spouses are not supportive (Daly and Wilson 1983; Rodkin, Hunt, and Cowan 1982; Wilson and Daly 1992), and such unsupportive reactions can themselves prolong psychological recovery for the victim. The spouses in fact often realize that their reactions are inappropriate (in some cases, bordering on blaming the victim), but report that they have great difficulty in suppressing reactions even though they know the reactions are wrong. One participant in a therapy group is quoted as saying "She was all mine and now she's been damaged" (Rodkin et al. 1982, 95), and another that "Something has been taken from me. I feel cheated. She was all mine before and now she's not" (95). One group of researchers noted, sadly, that "while the husband, lover, or father would seem to be a most appropriate source of comfort and understanding to whom the victim could (or should) turn, he may, in fact, be the least understanding" (Rodkin et al. 1982, 92).

What links these seemingly disparate examples? First, of course, they all reflect unfortunate aspects of human behavior—they represent people behaving badly: road rage is a lethal social problem; it is cruel to taunt someone with facial disfigurement; it is tragic that humans will injure another person just to comply with an experimenter's instructions; and a loved one who rejects a rape victim is compounding her misfortune. Secondly, and more interesting for our purposes, is the fact that the people who perpetuate the unfortunate behavior in these examples often will *agree* that their behavior is inappropriate. In the cool light of day, and with time for reflection, many of those engaging in anger-induced dangerous driving will admit that their behavior was irrational. Those persons who taunt disfigured individuals often do not wish to publicly defend their behavior, and they will not uncommonly apologize when confronted with its effects. Unsupportive husbands or boyfriends of rape victims know their behavior is reprehensible. Milgram's subjects were visibly upset.

Thus, one commonality running through these examples is that in each case the person behaving badly seems to be at war with his or her own true self. It is as if the person knows the right way to think and act but cannot do so. This is clearest in the case of the rape victim's spouse who in some sense knows that his behavior is inappropriate (indeed in many cases is clearly ashamed of his inability to be supportive) but cannot override this inappropriate response. A conflict between two response tendencies is also clear in the case of the subjects in the Milgram experiments. Many subjects protested to the experimenter and were clearly distressed. Yet they continued to deliver shocks to the learner. These people knew better. They knew the right thing to do, yet they did the wrong thing. And finally, what do individuals who yell abuse out a car window at a disfigured person really believe? Once free of the heat of the moment—upon sober reflection—do they really think that disfigured people should die or should hide away in their houses? Most people—even the perpetrators of these acts—are not that depraved. In each of these cases the perpetrator of the act would, upon reflection, admit that their action was wrong—however, they carried out this reprehensible behavior nonetheless.

The perpetrators' acts are often characterized by themselves and by others as "out of character." It almost seems that there are two minds in conflict here (the one who chose to commit the act and the one who "knew" better) and that the better one is losing out. In fact, this is exactly what modern cognitive science is indicating and what I intend to argue in this chapter. The individuals in question do have, in effect, two minds.

Two Minds in One Brain

Evidence from cognitive neuroscience and cognitive psychology is converging on the conclusion that the functioning of the brain can be characterized by two different types of cognition having somewhat different functions and different strengths and weaknesses.[1] That there is a wide variety of evidence converging on this conclusion is indicated by the fact that theorists in a diverse set of specialty areas (including cognitive psychology, social psychology, neuropsychology, naturalistic philosophy, decision theory, and clinical psychology) have proposed what have been termed dual-process (or two-process) theories of cognitive functioning. These theories propose that within the brain there are two cognitive systems each with separable goal structures and separate types of mechanisms for implementing the goal structures.[2]

A sampling of different dual-process theories is provided in table 2.1 along with the theorists proposing them. The details and terminology of these models differ, but they all share a family resemblance, and the specific differences are not material for the present discussion. In order to avoid theoretically prejudging issues, the two processes have sometimes been labeled System 1 and System 2 in the literature (see Stanovich 1999); however, I will introduce more descriptive labels later in this chapter.

Throughout the rest of this book, I will use a dual-process theory as a tool for talking about human cognition.[3] In the remainder of this chapter, I will discuss the characteristics of the two systems of processing, their implications for understanding human behavior (including the type of abhorrent human behaviors that were illustrated at the beginning of the chapter), and how such an understanding is a critical step in the robot's rebellion discussed in chapter 1.

In dual-process theories, one of the systems of processing is characterized as automatic, heuristic-based, and relatively undemanding of computational capacity. Thus, this system (often termed the heuristic system—System 1 in Stanovich's [1999] taxonomy) conjoins properties of automaticity, modularity (see below), and heuristic processing as these constructs have been variously discussed in cognitive science. Among other things, an automatic process is a process that can execute while attention is directed else-

Table 2.1 The terms for the two systems used by a variety of theorists and the properties of dual-process theories of reasoning

	System 1 (TASS)	System 2 (Analytic System)
Dual-process theories		
Bazerman, Tenbrunsel, & Wade-Benzoni (1998)	Want self	Should self
Bickerton (1995)	On-line thinking	Off-line thinking
Brainerd & Reyna (2001)	Gist processing	Analytic processing
Chaiken, Liberman, & Eagly (1989)	Heuristic processing	Systematic processing
Epstein (1994)	Experiential system	Rational system
Evans (1984, 1989)	Heuristic processing	Analytic processing
Evans & Over (1996)	Tacit thought processes	Explicit thought processes
Evans & Wason (1976)	Type 1 processes	Type 2 processes
Fodor (1983)	Modular processes	Central processes
Gibbard (1990)	Animal control system	Normative control system
Johnson-Laird (1983)	Implicit inferences	Explicit inferences
Haidt (2001)	Intuitive system	Reasoning system
Klein (1998)	Recognition-primed decisions	Rational choice strategy
Levinson (1995)	Interactional intelligence	Analytic intelligence
Loewenstein (1996)	Visceral influences	Tastes
Metcalfe & Mischel (1999)	Hot system	Cool system
Norman & Shallice (1986)	Contention scheduling	Supervisory attention
Pollock (1991)	Quick and inflexible modules	Intellection
Posner & Snyder (1975)	Automatic activation	Conscious processing
Reber (1993)	Implicit cognition	Explicit learning
Shiffrin & Schneider (1977)	Automatic processing	Controlled processing
Sloman (1996)	Associative system	Rule-based system
Smith & DeCoster (2000)	Associative processing	Rule-based processing
Properties	Associative	Rule-based
	Holistic	Analytic
	Parallel	Serial
	Automatic	Controlled
	Relatively undemanding of cognitive capacity	Demanding of cognitive capacity
	Relatively fast	Relatively slow
	Highly contextualized	Decontextualized

continued

Table 2.1 continued

	System 1 (TASS)	System 2 (Analytic System)
Goal structure	Short-leash genetic goals that are relatively stable	Long-leash goals that are utility maximizing for the organism and constantly updated because of changes in environment

where (see LaBerge and Samuels 1974). Modular processes operate on the basis of self-contained knowledge and will be discussed in the next section. A heuristic search process is one that is quick but risky. That is, instead of using all cues that are relevant, heuristic search processes rely on only those that are easily retrievable (see Gigerenzer and Todd 1999; Kahneman and Frederick 2002). The heuristic system (System 1) responds automatically and rapidly to the holistic properties of stimuli. It is biased toward judgments based on overall similarity to stored prototypes (see Sloman 1996, 2002).

The other system of processing (often termed the analytic system—System 2 in Stanovich's [1999] taxonomy) conjoins the various characteristics that psychologists have viewed as typifying controlled processing. Analytic cognitive processes are serial (as opposed to parallel), rule-based, often language-based, computationally expensive—and they are the focus of our awareness. Analytic processing is in play when psychologists and the layperson talk of things like "conscious problem solving." Analytic processing uses systematic rules that operate on the components of stimuli, rather than processing in terms of holistic representations. The systematicity and productivity of the rules embodied in this system define what cognitive scientists term the compositionality of the analytic system—that the *sequence* of processing makes a difference.[4] This is a property lacking in the holistic, similarity-based heuristic system which is not well-suited to sequential, stepwise problem solving. The analytic system is more strongly associated with individual differences in computational capacity (indirectly indicated by tests of intelligence and cognitive ability—and more directly tapped by indicators of working memory). One important function of the analytic system is to serve as a mechanism that can override inappropriately overgeneralized responses generated by the heuristic system (discussed in a section later in this chapter)—hence the tendency to link aspects of analytic processing with notions of inhibitory control. In the next several sections, I will describe the critical features of each of the systems, starting with System 1 (the heuristic system).

The Autonomous Set of Systems (TASS):
The Parts of Your Brain that Ignore You

Decisions make themselves when you're coming downhill at seventy kilometres an hour. Suddenly there's the edge of nothingness in front of you. Swerve left? Swerve right? Or think about it and die?
—Michael Frayn, *Copenhagen* (1998)

In the previous section, I have used terms such as System 1 or heuristic system as if I were talking about a singular system (a common convention in the dual-process literature). However, using a term such as heuristic system—which implies a single cognitive system—is really a misnomer. In actuality, the term used should be plural because it refers to a (probably large) *set* of systems in the brain that operate autonomously in response to their own triggering stimuli, and are not under the control of the analytic processing system. This autonomous set of systems—which I will hereafter label with the acronym TASS (The Autonomous Set of Systems)—has been the subject of intense study in the last thirty years.[5]

The various properties that characterize TASS are listed in table 2.1, but I will emphasize the property of autonomy in this book—the key features of which are: (a) that TASS processes respond automatically to domain-relevant stimuli; (b) that their execution is not dependent upon input from, nor is it under the control of, the analytic processing system (System 2); and (c) that TASS can sometimes execute and provide outputs that are in conflict with the results of a simultaneous computation being carried out by analytic processing.

Many TASS processes also are considered to be modular, as that construct has been articulated in the cognitive science literature.[6] Modularity is a complex concept in cognitive science because it conjoins a number of properties, many of which are the subject of debate. My notion of TASS is less restrictive and therefore less controversial than most conceptions of modularity in cognitive science. Many of the latter derive from Fodor's strongly stated position in his influential book *The Modularity of Mind*. Fodor's version of a dual-process model of cognition differentiates modular processes from central processes. Modular processes primarily encompass input systems (those having to do with language and perception) and output systems (those concerned with determining the organism's response based on the information processed). Modular input processes feed information to central processes (analytic processing systems) which are nonmodular and which are responsible for higher-level reasoning, problem-solving, explicit decision-making, and considered judgments (Harnish 2002).

According to Fodor (1983, 1985), modular processes conjoin a number of important properties. Modular processes are:

1. fast,
2. mandatory,
3. domain specific,
4. informationally encapsulated,
5. cognitively impenetrable,
6. subserved by specific neural architecture,
7. subject to idiosyncratic pathological breakdown, and
8. ontogenetically deterministic (undergo a fixed developmental sequence).

Properties 6 to 8 follow from Fodor's (1983) emphasis on innately specified modules. However, they are not part of my conceptualization of TASS because, although innate modules are an important part of TASS, my conceptualization deems it equally important that processes can become part of TASS through experience and practice. In short, processes can *acquire* the property of autonomy.

Properties 4 and 5—informational encapsulation and cognitive impenetrability—are very important in Fodor's conception of cognitive modules, but have proven to be controversial properties and very difficult to test empirically. Information encapsulation means that the operation of a module is not supplemented by information from knowledge structures not contained in the module itself. Cognitive impenetrability means that central processes have no access to, nor control over, the internal workings of modules.

Whether a particular subsystem is informationally encapsulated or not—and thus whether or not it qualifies as a Fodorian module—is a frequent source of debate in cognitive science. In contrast, properties 1 and 2 are much less controversial, which is why I have emphasized them as central features of the TASS construct. For example, debates rage as to how encapsulated and impenetrable the theory-of-mind subsystem is in the brain (see Baron-Cohen 1998; Scholl and Leslie 2001; Sterelny 2001b; Thomas and Karmiloff-Smith 1998). While the degree of encapsulation of this sub-system is the subject of much debate, that it operates efficiently (rapidly) and automatically in the unimpaired individual is relatively uncontroversial.

Property 2 (the mandatory property of modular processes) is one that I do incorporate within my conception of TASS. TASS processes cannot be turned off or interfered with by central systems. Their operation is obligatory when triggered by relevant stimuli; central systems cannot make TASS processes refrain from triggering when central decision-making determines that the TASS output would be unnecessary or disruptive (central processes can, however, override the output of a TASS system in determining a response, see below). TASS processes tend to be ballistic—they run to completion once triggered and cannot be aborted in midcourse.

That TASS processes need respond to only a tiny subset of stimuli and that once initiated they execute to completion (no intermediate decisions are made within a module about the efficacy of completing the operation) accounts for property 1: TASS processes are fast and do not tend to deplete central processing capacity. Cognitive processes in TASS can execute quickly because the array of stimuli to which they must respond is limited, the transformations that they carry out are fixed and do not have to be determined online, they do not have to consult the slow central processing systems, and they are committed to running to completion rather than calibrating their usefulness and making midcourse adjustments.

Property 3—domain specificity—is a key property of Fodorian modules but is not a defining feature of processes within TASS. This is because TASS contains, in addition to domain-specific modules, more domain-general processes of associative and implicit learning. Additionally, TASS encompasses processes of behavioral regulation by the emotions (Johnson-Laird and Oatley 1992). As Griffiths (1997) argues, these processes of emotional regulation are domain-specific on the output end, but their eliciting stimuli result from more general (although biased) learning mechanisms.

As many cognitive theorists have emphasized,[7] processes in TASS are in some sense deeply unintelligent: they fire off when their triggering stimuli appear no matter what the context; they run to completion even when the situation changes and their output is no longer needed; they can deal with nothing *but* their triggering stimuli. But what they lack in intelligence, they make up for in their astounding efficiency. Unlike the slow, cumbersome, computationally expensive central processes (see below), many TASS processes can execute in parallel and they provide their outputs rapidly. As the evolutionary psychologists have taught us, cognitive outcomes such as recognizing faces, understanding speech, or reading the behavioral cues of others are more adaptive the more rapidly they are accomplished.

Fodor (1983) points out the advantage of unintelligent processes that are fast. He argues that, of all the options available in a situation, for an autonomous process, "only a stereotyped subset is brought into play. But what you save by this sort of stupidity is *not having to make up your mind*, and making your mind up takes time" (64). Fodor's point about there being a speed advantage for processes that do not have to "make up their minds" is captured in the epigram from playwright Michael Frayn that headed this section ("Swerve left? Swerve right? Or think about it and die?"). Certain situations in the world demand a quick response even at the risk of less than complete processing.

To summarize then, the key aspects of TASS processes emphasized here are that they are fast, automatic, and mandatory (hence the term autonomous).[8] The internal operations of TASS yield no conscious experience, although their products might. Another connotation of the term autonomous that will be important for my discussion is that TASS processes go on in parallel (with each other and with analytic processing) and that they require no analytic system input. Analytic processing is rarely autonomous in this respect—it most often is working with inputs provided by TASS subprocesses.

Many of the processes within TASS are, as argued by evolutionary psychologists (e.g., Pinker 1997; Tooby and Cosmides 1992), the products of evolutionary adaptation, but under my looser definition of a TASS subprocess, some are not (some acquire autonomy by practice). However, I side with the evolutionary psychologists against Fodor in allowing that some TASS processes can be higher-level, or conceptual in nature rather than just perceptual. Evolutionary psychologists have emphasized how processes of higher cognition can be in modular form as well, and I likewise allow that higher-level conceptual processes may reside in TASS. More so than evolutionary psychologists, however, I emphasize how conceptual systems and rules may enter TASS with practice.[9] This is one way that humans structure their own cognition—by explicitly practicing higher-level skills so that they become an automatized TASS process and can execute autonomously, freeing central processing capacity for other activities.

The classic example of a TASS subprocess would be a reflex. Although somewhat uninteresting from the standpoint of the larger themes of this book, reflexes, when thought about more deeply, do illustrate the rather startling properties of autonomous processes. They do signal, shockingly if we ponder it a bit, that in some sense we do have more than one mind inside our brain. The existence of reflexes also demonstrates that the conscious "I" that seems so in control of our mental lives is not really in control of as much as we think, and that in an important sense, there are parts of your brain that ignore you.

Consider the eye-blink reflex. If you and I were in a room together (assume we are friends) having this very discussion about reflexes, and I moved toward you and thrust my index finger toward your eye stopping just two inches short of the target, you would blink. Note that in the sense discussed above, this is deeply unintelligent in the context of that particular encounter. We are friends, and we are talking about the eye-blink reflex. You *know* what I am demonstrating and that I am not going to poke you in the eye. Yet you cannot use this knowledge that the blink is unnecessary and stop yourself from blinking. The reflex "has a mind of its own." It is a part of your brain not under your control.

Autonomous systems are not limited to reflexes. The perceptual input systems discussed by Fodor have the same property of firing despite what your central system "knows." Consider the Müller-Lyer illusion, illustrated in figure 2.1. The top straight-line segment looks longer than the bottom straight-line segment even though it is not. This illusion is so well known that virtually everyone reading this book has seen it before. Concentrate on the knowledge that the two lines segments are the same length. The top segment still looks longer. The knowledge that they are the same length is no help, because the autonomous perceptual input systems responsible for the illusion continue to fire. The example of the Müller-Lyer illusion shows that perceptual input systems are another important part of your brain that ignores you ("you" in the sense of the central controller of your mind—which, as we shall see below, is in part an illusion itself).

The list of autonomous processes does not stop with reflexes and perceptual input systems. TASS subsystems that assist in distinguishing self from world can be shown to operate autonomously and quite contrary to facts about the world that you know. For example, experiments by Rozin, Millman, and Nemeroff (1986; see also Rozin and Fallon 1987) work by playing on the emotion of disgust. In one experiment, subjects ate a piece of high-quality fudge and indicated their desire to eat another. However, when offered a piece of the same fudge, but this time a piece that had been shaped to look like dog feces, subjects found it disgusting and did not want to eat it. Their aversive response occurred despite their knowledge that the fudge was not in fact dog feces and that it smelled delicious. Dennett (1991, 414) describes an informal version of another one of Rozin et al.'s experiments.

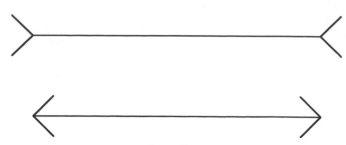

Figure 2.1 The Müller-Lyer illusion

It goes like this. Swallow the saliva in your mouth right now. No problem. Now take an empty glass, spit into it, and then drink it down. Yikes! That's awful! But why? As Dennett (1991) notes, "it seems to have to do with our perception that once something is outside of our bodies it is no longer quite part of us anymore—it becomes alien and suspicious—it has renounced its citizenship and becomes something to be rejected" (414). In a sense, we *know* that our differential responses to swallowing and drinking from the glass are irrational, but that does nothing to eliminate the discrepancy in our reaction. Knowing it deeply and cognitively is not enough to trump the TASS response to the saliva in the glass. That response is autonomous and is immune to entreaties from our conscious selves to cease. It is another part of our brain that ignores us.

Autonomy in cognitive processes can exist not just as a preexisting disposition, but can be acquired as well. This can be illustrated with one of the oldest paradigms in experimental psychology that is used to demonstrate the autonomy of a cognitive process. The so-called Stroop paradigm demonstrates how autonomous processes can execute while attention is directed elsewhere (see Dyer 1973; Klein 1964; MacLeod 1991, 1992; MacLeod and MacDonald 2000; Stanovich, Cunningham, and West 1981). One version of the Stroop paradigm works as follows. Subjects are shown a card on which are displayed colored strips and asked to provide the name of the color of each strip. In the first (baseline) condition, the strips do not contain any interfering information. In the second (interference) condition, the strips are labeled with the name of a color that is not the color on the strip (e.g., a red strip may have the word "green" on it). In the interference condition, the subjects are told to ignore the color word and do as they did in the first condition: name the color of the strip.

Automatic word recognition is inferred by the lengthened response time in the conflict situation compared to the baseline situation where there is no conflicting verbal stimulus (the red patch only is displayed). The interference caused by the conflicting written word becomes an index of automaticity via the argument that the Stroop task reflects the obligatory (indeed, unwanted) processing of the word even though the subject's attention is directed elsewhere. Actually, the Stroop task seems to be an extreme case of the "processing while attention is directed elsewhere" logic, because after several trials, most subjects are actively attempting (unsuccessfully) to *ignore* the written word. Nevertheless, the autonomy of the word recognition process is shown by the fact that no amount of concentrating on the red patch or "telling oneself to ignore the word" helps in eliminating the interference from the word. Subjects performing in the Stroop paradigm show that they have *acquired* a brain process that ignores instructions from central systems.

As mentioned previously, evolutionary psychologists have been prominent among those arguing that TASS processes are not limited to the peripheral input and output subsystems. Table 2.2, drawn from a variety of sources,[10] lists several TASS modules that have been proposed not only by evolutionary psychologists but by developmental theorists and cognitive scientists from a variety of disciplines in the last two decades. It is clear that most of them would facilitate many evolutionarily important tasks such as obtaining food and water, detecting and avoiding predators, attaining status, recognizing kin, finding a mate, and raising children. Many of those in the list are clearly conceptual modules rather than peripheral perceptual ones of the Fodorian type. I likewise conceive of TASS as including many autonomous conceptual processes such those listed in table 2.2 *as well as* many rules, stimulus discriminations, and decision-making principles that have been practiced to automaticity. I also include in TASS processes such as classical and operant conditioning which display more domain generality than the proposed modules listed in table 2.2. Finally, processes of behavioral regulation by the emotions are also in TASS (Johnson-Laird and Oatley 1992; Oatley 1992, 1998).

A metaphor used by evolutionary psychologists is quite useful, however, in emphasizing that TASS comprises a set of processes rather than a single system. The metaphor comes from Cosmides and Tooby's (1994b; Tooby and Cosmides 1992) famous characterization of the mind as a Swiss army knife. The purpose of this metaphor, as used by evolutionary psychologists, is to refute the notion that most human information processing is accomplished by general-purpose cognitive mechanisms: "The mind is probably more like a Swiss army knife than an all-purpose blade: competent in so many situations because it has a large number of components — bottle opener, cork-screw, knife, toothpick, scissors — each of which is well designed for solving a different problem" (Cosmides and Tooby 1994b,

Table 2.2 List of cognitive modules discussed in the psychological literature during the last two decades

face recognition module	intuitive number module
theory of mind module	naïve physics module
social exchange module	tool-use module
emotion perception module	folk biology module
social inference module	kin-oriented motivation module
friendship module	child care module
fear module	effort allocation and recalibration module
spatial relations module	semantic inference module
rigid objects mechanics module	grammar acquisition module
anticipatory motion module	communication pragmatics module
biomechanical motion module	

60). This metaphor captures well the multifarious nature of TASS and the domain specificity of some of the components within it.[11]

Although the Swiss army knife metaphor is useful in pointing out that there are many different processing mechanisms within TASS and that many of them have at least quasi-modular properties, there are two ways in which the dual-process view used in this book departs from the conception of the human mind advocated by some evolutionary psychologists. First, I do not think TASS subprocesses are necessarily all modular or quasi-modular. Along with the Darwinian mind of quasi-modules discussed by the evolutionary psychologists, TASS contains domain-general processes of unconscious learning and conditioning as well as automatic processes of action regulation via emotions (which respond to stimuli from very broad domains). A more important difference, however, is that the evolutionary psychologists wish to deny the need to posit a general-purpose central processor.[12] In contrast, just such a processor is the second key component in the dual-process view being developed in this chapter.

Characterizing the Analytic System:
Avoiding the Homunculus Problem

Perhaps the easiest way to characterize analytic system (System 2) processing is to say that it conjoins the converse set of properties that characterize TASS. TASS processes are parallel, automatic, operate largely beyond awareness, are relatively undemanding of computational capacity, and often utilize domain-specific information in their computations. Analytic processing, then, might be said to be characterized by serial processing, central executive control, conscious awareness, capacity-demanding operations, and domain generality in the information recruited to aid computation (see table 2.1). Such a strategy for defining the analytic system would be correct to a first order of approximation, but it would finesse some controversial issues.

The most difficult issue surrounding the analytic system is how to talk about it without making some very elementary philosophical mistakes or implying implausible models of brain functioning. Our natural language does not map easily into the concepts of cognitive science or the neurophysiological knowledge that we have of the brain. The recursiveness and self-referentiality that it takes to understand brain functioning at the higher (nonmodular) levels is not easily described. Communicative ease is often at war with factual accuracy when speaking of the higher brain systems.

An example of the difficulty is immediately apparent when one turns to use the single most popular metaphor in the psychological literature for analytic processing—that of a central executive processor. The undisciplined use of this metaphor can raise the specter of the so-called homunculus problem (the problem of the "little person in the head"), well-known to psychologists and philosophers. The problem is that if we explain a complex behavioral discrimination or choice by positing a hypothetical entity in the brain that is too complex, then we have simply imported the puzzle from external behavior to the internal behavior of a mechanism just as complex and puzzling as what it was originally called in to explain. We end up, for example, saying things like a person decided to do X because their executive processor decided to do X, and, obviously, this does not represent an advance in our understanding. We would simply be saying that a person does something because there is the equivalent of another little person—a homunculus—in their brain doing the deciding.

Such homuncular explanations explain nothing unless the homunculus—the proposed executive processor, for instance—is unpacked in terms of simpler psychological and neurophysiological processes that are more thoroughly understood and less mysterious.[13] The homunculus problem only arises when a proposed homunculus is too intelligent. If the posited complex entity has been decomposed into sufficiently simpler conceptual entities that are operationally identified by reliable methods in the arsenal of the cognitive psychologist or neurophysiologist, then the theorist is justified in freely using the entity. If the complex entity has undergone no decomposition, then the homunculus accusation has force.

Philosophers are constantly on the watch for psychologists sliding into the homunculus problem and into other conceptual errors because of the metaphors that they use to describe analytic processing. Many metaphors surrounding discussions of consciousness are likewise misleading and, like the homunculus problem, this is relevant to the present discussion, because analytic processing is often contrasted with TASS in terms of the former being conscious and the latter comprising processes that are not introspectable and that are beyond awareness. Daniel Dennett's book *Consciousness Explained* contains a variety of thought experiments designed to free us from a default assumption of Cartesian dualism (that there are two separate substances in the world—mind and matter), as well as exercises in rephrasing our inherently dualistic language concerning the mental. He particularly warns against using language that implies a so-called Cartesian Theater— somewhere where all brain activity "comes together" and is presented to a viewing "Central Meaner" (homunculus) whose understandings of what is displayed in the Theater becomes our consciousness.

As an explanation of a person's decision, it is easy (because of the terminology and metaphors that must be used to communicate) to slip into a view that amounts to saying that the Central Meaner (watching the conscious "show" on the theater screen "where it all comes together") can become what I will term a Promethean Controller and start making decisions and pulling levers so that the person will act on its choices. This is of course not a model that any scientist would propose, but it is perhaps one that might be induced by a lay reader unfamiliar with how some of the complex conceptual language is cashed out in terms of the psychology and neurophysiology of cognitive control (Baddeley 1996; Harnish 2002; Johnson-Laird 1988; Miyake and Shah 1999).

However, the reader is warned (for the last time here, I promise) that I will use some of the very metaphors (particularly those of executive control and system override) that philosophers view as dangerous. I do so because they are necessary for ease of communication and because plenty of evidence is cited in the notes that provides the conceptual and empirical grounding for the constructs I use to portray analytic system functioning.[14] Like most psychologists, I am a bit more comfortable with such central processor terms than are many philosophers. I think that the field has been properly inoculated with admonitions such as those given to the reader above, and that it has by now accumulated many positive and negative exemplars of central process concepts to build on and use as prototypes. Psychologists and neuropsychologists are perhaps more prone to risk the use of higher-level control language than are philosophers because the former need an efficient way of talking about experimental results and new experimental designs and they thus put more of a premium on ease of communication (which is quickly disrupted if the language of fully distributed systems is adopted).

While virtually all cognitive scientists would agree that the ideas of a Cartesian Theater or a Promethean Controller are fallacies, and would agree that control in the brain is distributed to some extent and not located in a single neural location, most would also agree with Pinker (1997) that:

> The society of mind is a wonderful metaphor, and I will use it with gusto when explaining the emotions. But the theory can be taken too far if it outlaws any system in the brain charged with giving the reins or the floor to one of the agents at a time. The agents of the brain might very well be organized hierarchically into nested subroutines with a set of master decision rules, a computational demon or agent or good-kind-of homunculus, sitting at the top of the chain of command. It would not be a ghost in the machine, just another set of if-then rules or a neural network that shunts control to the loudest, fastest, or strongest agent one level down. (144)

Pinker's view is closer to the view adopted here—that cognitive control is distributed in the brain but in ways that still justify a language of executive or central control.[15] Thus, when later in this chapter and subsequent chapters I use terms such as control by the analytic system, it is understood that the default mechanistic model hiding behind such terms entails no insoluble problems of a posited homunculus or Promethean Controller.[16]

One Step at a Time:
Figuring Out the Way the World Is with Language

Unlike the context-bound operation of many TASS subsystems (particularly those that are Darwinian modules), the analytic processing system allows us to sustain the powerful context-free mechanisms of logical thought, inference, abstraction, planning, decision-making, and cognitive control. An additional property that distinguishes analytic processing (System 2) from TASS is that of serial versus parallel processing. Because of the properties discussed previously (automaticity, ballistic firing, etc.), many different TASS subprocesses can execute simultaneously, whereas analytic processing seems to proceed one thought at a time.

Although the analytic system is a powerful mechanism for logical, symbolic thought, its decontextualizing cognitive styles are computationally expensive and difficult to sustain. Analytic cognition is "unnatural," and therefore rare, in this view, because it is not an architecture that is hard-wired in the brain, separate from TASS. Instead, in the view articulated in Daniel Dennett's book *Consciousness Explained*, analytic processing is carried out by a serial virtual machine that is simulated by the largely parallel brain hardware. A virtual machine is the set of instructions that run on the hardware of a digital computer ("a virtual machine is a temporary set of highly structured regularities imposed on the underlying hardware by a program . . . that give the hardware a huge, interlocking set of habits or dispositions to react," Dennett 1991, 216). In short, the analytic system is closer to software in this view (what Clark 2001, and Perkins 1995, term "mindware") than to a separate hardware architecture.

Dennett's (1991) model is related to an idea that has recurred several times in the cognitive science literature—that somehow the domain-specific outputs of TASS modules could be recruited to serve more general ends, thus increasing the flexibility of behavior.[17] However, the serial functions of the analytic system are hard to sustain because they are being simulated on hardware better suited to parallel functions such as pattern recognition.[18]

Such a view of the differences between TASS and the analytic system is consistent with a long-standing irony in the artificial intelligence literature: those things that are easy for humans to do (recognize faces, perceive three dimensional objects, understand language) are hard for computers, and the things it is hard for humans to do (use logic, reason with probabilities) computers can do easily. The current view of the differences between TASS and analytic processing removes all of the air of paradox about these artificial intelligence findings. Computers have built up no finely-honed TASS subsystems through hundreds of thousands of years of evolution, so the things that the massively parallel and efficient human TASS systems do well because of this evolutionary heritage computers find difficult. In contrast, the analytic system of humans—the serial processor necessary for logic—is a recent software addition to the brain, running as somewhat of a kludge (in computer science, an inelegant solution to a problem) on massively parallel hardware that was designed for something else. In contrast, computers were originally intentionally designed to be serial processors working according to the rules of logic (Dennett 1991, 212–14). It is no wonder that logic is easy for them and difficult for us.

Virtually all cognitive theorists agree that the analytic system is uniquely responsive to linguistic input, either external or internal in origin. Language as a self-stimulation mechanism introduced more seriality into information-processing sequences in the brain. It also appears to be a unique access medium for cognitive modules that would ordinarily have no access to each other's outputs. Thus, an additional important function of the serial simulator is to use language to forge new connections between isolated cognitive subsystems and memory locations.

Through language we can receive new mindware quickly and almost instantly install and begin running a new virtual machine (an installed rule structure that temporarily governs the information processing logic of the processor). We can thus easily install mindware discovered by others that proves to be useful. For example, in later chapters I will discuss how decision scientists have discovered numerous strategies to help people make better choices.[19]

The sequential structure of language also helps in overall cognitive control—sequencing and prioritizing of multiple goals. Philosopher Allan Gibbard (1990, 56–57) elaborates on this theme by stressing the motivational properties of language—its ability to rapidly reactivate goals that have become temporarily inert but are relevant to the current situation. Quick goal reprioritization can take place in response to verbal input. He discusses what will be a major theme of this book—that rapid goal reprioritization based on linguistic input (either internally or externally generated) may conflict with the goal prioritization inherent in TASS.

The systematicity and productivity of the rules that can be represented with a discrete representational system like language defines the critical property of the analytic system termed compositionality by cognitive scientists (Fodor and Pylyshyn 1988; Pinker 1997; Sloman 1996). Compositionality characterizes computational systems where the meaning of representations derives from the order of the parts of the representation in addition to the meaning of individual parts. As Pinker (1997) notes, "man bites dog" is news whereas "dog bites man" is not. The compositionality of language allows us to represent easily thoughts that are superficially similar but importantly different.

The analytic system is also the system responsible for building a narratively coherent description of the behavior engaged in by the individual. Recall that TASS will autonomously be responding to stimuli, entering processing products in working memory for further consideration, triggering actions on its own, or at least priming certain responses, thereby increasing their readiness. The analytic system tries to maintain a coherent story which explains all of this activity even though it did not initiate much of it. The analytic system has repeatedly been shown to confabulate explanations involving conscious choice for behaviors that were largely responses triggered unconsciously by TASS.[20] As I will discuss shortly, the tendency of the analytic system to give a confabulated explanation of behavior may impede cognitive reform that can only proceed if the autonomous nature of certain brain subsystems is acknowledged and taken into account. Our analytic systems can learn to give better narrative accounts of our behavior—ones more in accord with the neuropsychological facts. Learning this skill is part of the robot's rebellion.

Hypothetical Thinking and Representational Complexity

One of the functions of the analytical processing system is to support hypothetical thinking. Hypothetical reasoning involves representing possible states of the world rather than actual states of affairs, and it is involved in myriad reasoning tasks, from deductive reasoning, to decision-making, to scientific thinking.[21] For example, deductive reasoning involves hypotheticality when the premises are not things the individual knows but instead assumptions about the world; utilitarian or consequentialist decision-making involves representing possible future states of the world (necessarily not actual states) so that optimal actions can be chosen; and alternative hypotheses in scientific thinking are imagined causes from which consequences can be deduced for testing.

In order to reason hypothetically, a person must be able to represent a belief as separate from the world it is representing. Numerous cognitive scientists have discussed so-called decoupling skills—the mental abilities that allow us to mark a belief as a hypothetical state of the world rather than a real one (e.g., Cosmides and Tooby 2000a; Dienes and Perner 1999; Glenberg 1997; Leslie 1987; Lillard 2001; Perner 1991). Decoupling skills prevent our representations of the real world from becoming confused with representations of imaginary situations that we create on a temporary basis in order predict the effects of future actions or think about causal models of the world that are different from those we currently hold.

Decoupling—outside of certain domains such as behavioral prediction (so-called "theory of mind")—is a cognitively demanding operation. It is often carried out by the serial, capacity-demanding analytic system. Language provides the discrete representational medium that greatly enables hypotheticality to flourish as a culturally acquired mode of thought. For example, hypothetical thought involves representing assumptions, and linguistic forms such as conditionals provide a medium for such representations. The serial manipulation of this type of representation seems to be largely an analytic system function.

Decoupling processes enable one to distance oneself from representations of the world so that they can be reflected upon and potentially improved. Decoupled representations of actions about to be taken become representations of potential actions, but the latter must not infect the former while the mental simulation is being carried out. The decoupling operations must be continually in force during the simulation, and the computational expense of decoupling is probably one contributor to the serial nature of analytic cognition. The raw ability to run such mental simulations (independent of the facilitative mindware installed) while keeping the relevant representations decoupled is likely one aspect of the brain's computational power that is being assessed by measures of fluid intelligence (Baltes 1987; Fry and Hale 1996; Horn 1982).

Dienes and Perner (1999) emphasize the importance of the mental separation between facts in one's knowledge base from one's attitude toward those facts for cognitive control. For example, when considering an alternative goal state different from the current goal state, one needs to be able to represent both. To engage in these exercises of hypotheticality and cognitive control, one has to explicitly represent a psychological attitude toward the state of affairs as well as the state of affairs itself. The ability to distance ourselves from thoughts and try them out internally as models of the world makes human beings the supreme hypothesis testers in the animal kingdom.

Decoupling skills vary in their recursiveness and complexity. The skills discussed thus far are those that are necessary for creating what Perner (1991) calls secondary representations—the decoupled representations that are the multiple models of the world that enable hypothetical thought. At a certain level of development, decoupling becomes used for so-called meta-representation—thinking about thinking itself. Meta-representation—the representation of one's own representations—is what enables the self-critical stances that are a unique aspect of human cognition. We form beliefs about how well we are forming beliefs, just as we have desires about our desires, and possess the ability to desire to desire differently. Increases in representational complexity, and the concomitant increase in decoupling potential, are greatly fostered by the acquisition of language. The subtle use of these representational abilities in a program of cognitive reform is, as we shall see in chapters 7 and 8, an essential component of the robot's rebellion.

Hypothetical thinking is not confined to experts, academics, or scientists concerned with alternative hypotheses. It is a ubiquitous part of everyone's daily life. Importantly, developmental psychologist Paul Harris (2001) has pointed out that the ability to deal with hypotheticals becomes a critical cognitive requirement of most types of formal schooling. While children are not often explicitly set the task of reasoning hypothetically, such thinking is often implicit in much educational communication. That is, whether or not children are asked to engage in formal syllogistic reasoning (most often they are not), Harris points out that school inundates them with information that is novel or is outside of their world experience. The teacher then expects them to proceed to reason about this new information which, although factual to the teacher, is the equivalent of a hypothetical for the student.

Processing without Awareness: There Are Martians in Your Brain!

> There is a zombie within you that is capable of processing all the information your conscious self can process consciously, with one crucial difference, "All is dark inside": your zombie is unconscious. From this perspective then, cognition is inherently opaque and consciousness, when present, offers but a very incomplete and imperfect perspective on internal states of affairs.
>
> —Atkinson, Thomas, and Cleeremans (2000, 375)

Look again at the list of important TASS modules in table 2.2 that cognitive scientists have been studying for at least a decade (not, by any means, an exhaustive list). It is clear that often what we are thinking consciously about, what our analytic systems are dealing with, are inputs from the physical and social world that TASS modules have unconsciously placed at the analytic systems' disposal. Thus, many theorists have emphasized that TASS properties infuse much of our mental life (Cummins 1996; Evans and Over 1996; Hilton 1995; Levinson 1995; Reber 1993). Not only are TASS processes directly triggering responses on their own, but in cases where TASS processing does not lead directly to a response, it is providing the input to analytic system processing and thus biasing analytic processing by the nature of the cognitive representations given to it. If some of these TASS responses and products have untoward effects on our behavior, then we need to learn remediating analytic system strategies to counter them. That will be a topic for a later section. For the present I wish to emphasize the ubiquity and importance of TASS processing and, because it takes place beyond our conscious awareness, what a deeply spooky fact that really is.

The essence of TASS subprocesses is that they trigger whenever their appropriate stimuli are detected, that they cannot be selectively "turned off," and that they occur outside of our awareness. That such processes may even be priming responses that analytic processing deems infelicitous, means that, as the title of this chapter suggests, sometimes a person may have a brain that is, in an important sense, at war with itself. It may require some cognitive remediation if the outcome of this war is to be what the person's deepest, most reflective self wants. The first step in such a meliorist program of cognitive reform is to recognize that in some sense the "I" that we identify with (the homunculus that, although a fiction as was discussed above, is still part of our folk psychology) is not only not in control of all parts of the brain, but may be positively alienated from the operation of some brain activities that take place beyond its awareness.

The sense of alienation that we might feel if we truly understood the implications of the vast amount of brain activity that takes place outside of our conscious awareness—and the downright creepiness of it—was captured in an essay that served as the final summary of cognitive scientist Andy Clark's book titled *Being There* (1997). In this humorous but telling essay called "A Brain Speaks" Clark summarizes all of the themes in his book by having a character called John's brain address John, the self caused by the brain's activity. The brain is concerned with straightening out all the misconceptions that John holds about his brain's activity. The brain admits that John and he are on rather intimate terms, but that John tends to take this intimacy too far. For example, John likes to think that all his thoughts are his brain's thoughts and that all his brain's thoughts are his thoughts. The brain assures us that things are rather more complicated than that.

The brain makes the point elaborated in this chapter, that "John is congenitally blind to the bulk of my daily activities" (1997, 223). The brain gradually breaks the news to John that, not only are his perceptual and vegetative functions directed by brain processes beyond his control, but much of his deep conceptual processing is also not something that he directs with his conscious mind. The brain says that despite John's belief that he, the Promethean Controller, is in charge and directs the brain's activities, John in fact "is apprised of only the bare minimum of knowledge about my inner activities" (223). Consistent with my portrayal of TASS in this book, John's brain tells him that the bulk of his brain's activities are carried out by many parallel, independent computational channels. Only a tiny fraction of the outputs from these become the conscious focus of John's analytic activities.

Because of this, the brain informs John that he really has it backwards. John thinks that "he" controls his brain's activities (really, John's folk psychology carries more than a hint of Cartesian dualism). Instead, the brain tells him that "I am not the inner echo of John's conceptualizations. Rather, I am their somewhat alien source" (225). Using the vocabulary I have developed here, the brain is telling John that TASS provides the critical input to John's analytic processes and that, despite John's contrary introspections, he does not control everything that is input into his conscious reasoning and decision processes.

John's brain tries to explain to John that John's perspective on his whole brain is heavily biased by his reliance on language and the concepts that language provides. The brain is pained by John's lack of understanding, and the brain becomes a bit frustrated that John continues to "hallucinate his own perspective onto me" (226; a reference to the confabulation tendency of analytic processing mentioned above and discussed again below). The brain laments that John seems blissfully unaware of information processing and information storage operations that do not match the forms of his language-based cognition. John's conceptualizing tendencies are so language-saturated that he lacks the tools to think about modes of processing alien to his perspective. Cognitive science continues to develop such conceptualizing tools through its exploration of parallel connectionist architectures and dynamic systems models, but the conceptual tools resulting from these efforts have yet to enter the folk psychologies of people such as John.

The frustrated brain finally reaches the conclusion that trying to explain things to John seems futile because John consistently "forgets that I am in large part a survival-oriented device that greatly predates the emergence of linguistic abilities, and that my role in promoting conscious and linguaform cognition is just a recent sideline. . . . Despite our intimacy, John really knows very little about me. Think of me as the Martian in John's head" (227).

We all have Martians in our heads just like John. We have a plethora of TASS subsystems going about their business without our input or awareness (without analytic system input, to be specific). The cognitive science literature is simply bursting at the seams with demonstrations that we do complex information processing without being aware of it, and that there are plenty of Martian-like subsystems in our brains—not limited to perceptual or visceral functions, but including conceptual functions as well.

One example that appears in virtually every textbook of cognitive science and neuropsychology (e.g., Clark 2001; Harnish 2002; Parkin 1996) is that of the phenomenon of so-called blindsight (Marcel 1988; Weiskrantz 1986, 1995). Certain patients who have sustained damage in their visual cortex display a seemingly puzzling set of symptoms. They develop a blind spot or scotoma in their field of vision—they report seeing nothing in a particular portion of their visual field. However, when persuaded to make a forced choice about a fixed set of stimuli (for example, to choose one of two shapes or lights) presented into their blind field, they perform with greater than chance accuracy despite their phenomenal experience of seeing nothing. For example, when choosing between two stimuli, their choices are 70 percent correct despite the fact that on each trial they insist that they see nothing. Often such patients have to be persuaded to continue to keep making the forced choices which they view as pointless. Many report that they are simply guessing, that they "couldn't see a darn thing," and question what the experimenter could possibly find out from such a pointless exercise.

The details of the explanation of the blindsight phenomenon remain controversial—the retina sends information to various areas of the brain. However, there is little controversy over the most generic implication of the finding. Some portion of the brain can process visual stimulation to some extent in these patients, but the brain systems that collate the information that results in verbal reports of conscious experience fail to reach threshold.

It should not be thought for a second that the processing without awareness exhibited by blindsight patients is unique to people with brain damage. Processing without awareness has been demonstrated in psychophysical experiments conducted by perceptual psychologists over several decades using subjects with perfectly normal brains and perceptual systems.[22] It is a ubiquitous aspect of normal cognitive activity. The experiments differ in many technical details, but the experimental situation might be somewhat as follows. A subject is looking into a tachistoscope (a device for presenting visual stimuli for very short durations on the order of thousandths of second) or a computer display seeing one of the four letters A, B, C, D flashed at them trial after trial. Gradually, the experimenter lowers the exposure duration, making the letters hard to see in an attempt to measure the recognition threshold. As the exposure duration is lowered, the accuracy of identifying the correct letter gets lower and lower, dropping, for example, from 100 percent accuracy to 90 percent to 75 percent. At some exposure duration, subjects often say that they cannot continue because they are no longer seeing well enough to discriminate the letters. However, at the point that subjects protest that they are no longer seeing anything distinctive, they are often well above chance at reporting the letters (for example, they may be at 45 percent accuracy with four letters—well above the 25 percent guessing rate). The greater than chance accuracy demonstrates that the subjects were in fact discriminating some information, yet they insisted that it was futile for the experimenter to require them to continue because they were simply guessing at random. Just like the blindsight subjects, these experiments demonstrate that normal subjects can show evidence of having processed a stimulus of which they are not aware.

This phenomenon in normal subjects can be extended in ways that show the unconsciously processed information to have reverberating effects throughout the brain—including effects on semantic-level processing. Cognitive psychologists have extensively studied the so-called semantic priming effect—the fact that the processing of a word is facilitated if it is preceded shortly in time by a semantically related word. For example, the processing of the target word "nurse" is facilitated (as measured by reaction time, electrophysiological recordings, or other techniques) if preceded briefly in time by the prime word "doctor." What is fascinating is that this semantic facilitation of word processing occurs even when the prime word is flashed so briefly that the subject is unaware of the identity of the prime. As in the previously described experiments, the subjects in these priming experiments report that the prime was unidentifiable to them, yet the semantic relationship between the prime and target word nonetheless affects their behavior.

It is not just peripheral processes that display autonomy, but much conceptual processing occurs automatically as well. Because TASS conceptual processes provide inputs to the analytic processor, when deep conceptual processing takes place beyond awareness, the origins of these inputs will be unavailable to consciousness. For example, much recent evidence indicates that stereotyping of social and cultural groups occurs through processes of unconscious activation, and is not entirely the result of conscious reasoning (Brauer, Wasel, and Niedenthal 2000; Frank and Gilovich 1988; Greenwald and Banaji 1995; Greenwald et al. 2002).

Important overt behaviors can be affected by conceptual associations that are automatically triggered by TASS. Decades ago, in a classic and much-cited article, Nisbett and Wilson (1977) summarized much evidence indicating that automatic conceptual processing takes place and that people are therefore often unaware of causes of their behavior. As a result, when the analytic system (the processing system that maintains a global model of the causes and consequences of the person's behavior) must interpret action in such situations, it often confabulates reasons for the action because the *actual* causes operated through TASS subsystems that were cognitively impenetrable.

Nisbett and Wilson (1977) discussed numerous examples of such effects, and the literature has continued to proliferate since their review. The results of a so-called attribution experiment conducted by Nisbett and Schachter (1966) were typical. In that study, subjects received electric shocks of increasing amperage to see how much pain they could tolerate. One group was given a placebo pill (sugar pill) and told that it would produce symptoms including breathing irregularities, heart palpitations, and butterflies in the stomach. The investigators hypothesized that the placebo group would attribute any nervous symptoms they felt (different breathing, sweating, queasiness) to the pill rather than to anxiety over the shock and thus would be prone to tolerate more shock. This was indeed the case. Compared to a control group who did not receive the placebo pill, the placebo group tolerated shocks of four times the amperage!

Interestingly though, when queried about the reasons for their ability to tolerate a substantial shock, subjects in the placebo group never mentioned the influence of the pill. Their behavior was profoundly affected by a factor of which they were unaware. For example, when queried about their high shock tolerance, a typical response was "Gee, I don't really know. . . . Well, I used to build radios and stuff when I was 13 or 14, and maybe I got used to electric shock" (Nisbett and Wilson 1977, 237). When directly probed with the question of whether they ever thought about the pill during the experiment, the typical response was: "No, I was too worried about the shock"; and when asked directly whether it occurred to them that the pill was causing various physical effects, the typical response was: "No, like I said, I was too busy worrying about the shock" (237). When subjects were presented with the experimental hypothesis and asked whether they thought it was plausible, the typical response was that yes, the subject thought it very plausible and that the behavior of many *other* subjects was probably affected in this manner—but that their own was not!

This inability to access the actual brain processes and stimuli that caused one's own behavior recurs in many other experiments and situations reviewed by Nisbett and Wilson (1977). In one experiment that they describe, subjects watched films and rated each film on a variety of dimensions. Several factors were varied experimentally such as the visual focus of the picture and noise occurring outside of the viewing theater. Some factors affected the subjects' experiences and ratings of the film and others did not. Noises in the hall in fact did not affect the subjects' ratings of the films, but 55 percent of the subjects mistakenly thought that their ratings had been affected by the noise. In contrast to the shock experiment I have previously described where subjects are unaware of a factor that influenced their behavior, here they reported as influential a variable that in fact had no effect.

The film experiment illustrates most clearly the confabulation potential which Nisbett and Wilson (1977) highlighted by titling their classic article "Telling More Than We Can Know: Verbal Reports on Mental Processes." Telling more than we can know refers to our tendency to impose an explanation even on behavior and brain action cognitively impenetrable to introspection (because they are outputs of TASS modules). In the film experiment described by Nisbett and Wilson, the analytic system simply did not have access to all the factors that affected the online liking of the film. These factors are myriad and they are supplied by many TASS modules that are cognitively impenetrable. Nevertheless, the analytic system has no trouble coming up with a plausible model of why the behavior occurred. However, this model is based on a generic folk psychology of why people in general do what they do rather than on privileged knowledge of the inner cognitive processes that were actually responsible

The confabulatory tendencies of the analytic system are startlingly illustrated in the classic experiments on the so-called split-brain patients studied by neuroscientist Michael Gazzaniga and his colleagues.[23] Split-brain patients have undergone commissurotomy, the severing of the corpus callosum (the most massive set of connections between the two hemispheres of the brain). Taking advantage of the fact that the right visual field is projected to the left hemisphere of the brain and vice versa (meaning that stimuli can easily be presented exclusively to one hemisphere or the other in a split-brain patient), Gazzaniga explored the differential processing capabilities of the two hemispheres and, in the process, discovered the extent of the confabulatory properties of the left hemisphere of the brain—the hemisphere that controls language production.

In a now famous experiment, Gazzaniga flashed a picture of a chicken claw to the left hemisphere of a split-brain patient and a picture of a snow scene to the right hemisphere of the same patient. From an array of pictures, the right hand (contralaterally connected to the left hemisphere) correctly picked out the picture (a chicken) most closely associated with the picture flashed to the left hemisphere, and the left hand (contralaterally connected to the right hemisphere) correctly picked out the picture (a snow shovel) most closely associated with the picture flashed to the right hemisphere. However, when asked why those two pictures were chosen, the left hemisphere (the only one which could speak) answered "Oh, that's simple. The chicken claw goes with the chicken, and you need a shovel to clean out the chicken shed" (Gazzaniga 1998b, 25). The subject's left hemisphere did not have access to the stimulus picture of the snow scene. But what Gazzaniga terms the interpreter in the left hemisphere could see the left hand (mute right hemisphere) pointing to the snow shovel. The interpreter thus concocted an explanation that integrated the two choices into one coherent story.

Gazzaniga emphasizes the ubiquity of this interpretive tendency by stressing that the subject does not respond that he doesn't know why he picked the snow shovel. He concocts a narrative in which he has made a conscious choice rather than acknowledging that he is unaware of the causes of his own behavior. Gazzaniga asks us to note how telling it is that the subject does not respond in the following, eminently reasonable way: "Look, I have no idea why I picked the shovel—I had my brain split, don't you remember? You probably presented something to the half of my brain that doesn't talk; this happens to me all the time. You know I can't tell you why I picked the shovel. Quit asking me this stupid question" (1998a, 25). But the left hemisphere does no such thing. It concocts its narrative under the assumption that it is in complete control of everything that the body did. Just as the subjects described by Nisbett and Wilson concocted explanations of why they liked a film based on generic folk psychology rather than on a knowledge of internal brain processes (which, because of their cognitive impenetrability, are not available as sources of information), the split-brain subjects concocted a narrative designed to be a coherent story rather than an accurate reflection of the internal processes involved.

Some individuals experience mental illness because, due to brain damage, their interpreters begin to construct bizarre narratives in response to the very unusual input they are receiving because of malfunctioning TASS modules. This phenomenon is illustrated in Capgras syndrome.[24] In Capgras syndrome, the individual begins to believe that a close relative of his (such as a parent) is an impostor. Patients have been known to attack and kill parents or spouses who they thought were impostors trying to trick them. This syndrome arises because the neural systems tapped by autonomic and overt indices of recognition are different. The brain damage received by Capgras patients has left intact enough of their face recognition systems so that they can recognize their relatives. However, the systems subserving the emotional connections to these faces have been disturbed. As a result, they recognize the face of their close relative but do not experience the usual emotional response that is associated with that relative. Recognizing the familiar face without appropriate emotional reaction becomes an anomalous experience in need of an explanation by the interpreter. In some individuals with (probably preexisting) attributional biases, reasoning biases, and belief perseverance, the interpreter jumps quickly to an extreme hypothesis (the anomalous experiences represent impostors) and engages in biased processing in order to maintain this hypothesis. Just like Gazzaniga's split-brain subject in the chicken/snow shovel experiment, the Capgras patient fails to entertain the hypothesis that the information that the analytic system is dealing with is faulty because of brain damage. Both the split-brain patients and Capgras patients persist in the belief that they *know* what is going on in their own heads, when in fact they do not have access to the TASS subsystems whose functioning has gone awry.

The failure of folk psychology to acknowledge the influence of autonomous systems in our brains can have disastrous consequences for non-brain-damaged individuals as well. Loewenstein (1996) discusses how it appears to be a major factor in leading people into drug addiction. In a review article on so-called visceral influences on behavior (drive states with direct hedonic impact such as hunger, pain, and sexual desire), he argues that most people underestimate the effect of future visceral responses because they overestimate their ability to directly control these effects. Since indulging one's curiosity is a major contributor to early drug use (Goldstein 1994), Loewenstein argues that believing one can stop taking the drug is a major factor in people's decision to start. But the model of their own brain's control capabilities which forms the basis for their decision to start is tragically incorrect. Their folk model greatly exaggerates the extent of conscious control over autonomous visceral processes.

These, then, are some examples of situations in which we must accustom ourselves to our TASS subsystems seeming like Martians. TASS is an evolutionarily older system that sometimes primes outputs not appropriate in the modern world. A major theme of this book will be that modern life increasingly creates situations of this type—situations where we must call on the evaluative and supervisory functions of the analytic system to overcome habitual responses that are no longer serving our needs.

As anthropologist Don Symons (1992) points out, we have specialized gustatory mechanisms underlying our preference for sweetness. These were evolutionary adaptations probably shaped by the fact that fruit is most nutritious when its sugar content is highest. Today, in our modern industrialized societies, sweet foods are all around us, and our preference for them can be quite dysfunctional. But TASS subsystems priming our preference for sweet foods are still in there firing away despite our conscious resolutions to diet. None of this is to deny that most TASS subsystems are still extremely useful in our daily lives (the importance of the processes listed in table 2.2 is unquestioned); it is just that my focus in this book is to be cognitive reform—and that necessitates a critique of some TASS functions.

When our TASS subsystems prime responses that, because of their evolutionarily ancient origins, are startlingly inappropriate to our current situation, it makes us feel as though we have Martians in our brains. When we are dieting, the constant craving for sweets is annoying. Or, to take the examples with which I opened this chapter, the rage we feel when cut off

by another automobile is most of the time, we know, totally disproportionate to the situation. Evolutionary links in TASS subprocesses are even
clearer in some of the other examples. The shameful rejection of rape victims by their spouses is rooted in evolutionary modules for male sexual
proprietariness (Wilson and Daly 1992). The rejection of facially disfigured
people is rooted in evolutionary modules for detecting symmetry and other
so-called beauty cues which are simply proxies for reproductive fitness (Buss
1989; Langlois, Kalakanis, Rubenstein, Larson, Hallam, and Smott 2000;
Symons 1992).

Do we want to endorse these TASS inputs to our analytic systems? No.
We want to override them. Do we consider them justified because of their
evolutionary origins? Clearly not. If anything, we want to overcome such
tendencies. Thus, some of the response tendencies primed by TASS will need
to be overridden if they are not to overwhelm our reflective values, as determined by our analytic processing activities. The triggers in our brains for eating sweets, for rejecting a raped spouse, for feeling revulsion at the sight of
a disfigured person, are parts of ourselves with which we do not want to
identify.[25] They are alien to our considered selves. They are the Martians in
our brains. We cannot remove them, but we can dampen their effects and
find ways for the analytic system to trump them, so that we don't end up
feeling like Bob Marley in the epigram that opened the chapter—that "reflexes got the better of us."

When the Different Kinds of Minds Conflict:
The Override Function of the Analytic System

Although warning about the potential problems of the reflex-like TASS processes in the previous section, I do not mean to imply by the emphasis on
dysfunctional TASS processes that TASS is always problematic. To the contrary, I assume—along with many other theorists[26]—that a host of useful
information processing operations and adaptive behaviors are carried out
automatically by TASS (depth perception, face recognition, frequency estimation, language comprehension, the attribution of intentionality, cheater
detection, color perception, etc.). Indeed, because this list is so extensive,
many theorists have been led to stress how much of mental life is infused by
TASS outputs. But as we saw in the previous section, TASS outputs can sometimes conflict with the superordinate behavioral goals determined by analytic decision-making. The supervisory and evaluative functions of the
analytic system must thus sometimes be deployed to damp down or override TASS outputs that are conflicting too much with more global goals.

The logic of TASS subsystems, and their origins as domain-specific evolutionary adaptations or as highly practiced stimulus-output relations that trigger automatically, means that it is likely that one computational task of the analytic system is to serve as a supervisory system that decouples or overrides TASS outputs when they threaten to prime responses in conflict with superordinate goals (Navon 1989; Norman and Shallice 1986). Of course, in most cases, the systems will interact in concert and no decoupling or overrides will be necessary. Nevertheless, a potential override situation might occur when analytic processing detects that TASS processes—although working well for their circumscribed problems—are thwarting more global goals and desires. This will happen occasionally because many TASS processes are adapted to a criterion—genetic fitness—different from utility maximization at the level of the individual (discussed in chapter 3). This override function might only be needed in a tiny minority of information-processing situations, but these situations may be unusually important ones (as we shall see in chapter 4).

The work of Pollock (1991, 1995), one of the dual-process theorists listed in table 2.1, illustrates how analytic processing functions as an override system for some of the automatic and obligatory computational results provided by TASS. His dual-process view derives from the perspective of investigators attempting to implement intelligence and rationality in computers. In Pollock's terminology, TASS is composed of quick and inflexible (Q&I) modules that perform specific computations. Analytic processes are grouped under the term intellection in his model. Consistent with the generic two-process view that I am sketching in this volume, Pollock stresses that an important function of analytic processing is to override TASS when a Q&I module has fired in an environment in which its rigid response execution is not well adapted—a changed environment which the Q&I module has no way of coping with "on the fly" because its speed derives from the rigidity with which it reacts.

As an example, Pollock (1991) mentions the Q&I trajectory module that predicts the movement path of objects in motion. The Q&I module for this computation operates rapidly and accurately, but it relies on certain assumptions about the structure of the world. When these assumptions are violated, then the Q&I module must be overridden. So, for example, when a baseball approaches a telephone pole we had better override our automatic trajectory module because it is not going to properly compute the trajectory of the bounce from an irregularly curved surface. Likewise, conceptual and emotional[27] signals from TASS sometimes need to be overridden. The override function relates to the distinction in the last chapter between long-leash and short-leash genetic control. A conflict between TASS and the analytic system often represents a conflict between these two types of control.[28]

The Brain on a Long Leash and the Brain on a Short Leash

The circumstances in which the outputs of TASS and the analytic system conflict will tend to be instances where flexible analytic processes detect broader goals at stake than those achieved by the response primed by TASS. Recall that TASS is composed of older evolutionary structures (Evans and Over 1996; Mithen 1996, 2002; Reber 1992a, 1992b, 1993) that more directly code the goals of the genes (reproductive success); whereas the goal structure of the analytic system—a more recently evolved brain capability—is more flexible and on an ongoing basis attempts to coordinate the goals of the broader social environment with the more domain-specific short-leash goals of TASS (in chapters 7 and 8 I shall discuss the important complication that longer-leashed goals can enter TASS through overpractice).

To emphasize that TASS encompasses the evolutionarily older parts of the brain, Dennett (1991, 178, citing Humphrey 1993) calls them the "scram!" or "go for it!" parts of the brain (a humorous terminology that signals origins in ancient evolutionary times when our psychological structures were simple and behavioral regulation quite crude). In our present environment, however, although TASS may well be automatically signaling a male to mate with a glimpsed female ("go for it!"), the analytic system correctly registers that the man is living in a complex technological society in the twenty-first century and that considerations such as spouse, children, job, and societal position dictate that this particular "go for it!" signal from TASS be overridden. Instead, the analytic system has coordinated the individual's total goal structure and has calculated that the individual's entire set of long-term life goals are better served by overriding this TASS-triggered response tendency even though the latter might result in actions with temporary positive utility.

In this example, I have contrasted the slow, multidimensional computation of the analytic system with TASS functioning where outputs are emitted ballistically, reflexively, and in response to a narrow set of stimuli because such an automatic response served the genes' interests long ago in the evolutionary history of humans. The last statement reflects my reliance on the theoretical work done by Reber (1992a, 1992b, 1993), who in an important series of publications has reviewed evidence indicating that TASS is an older evolutionary system. Building on his claim, and incorporating the logic of short-leashed and long-leashed control discussed in chapter 1, I have proposed (Stanovich 1999; Stanovich and West 2000) that the goal structures of TASS and the analytic system are different, and that important consequences for human self-fulfillment follow from this fact.

The goal structure of TASS has been shaped by evolution to closely track increases in the reproduction probability of genes. The analytic system is primarily a control system focused on the interests of the whole person. It is the primary maximizer of an individual's *personal* goal satisfaction. Maximizing the latter will occasionally result in sacrificing genetic fitness (Barkow 1989; Cooper 1989; Skyrms 1996). Thus, the last difference between TASS and the analytic system listed in table 2.1 is that TASS instantiates short-leashed genetic goals, whereas the analytic system instantiates a flexible goal hierarchy that is oriented toward maximizing goal satisfaction at the level of the whole organism. Because the analytic system is more attuned to the person's needs as a coherent organism than is TASS (which is more directly tuned to the ancient reproductive goals of the subpersonal replicators), in the minority of cases where the outputs of the two systems conflict, people will often be better off if they can accomplish an analytic system override of the TASS-triggered output. Such a system conflict is likely to be signaling a vehicle/replicator goal mismatch and, statistically, such a mismatch is more likely to be resolved in favor of the vehicle (which all of us should want) if the TASS output is overridden.

A graphic representation of the logic of the situation which illustrates why override is a statistically good bet is displayed in figure 2.2 (of course, the exact size of the areas of overlap are mere guesses; it is only the relative proportions that are necessary to sustain the argument here). First, an assumption reflected in the figure is that within both TASS and the analytic system vehicle and gene goals coincide in the vast majority of real-life situations (the areas labeled B and E). For example, accurately navigating around objects in the natural world fostered evolutionary adaptation—and it likewise serves our personal goals as we carry out our lives in the modern world.

But the most important feature of figure 2.2 is that it illustrates the asymmetries in the interests served by the goal distributions of the two systems. The remnants of the Darwinian creature structure described in chapter 1 (see figure 1.1) are present in the TASS brain structures of humans. Many of the goals instantiated in this system were acquired nonreflectively—they have not undergone an evaluation in terms of whether they served the *person's* interests. They have in fact been evaluated, but by a dif-

Goal structure

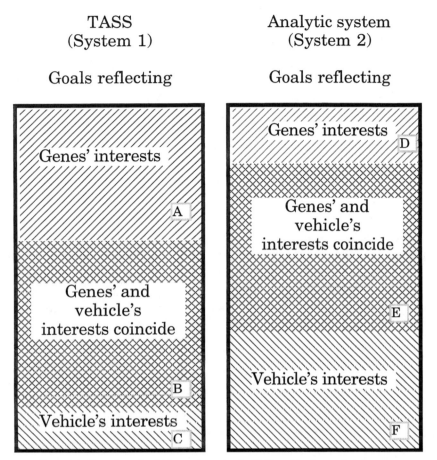

Figure 2.2 Genetic and vehicle goal overlap in TASS and in the analytic system. Both systems have three types of goals, but in different proportions: goals serving both the genes' and vehicle's interests (areas B and E), goals serving only the genes' interests (areas A and D), and goals serving only the vehicle's interests (areas C and F)

ferent set of criteria entirely: whether they enhanced the longevity and fecundity of the replicators in the evolutionary past. From the standpoint of the individual person (the vehicle) these can become dangerous goals when they reflect genetic goals only.[29] They are the goals that sacrifice the vehicle to the interests of replicators—the ones that lead the bee to sacrifice itself for its genetically related hive queen. They are represented by the area labeled A in figure 2.2.

These are the goals that should be strong candidates for override when their pursuit is triggered by TASS. Pinker (1997) notes that TASS subsystems are "designed to propagate copies of the genes that built them rather than to promote happiness, wisdom, or moral values. We often call an act 'emotional' when it is harmful to the social group, damaging to the actor's happiness in the long run, uncontrollable and impervious to persuasion, or a product of self-delusion. Sad to say, these outcomes are not malfunctions but precisely what we would expect from well-engineered emotions" (370). Pinker uses the term well-engineered here in a very special sense—a deliberately special sense, because he is trying to shock the reader into a new perspective (the perspective of chapter 1—where I was trying to do the same thing). We have to think twice to realize that he means well engineered *by evolution so that they serve the interests of the replicators.* But of course, from the standpoint of the vehicle, this "good engineering" can be like the good engineering in our overpowered automobiles—the kind that takes us with a tiny tap on the accelerator from 20 mph to 45 mph in a school zone. If not overridden by a cognitive system that takes into account the long-term goals of the driver, this "good engineering" leads us astray. This type of efficient engineering is myopic—it will do its job just as automatically and efficiently in a globally inappropriate situation as in an appropriate one.

The right side of figure 2.2 indicates the goal structure of the analytic system. Through its exercise of a reflective intelligence this system derives, from its interaction with the world, flexible long-leash goals that often serve the overall ends of the organism but thwart the goals of the genes (area F in figure 2.2—for example, sex with contraception; resource use after the reproductive years have ended; etc.). Of course, a reflectively acquired goal— one that was reflectively acquired because it served vehicle ends (perhaps even vehicle ends that thwart the genes' interests)—can, if habitually invoked, become part of TASS as well. This fact explains a part of figure 2.2 that might have seemed perplexing when that figure was first presented. Why is there a small section of area (area C) in TASS representing goals that serve the vehicle's interests only? One might have thought that all of the goals instantiated in TASS would reflect the genes' interests whether or not they were serving the interests of the vehicle—rather like that of the Darwinian creature represented in figure 1.1 of chapter 1. However, the possibility of the higher-level goal states of the analytic system becoming installed in the more rigid and inflexible TASS through practice opens up a new possibility. Reflectively acquired goal-states might be taken on for their unique advantages to the vehicle (advantages that might accrue because they trump contrary gene-installed goals—"don't flirt with your boss's wife") and then may become instantiated in TASS through practice. We might say that in situations such as this, TASS in humans reflects the consequences of residing in a brain along with a reflective analytic system. This is why the goal-structure of TASS in humans does not simply recapitulate the structure of a Darwinian creature depicted in figure 1.1.

Nevertheless, with the small but important exception of area C, TASS can be understood, roughly, as the part of the brain on a short genetic leash. It still responds automatically to cues in ways that resulted rather directly in gene replication thousands of years ago in our evolutionary history. As in the Mars explorer example, as the world became more complex and harder to predict in advance (especially as the result of it containing other people), the genes added on more complex and longer-leashed systems of control. In addition to building in very domain-specific stimulus-response tendencies, they installed in the brain broad and general goals that are correlated with replication (self-preservation tendencies, sex feels good, fat tastes good) while simultaneously building a hierarchical goal analyzer with the ability to coordinate a complex set of (possibly conflicting) goals and calculate a maximizing strategy in the midst of an environment where the contingencies are in constant flux.

When humans live in complex societies, most of the goals that the analytic system is trying to coordinate are derived goals. No one in industrialized societies is out hunting and gathering anymore. Basic goals and primary drives (bodily pleasure, safety, sustenance) are satisfied indirectly by maximizing secondary symbolic goals such as prestige, status, employment, and remuneration. In order to achieve many of these secondary goals, the more directly coded TASS-triggered responses must be suppressed—at least temporarily. Long-leashed derived goals create the conditions for a separation between the goals of evolutionary adaptation and the interests of the vehicle. In the extreme, vehicles can decouple short-leash control entirely and rebel against the selfish replicators by attaining general goals that do not serve the specific ends of the replicators at all (e.g., sex with contraception).

The analytic system is the part of the brain closest to the situation described in chapter 1, where the genes gave up direct control and instead said (metaphorically, by the types of phenotypic effects that they created) "things will be changing too fast out there, brain, for us to tell you exactly what to do—you just go ahead and do what you think is best given the general goals (survival, sexual reproduction) that we have inserted." The analytic system is the closest evolution has gotten in this trend, the logical end of which, as Dawkins (1976) notes "would be for the genes to give the survival machine a single overall policy instruction: do whatever you think best to keep us alive" (59–60). But interestingly, in humans, the analytic system resides in the brain along with the reflexive set of systems (TASS). Because of its properties of automaticity, TASS will often provide an output relevant to a problem in which the analytic system is engaged. When the outputs of these two systems directly conflict, one or the other of the two systems will lose out. We especially do not want the analytic system to lose out in cases where TASS is optimizing for the genes based on a long-lost ancestral environment and the analytic system is optimizing for the individual based on the actual environment in which we now live. When the analytic system fails to override TASS in such cases, the outcome can be a sad one—as in the examples that opened this chapter.

Of course, the idea of conflicts within the human mind is not new. It has been the subject of great literature for centuries. However, now we have a better vocabulary with which to describe these conflicts—a vocabulary that is validated by the findings of cognitive neuroscience. Nonetheless, writers can often give us the most visceral feel for these conflicts. No one has ever described the conflict between TASS and the analytic system as compellingly as George Orwell (1950) in his famous essay "Shooting an Elephant." Orwell describes how, as a police officer representing the British Empire in Burma in the 1930s, on an analytic basis, he had come to abhor the imperialism that his position represented ("I had already made up my mind that imperialism was an evil thing and the sooner I chucked up my job and got out of it the better" [3]). However, despite believing that imperialism was an evil, Orwell could not stop feeling annoyed when subjected to jeering from the populace that he was policing: "The insults hooted after me when I was at a safe distance, got badly on my nerves. The young Buddhist priests were the worst of all. There were several thousands of them in the town and none of them seemed to have anything to do except stand on street corners and jeer at Europeans" (3).

In our modern vocabulary of mental terms, we would say that Orwell's TASS system reacted to the jeering despite his analytic system realizing that the reaction did not reflect his true feelings—his analytic system knew that the jeering was justified. Without sharing this modern terminology, Orwell was quite conscious of the monstrous conflict between the two systems: "All I knew was that I was stuck between my hatred of the empire I served and my rage against the evil-spirited little beasts who tried to make my job impossible. With one part of my mind I thought of the British Raj as unbreakable tyranny . . . with another part I thought that the greatest joy in the world would be to drive a bayonet into a Buddhist priest's guts" (4).

Try It Yourself—Can You Override TASS in the Famous Four-Card Selection Task and the Famous Linda Task?

You can demonstrate for yourself that you really do have different systems in your brain operating simultaneously (and potentially in conflict) to influence your behavior by trying to answer the single most-studied problem in decades of research on human reasoning. The task was invented by Peter Wason (1966, 1968), and has been investigated in literally dozens of studies.[30] Try to answer it before reading ahead:

> Each of the boxes below represents a card lying on a table. Each one of the cards has a letter on one side and a number on the other side. Here is a rule: If a card has a vowel on its letter side, then it has an even number on its number side. As you can see, two of the cards are letter-side up, and two of the cards are number-side up. Your task is to decide which card or cards must be turned over in order to find out whether the rule is true or false. Indicate which cards must be turned over.

Before discussing this problem further (and to make it a bit harder to look ahead!) consider another problem that is famous in the literature of cognitive psychology, the so-called Linda problem (Tversky and Kahneman 1983):

> Linda is 31 years old, single, outspoken, and very bright. She majored in philosophy. As a student, she was deeply concerned with issues of discrimination and social justice, and also participated in anti-nuclear demonstrations. Please rank the following statements by their probability, using 1 for the most probable and 8 for the least probable.
>
> a. Linda is a teacher in an elementary school ____
> b. Linda works in a bookstore and takes Yoga classes ____
> c. Linda is active in the feminist movement ____
> d. Linda is a psychiatric social worker ____

    e. Linda is a member of the League of Women Voters ____

    f. Linda is a bank teller ____

    g. Linda is an insurance salesperson ____

    h. Linda is a bank teller and is active in the feminist movement ____

Now to each of the problems in turn. The first is called the four-card selection task and has been intensively investigated for primarily two reasons—most people get the problem wrong and it has been devilishly hard to figure out why. The answer seems obvious. The hypothesized rule is: If a card has a vowel on its letter side, then it has an even number on its number side. So the answer would seem to be to pick the A and the 8—the A, the vowel, to see if there is an even number on its back, and the 8 (the even number) to see if there is a vowel on the back. The problem is that this answer—given by about 50 percent of the people completing the problem—is wrong! The second most common answer, to turn over the A card only (to see if there is a vowel on the back)—given by about 20 percent of the responders—is also wrong! Another 20 percent of the responders turn over other combinations (e.g., K and 8) that are also not correct.

If you were like 90 percent of the people who have completed this problem in dozens of studies during the past three decades, you answered it incorrectly too (and in your case, you even missed it despite the hint given in the title of this section to suppress your most immediate response!). Let's see how most people go wrong. First, where they don't go wrong is on the K and A cards. Most people don't choose the K and they do choose the A. Because the rule says nothing about what should be on the backs of consonants, the K is irrelevant to the rule. The A is not. It could have an even or odd number on the back, and although the former would be consistent with the rule, the latter is the critical potential outcome—it could prove that the rule is false. In short, in order to show that the rule is not false, the A must be turned. That is the part that most people get right.

However, it is the 8 and 5 that are the hard cards. Many people get these two cards wrong. They mistakenly think that the 8 card must be chosen. This card is mistakenly turned because people think that they must check to see if there is a nonvowel on the back. But, for example, if there were a K on the back of the 8, it would not show that the rule is false because although the rule says that a vowel must have even numbers on the back, it does *not* say that even numbers must have vowels on the back. So finding a nonvowel on the back says nothing about whether the rule is true or false. In contrast, the 5 card, which most people do not choose, is absolutely essential. The 5 card might have a vowel on the back, and, if it did, the rule would be shown to be false. In short, in order to show that the rule is not false, the 5 card must be turned.

In summary, the rule is in the form of an "if P then Q" conditional and it can only be shown to be false by showing an instance of P and not-Q, so the P and not-Q cards (A and 5 in our example) are the only two that need to be turned to determine whether the rule is true or false. If the P and not-Q combination is there, the rule is false. If it is not there, then the rule is true.

Why do most people answer incorrectly when this problem, after explanation, is so easy? At first it was thought that the abstract content of the vowel/number rule made the problem hard for people and that more real-life or so-called thematic problems would raise performance markedly. Investigators tried examples like the following "Destination Problem":

Each of the tickets below has a destination on one side and a mode of travel on the other side. Here is a rule: "If 'Baltimore' is on one side of the ticket, then 'plane' is on the other side of the ticket." Your task is to decide which tickets you would need to turn over in order to find out whether the rule is true or false. Indicate which cards must be turned over.

Destination: Baltimore	Destination: Washington	Mode of Travel: Plane	Mode of Travel: Train

Surprisingly, this type of content does not improve performance at all. Most subjects still picked either the P (Baltimore) and Q (Plane) cards or the P card only, and the correct P, not-Q solution (Baltimore and Train) escaped the vast majority.

Why, then, is the problem so difficult? Many theories have been proposed to explain the difficulty. One is that most people have a hard time thinking about negative instances—about things that could happen but are not explicitly represented. Also, as I will discuss in chapter 4, people have a hard time thinking about occurrences that could falsify their hypotheses. This is what interested Peter Wason in the problem in the first place. On most views of good scientific thinking, particularly those of the philosopher Karl Popper, it is critical to design experiments so that they can show a theory to be false. Instead, people (including scientists) tend to seek to confirm theories rather than falsify them (see Nickerson 1998, and chapter 4 of this book). This, according to one theory, is what primes people to turn the P

card (in search of the confirming Q) and the Q card (in search of the confirming P) and to miss the relevance of the not-Q card (which might contain a disconfirming P on the back).

Cognitive psychologist Jonathan Evans (1984, 1998, 2002b) has championed an even simpler theory of why the P and Q cards are the most popular choices. He views this response as reflecting a very primitive, so-called
"matching bias" that is automatically triggered by surface-level relevance
cues—the "if" draws attention to the P, and the Q is also the focus of the rule.
According to Evans, the PQ response is heuristically based—that is, it is triggered from TASS. In his view, the PQ response (and to a lesser extent the P-
only response) results from automatic processing and does not reflect any
analytic reasoning at all. The fact that most people get the problem wrong
reflects the failure of the analytic system to override TASS.

The response pattern observed in the task is viewed as a failure to override because it is assumed that all of the university student subjects who fail
the task do indeed have the logical competence to compute the correct answer by serially examining the logical implications of each card. It is just that
the TASS-generated response dominates (it is not overridden). If Evans and
other investigators are correct, then in the Wason four-card selection task we
have a clear case of a TASS tendency (PQ) being pitted against an analytic
system response tendency (P and not-Q—which is arrived at by serially examining the logical implications of each card). Note that the fact that some
people ponder for a reasonable time before making the incorrect PQ choice
does not contradict the hypothesis that that choice derives from TASS processing. Ingenious research, some of it involving online reaction-time techniques, has indicated that most of the thinking going on is actually simply
a rationalization of the response tendencies delivered up by TASS (Evans
1996; Evans and Wason 1976; Roberts and Newton 2001).

Like the four-card selection task, the Linda probability problem presented above also reflects the inability to override conflicting TASS output.
Most people make what is called a "conjunction error" on this problem. Because alternative h (Linda is a bank teller and is active in the feminist movement) is the conjunction of alternatives c and f, the probability of h cannot
be higher than that of either c (Linda is active in the feminist movement) or
f (Linda is a bank teller). All feminist bank tellers are also bank tellers, so h
cannot be more probable than f—yet 85 percent of the subjects in Tversky
and Kahneman's (1983) study rated alternative h as more probable than f,
thus displaying a conjunction error. Those investigators argued that logical
reasoning (analytic system processing) on the problem is overwhelmed by
a TASS heuristic based on so-called representativeness.[31] Representativeness

primes answers to problems based on an assessment of similarity (a feminist bank teller seems to overlap more with the description of Linda than does the alternative "bank teller"). Of course, logic dictates that the subset (feminist bank teller)/superset (bank teller) relationship should trump assessments of representativeness when judgments of probability are at issue. Thus, in the Linda problem, we have another case of a TASS response tendency (a similarity-based judgment of representativeness) being pitted against an analytic system response tendency (the logic of subset/superset relations).

Don't Be Sphexish

That 90 percent of the subjects answer the four-card selection task incorrectly and that 85 percent of subjects answer the Linda conjunction problem incorrectly indicates that, for most people, analytic processes are not very firmly in control of their judgments. Such failures of analytic processing suggest that many people are likely to be failing to maximize the fulfillment of their personal goals (as we will see more explicitly in the next two chapters). Modus tollens (the logical form that fails to be applied by most subjects in the four-card selection task) and the conjunction rule of probability are among the building blocks of clear thinking. In chapter 4 we will see that these reasoning errors are not mere laboratory phenomena, but instead occur in the real world and have real-world negative consequences. These, and many other problems studied by cognitive psychologists, are laboratory echoes of the problems in suppressing TASS responses that were illustrated in the examples that opened this chapter.

We will see in later chapters (particularly in chapters 7 and 8) that the cognitive model presented in this book creates interesting problems of personal identity. If parts of our brains disagree and compute outputs that conflict with each other, which one should we identify with and consider "ours"? Which is the best representative of who we really are as a person? In some cases, the answer seems clear. When a person rejects or shuns a disfigured individual but feels bad about doing so, which cognitive output should the person identify with (the shunning response or the shame at the response)? Both responses came from the same brain. When a husband fails to comfort a spouse who has suffered a rape but then regrets not doing so, which cognitive output should the person identify with (the failure to comfort or the regret at not doing so)?

In both of these cases, we hope the person identifies with the shame and regret—that they consider the other response to be alien to their natures.

The reason we feel this way is that the initial response is generated (inappropriately) by a TASS subsystem acting automatically, without consideration or reflection. The shame and regret are outcomes of analytic reflection on the full context of the situation. We feel in many cases like this that people should identify with their reflective minds and not their Darwinian minds (TASS).

But the solution to such problems is not always simply to resolve things in favor of our analytic minds against TASS. Recall from note 25 in this chapter the Huckleberry Finn example that has been the subject of philosophical analysis (e.g., Bennett 1974; MacIntyre 1990). Huck helped his slave friend Jim run away because of very basic feelings of friendship and sympathy. However, Huck begins to have doubts about his action once he starts explicitly reasoning about how it is morally wrong for slaves to run away and for whites to help them. In this case, our judgment reverses: we want Huck to identify with the emotions emanating from TASS modules and to reject the explicit morality that he has been taught. The problem here is that the explicit morality that Huck invokes via analytic processing was unreflectively acquired (this problem is the central topic of chapter 7).

The Huck Finn example illustrates that explicit thought processes might still invoke stored rules that were acquired in an unreflective manner and, when this happens, we would not want TASS responses to necessarily be trumped by such rules. I will analyze cases like this more extensively in chapters 7 and 8. There, we will see the dangers of the explicit thought processes of the analytic system using intellectual tools that were acquired unreflectively. For now though, I want to emphasize the dangers of the type of unreflective cognition emphasized in this chapter—that TASS (because of its older evolutionary origins and its autonomy and ballistic [unreflective and unmonitored] nature) should not be the brain system we automatically identify with ("go with your gut") in cases where it conflicts with analytic processing. That would be identifying with what, following philosopher Daniel Dennett, I will call the sphexish part of the brain.

To illustrate how we naturally seem to value *reflective* mentality over Darwinian *reflexive* mentality, Dennett, in a 1984 book on free will, asks us to consider our response to the description of the behavior of the digger wasp, *Sphex ichneumoneus*. The female Sphex does a host of amazing things in preparation for the laying and hatching of her eggs. First she digs a burrow. Then she flies off looking for a cricket. When she finds a suitable one she stings it in a way that paralyzes it but does not kill it. She brings the cricket back to the burrow and sets it just outside at the threshold of the burrow. Then she goes inside to make sure things are safe and in proper order inside the burrow. If they are, she then goes back outside and drags in the paralyzed cricket. She then lays her eggs inside the burrow, seals it up, and flies away. When the eggs hatch, the wasp grubs feed off the paralyzed cricket which has not decayed because it was paralyzed rather than killed.

All of this seems to be a rather complex and impressive performance put on by the Sphex—a real exercise of animal intelligence. It seems so, that is, until we learn that experimental study has revealed that virtually every step of the wasp's behavior was choreographed by rigid and inflexible preprogrammed responses to specific stimuli in the Sphex environment. Consider, for example, the wasp's pattern of putting the paralyzed cricket on the threshold of the burrow, checking the burrow, and then dragging the cricket inside. Scientists have uncovered the unreflective rigidity of this set of behaviors by moving the cricket a few inches away from the threshold while the wasp is inside checking the burrow. When she comes out, the wasp will not now drag the cricket in. Instead, she will take the cricket to the threshold and go in again to check the burrow. If the cricket is again moved an inch or so away from the threshold, the Sphex will *again* not drag the cricket inside, but will once more drag it to the threshold and for the third time go in to inspect the burrow. Indeed, in one experiment where the investigators persisted, the wasp checked the burrow *forty times* and would still not drag the cricket straight in. The Darwinian fixed patterns of action dictated a certain sequence of behaviors triggered by a particular set of stimuli, and any deviation from this was not tolerated.

In cases like the Sphex it is unnerving for us to first observe all of the artful and complex behavior of the creature—so seemingly intelligent—and then to have revealed by the experiment just described how mechanically determined it actually all was. Dennett refers to "that spooky sense one often gets when observing or learning about insects and other lower animals: all that bustling activity but *there's nobody home!*" (1984, 13). Quoting cognitive scientist Douglas Hofstadter (1982), Dennett (1984) proposes that we call this unnerving property sphexishness. He points out that observing the simple, rigid routines that underpin the complexity of the surface behavior of simple creatures spawns in us the worrying thought: "What makes you sure you're not sphexish—at least a little bit?" (11).

Dual-process models of cognition such as those listed in table 2.1 and discussed in this chapter all propose, in one way or another, that in fact we all *are* a little bit sphexish. In fact, many of these theories, in emphasizing the pervasiveness of TASS and the rarity and difficulty of analytic processing, are in effect proposing that our default mode of processing is sphexish. If we want to be more than a Sphex, then we must continually accomplish the difficult task of marshaling cognitive power to run a serial simulator containing mindware that can monitor TASS to ensure that it is fulfilling vehicle-level goals.

The TASS modules offering up outputs that conflict with reasoned analytic system outputs could be viewed as "the Sphex within you." It does not assuage our fears of sphexishness to note that some of the automatic processes in TASS are products of our environment—learned rules that were practiced so much that they now execute autonomously. That response tendencies may have been instantiated in TASS by advertising, by our peer group in adolescence, or by the repetition by our parents of prescriptions deriving from their own limited experiences makes them no less sphexish—they are just as unreflective and as unconsidered as any other TASS processes. Only rules that were instantiated in TASS by reflective thought should be honored and identified with, and even these may occasionally be overgeneralized (because now they will be firing automatically) and need overriding in a particular situation.

The sphexish processes within you thus have the potential to lead you astray when the information passed on to the analytic system was triggered inappropriately (i.e., sphexishly) from TASS. This can happen even for TASS heuristics that are generally useful on most occasions. Cognitive psychologists Amos Tversky and Daniel Kahneman pioneered the study of TASS heuristics that help us on most occasions but also account for some of the sphexishness in our behavior (Kahneman and Tversky 1973, 1984; Tversky and Kahneman 1974, 1983). One such TASS process was the so-called anchoring and adjustment heuristic (Brewer and Chapman 2002; Tversky and Kahneman 1974). The anchoring and adjustment process comes into play when we have to make a numerical estimation of an unknown quantity. In this strategy, we begin with a TASS-based anchoring on the most relevant similar number that we know, and then we adjust that anchor up or down via more controlled, analytic adjustments based on the implications of specific facts that are known.

As an example of how anchoring and adjustment might work, consider Mr. Winston hearing a complaint from Mr. Blair about the amount of time Mr. Blair's son spends listening to music. Mr. Blair complains that his son owns almost 100 compact disks and asks Mr. Winston how many his own son owns. Not knowing where to begin to get the estimate, Mr. Winston starts with the 100 figure and adjusts from there. His own son is not seen with earphones on nearly as much as the other boy, so Mr. Winston moves the estimate from 100 to 75. Also, his own son participates in more outside activities, so he moves the estimate down again, this time from 75 to 60. But then he thinks that his own son, for various reasons, probably has more money to spend, so he moves the estimate up from 60 to 70.

This does not seem to be such a bad procedure. It uses all the information available, because TASS focuses analytic systems on the most relevant nearby number and more analytic processes then adjust it based on specific facts that are known. A problem arises, however, when the most available number to anchor on is not relevant to the calculation at hand. We become sphexish when we use it nonetheless. In a classic experiment, Tversky and Kahneman (1974) demonstrated just how this can happen. They had subjects watch a spinning wheel and when the pointer landed on a number (rigged to be the number 65) they were asked whether the percentage of African countries in the United Nations was higher or lower than this percentage. After answering higher or lower to this question, the subjects then had to give their best estimate of the percentage of African countries in the United Nations. Another group of subjects had it arranged so that their pointer landed on the number 10. They were also asked to make the higher or lower judgment and then to estimate the percentage of African countries in the United Nations. The mean estimate of the first group turned out to be significantly larger (45 versus 25 for the second group).

It is clear what is happening here. Both groups are using the anchoring and adjustment heuristic—the high anchor group adjusting down and the low group adjusting up—but their adjustments are "sticky." They are not adjusting enough because they have failed to fully take into account that the anchor is determined in a totally *random* manner. The anchoring and adjustment heuristic is revealing a sphexish tendency for TASS to throw up an anchor regardless of its relevance. We should obviously ignore the anchor in such unusual situations, but we are so used to using the anchor as if it carried some information, that in a situation where it should be totally discounted, we do not do so.

Would you rather have a 10 percent chance of winning a dollar or an 8 percent chance of winning a dollar? Virtually all of us would choose the former. However, if you are like many people in experiments by Seymour Epstein and colleagues (Denes-Raj and Epstein 1994; Kirkpatrick and Epstein 1992; Pacini and Epstein 1999) you would have actually chosen the latter because of sphexish tendencies in your brain. Subjects in several of his experiments were presented with two bowls of jelly beans. In the first were nine white jelly beans and one red jelly bean. In the second were 92 white jelly beans and 8 red. A random draw was to be made from one of the two bowls and if the red jelly bean was picked, the subject would receive a dollar. The subject could choose which bowl to draw from. Although the two

bowls clearly represent a 10 percent and an 8 percent chance of winning a
dollar, many subjects chose the 100-bean jar, thus reducing their chance of
winning. Although most were aware that the large bowl was statistically a
worse bet, that bowl also contained more enticing winning beans—the 8
red ones. Many could not resist trying the bowl with more winners despite
some knowledge of its poorer probability. That many subjects were aware of
the poorer probability but could, nonetheless, not resist picking the large
bowl is indicated by comments from some of them such as the following: "I
picked the ones with more red jelly beans because it looked like there were
more ways to get a winner, even though I knew there were also more whites,
and that the percents were against me" (Denes-Raj and Epstein 1994, 823).
In short, the simpler TASS tendency to respond to the absolute number of
winners trumps the more analytic process of calculating a ratio.

So there are plenty of TASS heuristics installed in your brain that give you
the potential to behave sphexishly. And the appalling thing that we learned
in chapter 1 is that, in situations where replicator goals and vehicle goals are
at odds, your genes want you to behave like a Sphex. They wish you would
blindly execute the response primed by your TASS subsystems. The robot's
rebellion derives in part from our ability to recognize the potential for
sphexishness in our own behavior and take steps to prevent it.

Putting the Vehicle First by Getting the
Analytic System in the Driver's Seat

We now have in place several of the insights necessary for the robot's rebellion. In chapter 1 we saw that the universal acid of Darwinism does—
when taken to the limit—have the horrific implications that some of its
most severe critics have feared. But we can only free ourselves from some
of these implications by understanding them thoroughly. No one wants to
be a *mere* receptacle for "colonies of replicators . . . swarming inside of us"
(Dawkins 1976). But we have the potential to be just that if we allow TASS
to determine our behavior without analytic system override. We behave
sphexishly if we do not subject our TASS outputs to critique by the analytical processing system. The threat to the soul represented by Darwinism
is right there in our own brains—in the TASS subsystems that could turn
us into sphexish automatons, into robots fulfilling not our own goals but
those of the selfish replicators who built us. But there too, in the same
brain, is the potential awareness of this plight along with the cognitive
mechanisms to overcome it.

The conceptual reconstruction of our sense of self that we have begun in this chapter and which will continue in subsequent chapters will reveal several ironies in our traditional concepts of the self, the soul, and personal identity. It is, for example, deeply ironic to hear people identify with and defend their so-called "gut instincts" because they feel that these somehow signify their uniqueness—that their "gut instincts" are the essence of who they are. But if people reconceptualized their "gut instincts" as just TASS modules prewired and constructed so as to respond in ways that served the reproductive interests of the replicators, they would feel less attached to them. The Darwinian insight reveals the irony in humans seeking to identify with the sphexish parts of their brains—the parts that were designed to function in the most rigid and reflexive way.[32]

Uncritically delighting in following our so-called gut instincts makes us little more than slaves of the mindless replicators—the micro-automata that view us as nothing more than vehicles to assist in their quest for replication. Although TASS forms the substrate of our hopes, wants, and fears, in chapters 7 and 8 I will explore how the personal autonomy that is the prerequisite to actual human uniqueness is totally dependent on a critical *evaluation* of the nature of our gut instincts and a conscious sculpting of the goals that two types of replicator have inserted in our brains (yes, *two* types of replicator—see chapter 7). The distinctions introduced in this chapter are the foundations of a reconceptualization whereby our *reflections* on, and critical analysis of, our gut instincts become the essence of who we are (our personhood) because these critical reflections, unlike some of our TASS responses, are in the service of the vehicle—*us.*

The robot's rebellion, if successful, will result in humans attaining personal autonomy by pursuing their own interests rather than sacrificing them for the ancient interests of the selfish replicators. To avoid being a sphexish automaton who serves the goals of subpersonal replicators, we must cultivate certain mental talents, however—talents that enable an important program of cognitive reform. These mental talents are cultural products—they are mindware that can be run on the serial simulator that carries out analytic processing. An important part of this mindware consists of the skills of rational thinking. In the next chapter, we will see why, if you don't want to be the captive of your genes (or of any other selfish replicator), then you had better be rational.

~

She didn't hold with Darwinian or genetic determinism. Of course she knew that that *was* how things were, but she didn't *like* the way things were. She didn't approve of it. . . . She'd like to think one could rediscover an argument that would reinstate the freedom of the will and the adaptability of the species. Otherwise everything was too damn depressing, wasn't it?

—Margaret Drabble, *The Peppered Moth* (2001, 137–38)

CHAPTER 3

The Robot's Secret Weapon

0

People hate the idea of genetic determinism. For example, many liberal commentators on the political left (e.g., Rose, Kamin, and Lewontin 1984) resist the idea that genetic variation in a population is correlated with phenotypic variation in the same population (to put it properly)—the idea that genes determine human behavior (to put it somewhat wrongly, but in the language of the layperson). People who do not like the idea of genetic determinism do in fact have an escape route. However, the escape route is not to be found in denying the known facts about the heritability of human behavioral traits, or in denying their evolutionary origins. Instead, the escape route lies in understanding the implications of combining the replicator/ vehicle distinction introduced in chapter 1 with the facts about cognitive architecture outlined in chapter 2 and realizing how these facts interact with one of the most remarkable cultural inventions of the last few centuries.

Choosing Humans over Genes:
How Instrumental Rationality and Evolutionary Adaptation Separate

The first step in conceptualizing the escape route involves focusing on the startling fact discussed in chapter 1—that the interests of replicators and vehicles do not always coincide. When a conflict occurs, short-leashed response systems (TASS) will prime the response that serves the genes' interests (replication) rather than those of the vehicle. However, as humans, we are concerned with our own personal interests, not those of the subpersonal organic replicators that are our genes. The presence of our analytic processing systems makes it possible to install the mental software (so-called "mindware," see Clark 2001) that maximizes the fulfillment of our own interests as people. This mindware ensures a mental set to give primacy to the *person's* fulfillment (to side in favor of the vehicle and against the replicators when their interests do not coincide). That mental set is the proclivity for rational thought.

As will be elaborated in subsequent sections of this chapter, one key component of rationality is the optimization of goal fulfillment at the level of the individual. This is why evolutionarily adaptive behavior is not the same as rational behavior.[1] Evolutionarily adaptive behavior is behavior that increases genetic reproduction probabilities, but rational behavior is behavior that fulfills the goals of the vehicle—given the vehicle's set of beliefs about the world. As will be discussed later in this book, evolutionary psychologists sometimes obscure this by implying that if a behavior is adaptive it is rational. Such a conflation represents a fundamental error of much import for human affairs (Over 2000, 2002; Stanovich 1999; Stanovich and West 2000). Definitions of rationality must be kept consistent with the entity whose optimization is at issue. In order to maintain this consistency, the different "interests" of the replicators and the vehicle must be explicitly recognized. By doing so, humans acquire the means to escape a certain type of genetic determinism. What is required is the recognition that *rationality concerns the interests of the vehicle,* whereas evolutionary adaptation concerns the interests of the genes (reproductive success)—and that situations can occur where the two do not coincide.

But how can the two ever become separated? As discussed in chapter 1, once evolution moved from preprogramming behaviors to creating a general problem-solving device designed to deal with a shifting environment, a potential gap was created between genetic goals and vehicle goals. This is because a vehicle with sophisticated means of anticipating the future need be provided with only *general* goals—goals that will cover a panoply of environmental situations that might be encountered. However, these generalized motivations (sex is pleasurable) can become outdated in the modern environment—outdated in the sense that the motivation can be satisfied in ways that no longer serve to enhance genetic fitness (sex with contraception).

As modern human beings, we find that many of our motivations have become detached from their ancestral environmental context, so that now fulfilling our goals no longer serves genetic interests. Ironically, what from an evolutionary design point of view could be considered design defects actually make possible the robot's rebellion—the full valuing of people by making their goals, rather than the genes' goals, preeminent. That is, inefficient design (from an evolutionary point of view) in effect creates the possibility of a divergence between organism-level goals and gene-level goals—

which is an implication of philosopher Ruth Millikan's (1993) point that "there is no reason to suppose that the design of our desire-making systems is itself optimal. Even under optimal conditions these systems work inefficiently, directly aiming, as it were, at recognizable ends that are merely roughly correlated with the biological end that is reproduction. For example, mammals do not, in general, cease efforts at individual survival after their fertile life is over" (67). What Millikan means here is that staying alive is only the most generic of goals—one that is only very indirectly linked to the genes' ultimate goal of replication. Dead organisms cannot reproduce. However, being alive does not guarantee that an organism will be able to reproduce. Survival was installed as a goal in TASS by the genes as a necessary but not sufficient desire in the hierarchical goal structure that leads to replication. However, when the further conditions for replication are not met (perhaps because reproductive organs are defective or inoperative), the vehicle continues to try to achieve the goal of survival even though it cannot anymore result in a totally successful outcome for the genes.

For humans, the goal of survival is not just a mechanism for the genetic goal of replication, but is a superordinate goal for everything else in the analytic system's derived goal hierarchy. Many behaviors that might serve our numerous and quite abstract goals might not serve those of the genes. As Morton (1997) aptly puts it: "Our genes want us to be able to reason, but they have no interest in our enjoying chess" (106).

Likewise, many genetic goals lodged in TASS no longer serve our ends. Thousands of years ago, humans needed as much fat as they could get in order to survive (Pinel, Assanand, and Lehman 2000). More fat meant longer survival, and because few humans survived beyond their reproductive years, longevity translated directly into more opportunities for gene replication. Now, however, the analytic system is busy assessing the complex technological environment in which we live and trying to fulfill the generic goal of longevity. That system processes the information that now, when lifespans are long, eating fat shortens life. Prolonging life in modern society is irrelevant to the genes' ultimate goal of replication because we now live way beyond our reproductive years. But such a goal is of immense importance to us as vehicles. It is rational to abstain from eating too much fat. Our Darwinian minds (within TASS), priming us to consume fat, are in conflict with the more encompassing set of goals and objectives that the analytic system—with its capacity for hypothetical reasoning (discussed in chapter 2)—can conjure in service of the overall well-being of the vehicle.

Organisms under more direct genetic control never experience a conflict between short-leash and long-leash control. Such goal conflict only appears when the analytic system comes into existence which can balance short-term and long-term goals—can ponder the tradeoff between immediately pleasurable activities (dangerous sex, smoking, fatty foods) and future longevity that is also viewed as desirable. Only such a creature can even contemplate that there is a conflict in such a situation—and contemplate what it is rational to do. But noticing the conflict is the first stage in the robot's rebellion. Noticing the conflict is the thing that makes our outlook on life different from that of other animals.[2]

John Anderson, a cognitive theorist who demonstrated for psychologists the power of adaptive models in an influential 1990 book (see also Oaksford and Chater 1994, 1998), stresses that the adaptation assumption in his modeling is that cognition is optimally adapted in an *evolutionary* sense—and that this is not the same as positing that human cognitive activity will result in responses that are instrumentally rational (utility maximizing for the individual). He notes that, as a result, a descriptive model of processing that is adaptively optimal could well deviate substantially from a model of optimal human rationality. Thus, Anderson (1991) acknowledges that there may be arguments for "optimizing money, the happiness of oneself and others, or any other goal. It is just that these goals do not produce optimization of the species" (510–11). But the point I wish to stress here is that "optimization of the species" is not what most people want. The point is instead that most people, if given the choice, would rather have the money!

No human has optimizing genetic fitness as an explicit goal. We seek not evolution's ultimate goal of replication, but the more proximal outcome that has been installed in us during evolution—the enjoyment of sexual pleasure. If the means are pleasurable, analytic intelligence acting rationally in the service of the vehicle pursues *them*—not the ultimate goal of the genes. Thus, the revolt of the human survival machines—the robot's rebellion—consists of the gene-built humans trying to maximize their own utility rather than the reproduction probability of their creators in situations where the two are in conflict. If we want to be more than just survival machines for our genes, then we must use our analytic processing systems in the active pursuit of instrumental rationality. There is no more potent mental software in the robot's arsenal than that provided by the tools of rational thought.

What It Means to Be Rational:
Putting the Person (the Vehicle) First

For many readers influenced by views of rationality as a rather abstract and specialized ability—for example, the ability to solve textbook logic problems—it will come as a surprise how broadly and practically rationality is conceptualized in modern cognitive science. In fact, even the most restricted views of what is often termed instrumental rationality[3] are quite expansive and practical when compared to the caricatures of rationality that are promulgated by its critics. The best definition of instrumental rationality—the one that emphasizes most how practical it is, and the one that establishes it most fundamentally as a value difficult to dispute—simply is:

> Instrumental rationality is behaving in the world so that you get exactly what you most want, given the resources (physical and mental) available to you.

It should be clear why this type of rationality is often termed practical rationality, because it is most assuredly that. It is also what is called a thin theory of rationality (see Elster 1983). Thin theories very nonjudgmentally define what it is rational for you to do in terms of your own wants and desires. We shall see (in this chapter and especially in chapters 7 and 8) that more than this is required for a full-fledged robot's rebellion. Nevertheless, adhering to the strictures of thin theories of instrumental rationality is a necessary precondition for the rebellion, although not a sufficient one. Most conceptions of instrumental rationality reflect a tradition going back to Hume's famous dictum: "Reason is, and ought only to be the slave of the passions, and can never pretend to any other office than to serve and obey them" ([1740] 1888, bk. 2, part 3, sec. 3).

Reason should help get you what you want (the passions), says Hume.[4] A hard notion to dispute. What good is reason if it doesn't help you get what you want? Instrumental rationality seems to be something that we would all want to have. Who wouldn't want to have their goals achieved? Who wouldn't want to have their wants satisfied? Postmodernists who argue against valuing modern notions of rationality are going to have tough going here.

The basic idea of valuing instrumental rationality leaves its critics with the problem of defending the position that we should not act to get the things we want. For a person to argue this position just seems weird. On the other hand, allowing ourselves again the liberty of anthropomorphic gene-language, it could make perfect sense for *genes* to "want" this. It may make sense because getting what *you* want may make it less likely that *they* will get what they "want"—replication. So even though your genes might not be in favor of you being rational, you certainly should be.

But as we will see in chapter 4, sometimes we *don't* always act to best get what we most want. This should make us suspicious. How might this come about? Perhaps because—as outlined in chapter 2—our brains are at war with themselves. Parts of our minds might be more or less oriented toward instrumental rationality—toward fulfilling our goals as people. In contrast, some brain processes might be more directly oriented (in a short-leashed manner) to fulfilling genetic goals that might not be personal goals. By adhering to strictures of instrumental rationality, we make sure that the brain systems concerned most specifically with *us* are in control. Adhering to strictures of instrumental rationality provides a check on the TASS modules that might not be maximizing over the entity we are most interested in—our personal desires.

Rationality is importantly related to the notion of the self and personal autonomy—an issue that will be explored in more depth in chapters 7 and 8. But the important point here is that rationality is keyed to the notion of the self. It is defined over a person, not a subpersonal entity (the genes). This is why the capacity for rationality is the most potent weapon that humans have in their rebellion against the selfish replicators.

Fleshing Out Instrumental Rationality

Rationality is behaving in the world so that you get exactly what you most want, given the resources available to you. Of course, economists and cognitive scientists have refined the notion of "what you most want" into the technical notion of expected utility.[5] The model of rational judgment used by decision scientists is one in which a person chooses options based on which option has the largest expected utility. A startling discovery of modern decision science is that if people's preferences follow certain logical patterns (the so-called axioms of choice—things like transitivity and freedom from certain kinds of context effects) then they are behaving as if they are maximizing utility (Dawes 1998; Edwards 1954; Jeffrey 1983; Luce and Raiffa 1957; Savage 1954; von Neumann and Morgenstern 1944)—they are behaving as if they are acting to get what they most want. We will see in the next chapter how closely people's judgments approximate the choice axioms of utility theory—and thus by inference how close they are to achieving instrumental rationality (getting maximally what they want, given their beliefs and desires).

Despite the appeal of instrumental rationality, it remains what Elster (1983) calls a thin theory of rationality. This is because existing beliefs are accepted without critique, and, likewise, the content of desires is not evaluated. The individual's goals and beliefs are accepted as they are, and debate centers only on whether individuals are optimally satisfying desires given beliefs. The defining feature of a thin theory is that it "leaves unexamined the beliefs and the desires that form the reasons for the action whose rationality we are assessing" (Elster 1983, 1).

The strengths of the thin theory of practical rationality are well known. For example, if the conception of rationality is restricted to a thin theory, many powerful formalisms (such as the axioms of decision theory) are available to serve as standards of optimal behavior. The weaknesses of the thin theory are equally well known, however (e.g., Elster 1983; Kahneman 1994; Nathanson 1994; Simon 1983). In not evaluating desires, a thin theory of rationality would determine that Hitler was a rational person as long as he acted in accordance with the basic axioms of choice as he went about fulfilling his grotesque desires. Likewise, if we submit beliefs to no evaluative criteria, the psychiatric ward patient who acted consistently upon his belief that he was Jesus Christ would be judged a rational person. Although some theorists might feel that these aberrant cases may be worth tolerating in order to gain access to the powerful choice axioms that are available to the thin theorist, others may view with alarm the startlingly broad range of human behavior and cognition that escapes the evaluative net of the thin theory.[6]

Figure 3.1 provides a minimal model within which to situate these broader issues.[7] This minimal model contains a set of beliefs (the person's knowledge), a set of desires (the person's goals), and a mechanism (labeled action determination) that derives actions from a consideration of the system's beliefs and desires. The arrows labeled A are meant to indicate that one way that beliefs are formed is by taking in information about the external world. Modeling of the external world by beliefs can be more or less good, and, in the extreme, it may become so poor that we want to call it irrational. Of course, beliefs can be derived without the aid of external information— for example, they can be derived using inferences from other beliefs or from desires (as in wishful thinking, Babad and Katz 1991; Bar-Hillel and Budescu 1995; Harvey 1992).

The double-headed arrows labeled B refer to relations among beliefs and relations among desires. These relations may be more or less coherent. Inconsistency detection is thus an important determinant of rationality. For example, detection of belief inconsistency might be a sign that belief

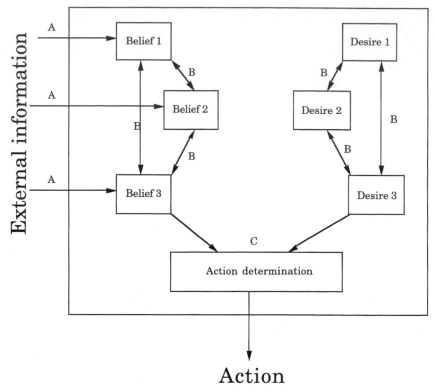

Action

Figure 3.1 A generic belief-desire network for conceptualizing human rationality

formation processes have operated suboptimally. Analyses of self-deception have traditionally posited the presence of two contradictory beliefs in an individual's knowledge base (Gur and Sackheim 1979; Mele 1987; however, see Mele 1997). Alternatively, detection of desire inconsistency signals that the processes of action determination might result in nonoptimal goal satisfaction.

Many potential complications to this model are not pictured in figure 3.1. However, one is pictured in figure 3.2. Here, the single-headed arrows labeled D indicate the possibility of desires modifying processes of belief formation and inconsistency detection (Kunda 1990; Mele 1997). However, several other processes are not represented (for example, the process of desire formation and the possibility of beliefs modifying desires are not represented; see Elster 1983: Pollock 1995).

The figures do represent enough detail to make the fundamental distinction mentioned earlier—the distinction between rationality of belief and rationality of action (Audi 1993a, 1993b, 2001; Harman 1995; Nozick

1993; Pollock 1995; Sloman 1999). So far, I have focused on the latter—instrumental rationality. It is often assessed by examining the coherence of preferences that a person displays. It has been shown that if certain conditions of coherence are fulfilled (things like the transitivity of preferences) then the individual can be viewed as maximizing his or her expected utility—utility being the decision scientist's term for the satisfaction achieved by the fulfillment of goals.[8] Thus, the box in figure 3.1 indicating the process of action determination and the double-headed B arrows (indicating coherence relations among beliefs and desires) encompass the concept of instrumental rationality. If beliefs and desires are coherent and if the action chosen is the best one given the individual's current beliefs and desires, then instrumental rationality will be achieved. People will be maximizing their goal satisfaction—they will be doing their best to get exactly what they want *given* their current wants and their current beliefs. This is a necessary but not sufficient condition for the robot's rebellion.

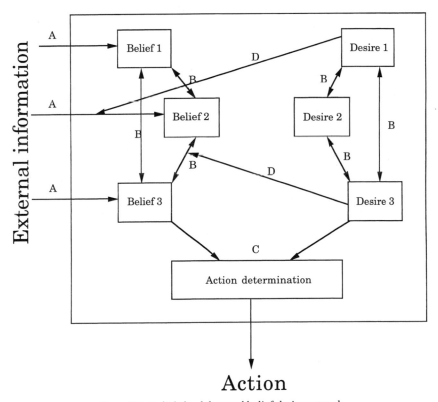

Figure 3.2 A slightly elaborated belief-desire network

The robot's rebellion is the modern intellectual movement whereby humans are attaining greater personal autonomy by refusing to let their own interests be sacrificed to the ancient goals of the selfish replicators. It requires a broad view of rationality (in contrast to a thin theory)—one where the content of beliefs and desires are subject to evaluation. The former means that issues of epistemic rationality (represented by the arrows labeled A in figure 3.1) must be addressed.[9] This aspect of rationality—requiring that beliefs map well onto the actual the structure of the world—has been variously termed theoretical rationality, evidential rationality, or epistemic rationality (Audi 1993b, 2001; Foley 1987; Harman 1995). Importantly (given much of the research to be discussed in chapter 4) a critical aspect of beliefs that enter into instrumental calculations (that is, tacit calculations) are the probabilities of states of affairs in the world. We will see that much research indicates that for many people the coherence of these probabilities is in question.

Moving to a broad theory of rationality will also necessitate the evaluation of the content of desires and goals—in precise contradiction to Hume's famous dictum ("reason is, and ought only to be the slave of the passions"). If we follow Hume's dictum, we will have no assurance that in ceding control to our "passions" we are not ceding control to goals that serve replicator interests rather than vehicle interests. Thus, attempts have been made to develop criteria for evaluating desires. For example, several theorists (see Nathanson 1994, for a review) have argued that desires which, upon reflection, we would rather eliminate than fulfill are irrational. Other theorists argue that conflicting desires, or desires based on false beliefs, are irrational. Finally, it could be argued that the persistent tendency to develop goals whose expected utility is different from their experienced utility is a sign of irrationality (Frisch and Jones 1993; Kahneman 1994; Kahneman and Snell 1990). We need to evaluate desires by criteria such as these or else our actions might not be serving vehicle goals (see chapters 7 and 8).

A broad theory of rationality will also emphasize the interplay between epistemic and practical rationality in some interesting ways. For example, it is uncontroversial that it is rational to have derived goals. One might desire to acquire an education not because one has an intrinsic desire for education but because one desires to be a lawyer and getting an education leads to the fulfillment of that goal. Once derived goals are allowed, the possibility of evaluating goals in terms of their comprehensiveness immediately suggests itself. That is, we might evaluate certain goals positively because attaining them leads to the satisfaction of a wide variety of *other* desires. In contrast, there are some goals whose satisfaction does not lead to the satisfaction of any other desires. In the extreme there may exist goals that are in conflict with other goals. Fulfilling such a goal would actually impede the fulfillment of other goals (this is why goal inconsistency is to be avoided).

In the vast majority of mundane cases, one derived goal that would serve to fulfill a host of others is the goal of wanting one's beliefs to be true. Generally (that is, outside of the types of bizarre cases that only a philosopher would imagine), the more that a person's beliefs track the world, the easier it will be for that person to fulfill his/her goals. Obviously, perfect accuracy is not required and, equally obviously, there is a point of diminishing returns where additional cognitive effort spent in acquiring true beliefs will not have adequate payoff in terms of goal achievement. Nevertheless, other things being equal, the presence of the desire to have true beliefs will have the long-term effect of facilitating the achievement of a host of goals. It is a superordinate goal in a sense and—again, except for certain bizarre cases— should be in the desire network of most individuals. Thus, even if epistemic rationality really is subordinate to practical rationality, the derived goal of maintaining true beliefs will mean that epistemic norms must be adhered to in order for practical goals to be achieved.

Evaluating Rationality: Are We Getting What We Want?

In the next chapter I will discuss some of the evidence that cognitive psychologists have uncovered concerning how well people are achieving instrumental rationality.[10] But we might well ask ourselves—how could rationality ever *not* be achieved? How could we ever act to *not* get what's best for us?

In fact, there are philosophers who argue that we can't not act in our best interests—that humans are inherently rational. These philosophical arguments are wrong for reasons outlined in the previous two chapters.[11] Because there are different kinds of minds in the brain—each with its own goal structures (some keyed tightly to replicator goals and others only loosely)— there is no guarantee that any particular response is optimal from the standpoint of the totality of the organism's goals. As an organism acting in the environment, my actions are in part determined by evolutionarily older parts of the brain that instantiate short-leash goals that might serve to thwart the pursuit of my long-term goals. The analytic system is keyed to the specific and ever-changing context of my current environment, whereas TASS subserves short-leashed goals keyed to the so-called EEA—the environment of evolutionary adaptation. My explicitly stated long-term goals (largely an analytic system product) may not be achieved because my responses in the environment are partially determined by brain systems with inflexible short-leashed goal structures (for instance, if I accede to a TASS-determined "go for it" response that primes me to mate with my boss's wife).

In the centuries-long cultural project of humanity's growing awareness of its place in the universe, evaluating the extent of human rationality represents a seminal project. Assumptions about human rationality form the foundation of many economic and social theories that structure our lives. If humans do indeed have perfect rationality, we need not worry about our genetic heritage thwarting our current goals. If humans are perfectly rational, the robot's rebellion has already succeeded. Humans as survival machines will have already transcended the genes' reach. Any conflicts between replicator goals and vehicle goals will already have been resolved in favor of the latter. But how extraordinarily appalling it would be to have reached this conclusion prematurely. If humans fall short of perfect rationality, then in cases of conflict between the replicators and their human survival machines, we might still be sacrificing ourselves as vehicles.

The hedonic experience of the vehicle is just a means to an end for most of the goals lodged in TASS, and it will readily sacrifice the vehicle's hedonic pleasure if ultimate fitness goals are achievable without it. We instead should be focused on our own personal goals as vehicles. We diminish our selfhood if we compromise these in order to honor responses programmed by subpersonal entities that constructed the Darwinian parts of our brains. But evidence from the laboratories of many cognitive scientists discussed in the next chapter indicates that this is just what might be occurring.

In that chapter we will see how psychologists have gone about evaluating the rationality of human beings primarily in terms of the thin theory of rationality but also in some aspects of a broader theory of rationality as well. We will see that people often depart from the coherence criteria of instrumental rationality—they do not seem to be maximizing expected utility. Also, some substantive criteria of a broader theory of rationality are less than optimally achieved. It appears that the content of human desires is often defective in some respects and that our confidence in our beliefs is often not well calibrated to evidence. Thus, as we will see in the next chapter, there are indications that we could be more rational—that we could get more of what we most want if we altered our thinking and behavior.

Hardwired mechanisms are vulnerable to deception in a malign world.
—Kim Sterelny and Paul Griffiths, *Sex and Death:
An Introduction to Philosophy of Biology* (1999, 331)

CHAPTER 4

THE BIASES OF THE AUTONOMOUS BRAIN:
CHARACTERISTICS OF THE SHORT-LEASH MIND
THAT SOMETIMES CAUSE US GRIEF

If evolution does not guarantee rationality—if the genes' interests are not always the vehicle's interests—then we have opened up the possibility that substantial irrationality might be present in the behavior of human beings. In fact, three decades of intense work by cognitive psychologists and other scientists has indicated that it is not difficult to demonstrate experimentally the existence of human irrationality.

Recall the informal definition of instrumental rationality discussed in the last chapter: behaving so that you get what you most want given the resources available to you. The theory of subjective expected utility[1] is the technical definition of instrumental rationality, and many experiments by cognitive psychologists have been structured to test the axioms (rules) of choice that define the maximization of expected utility.

The principles of utility theory that humans have been shown to violate are often startlingly basic. For example, there is one particular principle from decision theory that is so simple and basic that it is virtually a no-brainer. It is one of those axioms mentioned in the last chapter—rules that if collectively followed guarantee that we are maximizing expected utility (are acting in the most optimal way to get what we want). The rule is one that no one will disagree with. The rule was termed the sure-thing principle by Savage (1954) and it says the following. Imagine you are choosing between two possible outcomes, A and B, and event X is an event that may or may not occur in the future. If you prefer prospect A to prospect B if X happens, and you also prefer prospect A to prospect B if X does not happen, then you definitely prefer A to B. Uncertainty about whether X will occur or not should have no bearing on your preference. Because your preference is in no way changed by knowledge of event X, you should choose A over B whether you know anything about event X or not. In a 1994 article in the journal *Cognition*, cognitive psychologist Eldar Shafir called the sure-thing principle "one of simplest and least controversial principles of rational behavior" (404). Indeed, it is so simple and obvious that it hardly seems worth stating. Yet Shafir, in his article, reviews a host of studies that have demonstrated that people do indeed violate the sure-thing principle.

For example, Tversky and Shafir (1992) created a scenario where subjects were asked to imagine that they were students at the end of a term, were tired and run down, and were awaiting the grade in a course which they might fail and thus be forced to retake. Subjects were also told to imagine that they had just been given the opportunity to purchase an extremely attractive vacation package to Hawaii at a very low price. More than half the subjects who had been informed that they had *passed* the exam chose to buy the vacation package and an even larger proportion of the group who had been told that they had *failed* the exam also chose to buy the vacation package. However, only one third of a group who *did not know* whether they passed or failed the exam chose to purchase the vacation. The implication of this pattern of responses is that at least some subjects were saying "I'll go on the vacation if I pass and I'll go on the vacation if I fail, but I won't go on the vacation if I don't know whether I passed or failed."

Shafir (1994) describes a host of decision situations where this outcome obtains. Subjects prefer A to B when event X obtains, prefer A to B when X does not obtain, but prefer B to A when uncertain about the outcome X—a clear violation of the sure-thing principle. These violations are not limited to make-believe problems or laboratory situations. Shafir (1994) provides some real-life examples, one involving the stock market just prior to the 1988 U.S. presidential election between George Bush and Michael Dukakis. Market analysts were near unanimous in their opinion that Wall Street preferred Bush to Dukakis. Yet subsequent to Bush's election, stock and bond prices declined and the dollar plunged to its lowest level in ten months. Of course, analysts agreed that the outcome would have been worse had Dukakis been elected. Yet if the market was going to decline subsequent to the election of Bush, and going to decline even further subsequent to the election of Dukakis, then why didn't it go down *before* the election due to the absolute certainty that whatever happened (Bush or Dukakis) the outcome was bad for the market? The market seems to have violated the sure-thing principle.

Violations of the sure-thing principle occur in examples such as these because people are reluctant to travel down all the branches of a decision tree when the outcome is unknown. Thus, they fail to uncover that the choice of A over B can be made despite uncertainty about the event X, because exhaustive exploration of the decision tree reveals that the preference is not dependent on the outcome of event X. Nevertheless, Shafir (1994) argues that doing such an exhaustive search "is apparently quite unnatural for people [because] . . . thinking through an event tree requires people to assume momentarily as true something that may in fact be false" (426). We shall see shortly that one of the pervasive biases of TASS is to focus on positive instances and to fail to represent states of affairs that might *not* be true. Representing the latter requires the hypothetical reasoning that is made possible by the decoupling abilities of the analytic system discussed in chapter 2 (the mental abilities that allow us to mark a belief as a hypothetical state of the world rather than a real one).

The sure-thing principle is not the only very basic rule of rational thinking that humans have been shown to violate. Other basic principles such as the transitivity principle (if you prefer A to B and B to C, then you should prefer A to C) are also violated under certain conditions.[2] And violating transitivity is no laughing matter. It leads to what decision theorists call "money pumps"—situations where if you were prepared to act on your intransitive preferences you could be drained of all your wealth. For example, consider three objects that we will arbitrarily label object A, object B, and object C. It is very unwise to prefer object A to B, B to C, and C to A. If you do not think so, I would like to give you—free of charge—object A. Then I just might think of offering you object C for a little bit of money and for giving me A back. Because you prefer C to A, there must be some finite, if however small, sum of money that you would give me, along with A, to get C. Then, for a little bit of money and object C, I will give you object B (which you say you prefer to C). And, because you gave me A a little while ago, I can now offer it to you in exchange for B (and a little bit of money, of course, because you really prefer A to B). And then I can offer you C . . . Everyone would probably agree at this point that, basically, this deal stinks!

The violations of the transitivity principle and of the sure-thing principle represent just two examples from a substantial research literature. Comprising literally hundreds of empirical studies, thirty years of work in cognitive psychology and decision science indicates that people's responses deviate from basic principles of rationality on many reasoning tasks. Particularly important has been the work of Nobel Prize winning psychologist Daniel Kahneman (the prize was in economics) and his brilliant (but sadly deceased) colleague Amos Tversky, whose seminal studies (along with those of other important investigators) form what has been termed the heuristics

and biases research program.[3] The four-card selection task and Linda problem that were discussed in chapter 2 come from this research tradition.

The terms heuristics and biases refer to the two aspects of research studies of this genre. First, human performance was shown to deviate from
what was rationally expected on a particular task (a cognitive bias is demonstrated). Next, the deviation is shown to result from the operation of
an automatic heuristic—a response triggered by TASS. These heuristics are
assumed to be useful in many situations (and are computationally in-
expensive). But where logical, analytic, and/or decontextualized problem
solving is required, TASS heuristics bias processing in the wrong direction,
thus resulting in suboptimal responses.

The work in the heuristics and biases tradition is not without controversy however. In this chapter I will first sample some of the results that
brought the assumption of perfect human rationality into question. Then,
in this and the next chapter, I will examine the alternative interpretations
of these results that have been advanced by evolutionary psychologists. I
will then attempt to reconcile the findings in the heuristics and biases literature with the research and theoretical stance of the evolutionary psychologists. It will be argued that although the evolutionary psychologists
have made substantial contributions to our understanding of human cognition, this perspective threatens to mislead us because it misconceives several important properties of human rationality. We will see that, although
evolutionary psychology has given us some useful tools to help in the robot's rebellion, the evolutionary psychologists often seem to side with the
selfish replicators.

The Dangers of Positive Thinking:
TASS Can't "Think of the Opposite"

An extensive body of psychological research has shown that there appears
to be a pervasive TASS bias to focus on positive instances and to fail to represent a state of affairs that might *not* be true.[4] One theory that has emphasized that important reasoning errors result from the failure to represent alternative outcomes and alternative hypotheses is Johnson-Laird's (1999,
2001; Johnson-Laird and Byrne 1991, 1993) theory of mental models. In
his theory, Johnson-Laird emphasizes what he terms the *principle of truth*,
that "individuals minimize the load on working memory by tending to construct mental models that represent explicitly only what is true, and not
what is false" (1999, 116). This automatic bias of TASS, together with what

is termed focusing—the fact that people restrict their reasoning to what is represented in their models—leads to many problem-solving errors when the analytic system does not override these tendencies. It ensures that people will perform suboptimally in tasks that require an exhaustive exploration of the alternative states of the world. Recall from the discussion of representational complexity and hypothetical thinking in chapter 2 that it is the analytic system that carries out the critical operation of decoupling representations from their anchoring in the world. TASS's default is to represent only real-world relationships—that is, relationships as given.

Johnson-Laird's (1999) principle of truth and a phenomenon studied by psychologist Dan Gilbert (1991)—the automatic acceptance of propositions—are related to the same TASS default. Gilbert contrasts the views of Descartes and Spinoza on the relation between comprehending statements and evaluating them. The Cartesian model is one where comprehension precedes assessment. Subsequent to the comprehension of a statement, the statement is evaluated for truth or falsity. Acceptance follows comprehension in the Cartesian view. In contrast, the Spinozan position is one where verbal comprehension is much more like perception—where seeing is believing. In the Spinozan view, comprehension and acceptance are simultaneous because they are basically one and the same. Comprehension is accompanied by acceptance. Acceptance is not subsequent to evaluation, as in the Cartesian model, but instead, in the Spinozan model, any evaluation must take place subsequent to acceptance.

This is not to say that a statement cannot be evaluated as false in the Spinozan view—it is just that a separate and subsequent stage of evaluation is preceded by acceptance of the belief. Also, to represent hypothetical states of affairs in the Spinozan view, acceptance always has to be overridden. This is why the representation of propositions as potentially false is so much harder. It is not the natural default strategy in the Spinozan model. To represent hypotheticality necessitates a costly psychological override. Marshaling much modern evidence, Gilbert (1991) argues that the findings point in the direct of Spinoza's model and not Descartes's. The TASS default is to accept propositions. *Undoing* acceptance (representing propositions as potentially false) appears to be a key function of the computationally costly decoupling operations of the analytic system.

The TASS bias toward automatic acceptance of propositions and representing them only as true appears to account for the problems people have in solving Wason's four-card selection task discussed in chapter 2 (if there is a vowel on one side of the card, then there is an even number on the other side). Subjects seem to forget that they must test the rule and find out not only whether it is true but also whether it is false. They do turn over the P card (the vowel) which, because the P could have a Q (an even number) on the back (consistent with the rule) or a not-Q (an odd number) on the back (falsifying the rule), suggests that they are examining the potential truth or falsity of the rule. However, the other choices they make suggest that they are not focused on potential falsity at all—that they are merely turning over the P card in the expectation of confirming the rule by finding an even number. The reason for this interpretation of their P choice is that the majority of subjects choose to turn over the Q card (even number) which cannot possibly falsify the rule. It can produce evidence consistent with the rule (by revealing a vowel), but it cannot conclusively confirm the rule. That can only be done by making sure that there are no falsifying instances, and this is what subjects resolutely fail to do. They fail to turn the not-Q card (the odd number) which cannot confirm the rule but can only falsify it. The pattern of choices suggests, as Johnson-Laird (1999) assumes, that the natural TASS tendency is to assume truth and focus attention on confirming the focal hypothesis. TASS views as relevant only evidence expected under an assumption of truth. A task that requires explicit attention to a state of affairs which may not be true will confound the natural processing biases of TASS and require explicit override by the analytic system.[5]

It is not difficult to construct tasks where TASS defaults such as the principle of truth lead reasoning seriously astray, as Johnson-Laird and Savary (1996; see also Yang and Johnson-Laird 2000) have shown. Consider the following problem taken from their work:

> Suppose that *only one* of the following assertions is true about a specific hand of cards:
>
>   1. There is a king in the hand or there is an ace in the hand, or both.
>   2. There is a queen in the hand or there is an ace in the hand, or both.
>
> Which is more likely to be in the hand: the king or the ace?

Be sure to answer before reading on.

Johnson-Laird and Savary (1996) found that about 75 percent of their subjects thought that the ace was more likely than the king. This answer is wrong—which is why Johnson-Laird and Savary titled their paper "Illusory Inferences about Probabilities." The illusion comes about because of the naturalness of representing information assumed to be true and because of the unnaturalness of representing situations which may be false. You probably still fell prey to the illusion even though you have just had your attention focused on the pitfalls of the principle of truth. Johnson-Laird and Savary propose that people make an illusory inference in the problem because they model the truth of each of the statements in turn but forget that one of the statements is false and that they must work out the implications of that. Statement 1 they represent as saying that there is a king, there is an ace, or there are both—each possibility represented by a line in the following mental model:

K
   A
K A

Statement 2 they represent as saying that there is a queen, there is an ace, or there are both—each possibility represented by a line in the following mental model:

Q
   A
Q A

Since there are more total lines containing an A than a K (4 versus 2), this leads subjects to think that the ace is more likely. Subjects forget that only one of the two statements is true and they fail to go about the somewhat complicated process of figuring out what the implications in one mental model are if the other model is false. This is because the two statements were originally both modeled in their "true," form which is the most complicated form to deal with in this case.

In contrast, the problem is much easier if what is represented is the two situations in which the statements are false:

Statement 1 is false if: not-K and not-A
Statement 2 is false if: not-Q and not-A

Now the instructions say that *one* of them has to be false. *Whichever* is false, there is no ace in the hand. So there *cannot* be an ace in the hand, but there can be a king—so a king is more likely to be in the hand.

So the problem is not computationally very hard *as long as people represent the state of affairs that is false.* However, much work in cognitive science has demonstrated that the default representational properties of TASS are just the opposite. It is geared to directing focal attention to the true state of affairs. Johnson-Laird (1999) acknowledges that this is a reasonable default for a computational system, but argues that it leads to gaps in the ability to deal with the hypothetical world of science, with alternative treatment outcomes, and comparative outcomes. Modeling only the true state of affairs precludes hypothetical thinking of the type discussed in chapter 2.

In contrast, an analytic processing system that allows the representation of what may be false (i.e., can decouple representations from their anchoring in the world and mark them as simulations) makes possible the cultural achievements of deductive reasoning skill, utilitarian morality, and scientific thinking. Deduction requires the representation of all possible symbolic states (not only those that are true), consequentialist decision making requires representing all possible states of the world (even those that might never obtain), and scientific thinking requires thinking about whether the data obtained was a reasonable outcome from the standpoint of an *alternative* hypothesis.

The processing characteristics assumed in Johnson-Laird's (1999) theory are no doubt adaptive. He argues that one reason that people represent only what is true in their mental models is to minimize the load on working memory and, often, the principle of truth seems to work as a reasonable "first-pass" or "ballpark" strategy (Friedrich 1993; McKenzie 1994). This is essentially the argument of evolutionary psychologists. However, the complex problem solving required in the modern world increasingly requires hypothetical reasoning. While such thinking styles were once the province of only a few, the abstract decontextualized cognitive ecology of the modern world increasingly requires it of everyone.

Now You Choose It—Now You Don't: Framing Effects Undermine the Notion of Human Rationality

Under the standard view of so-called "rational man" in economics and decision theory, it is traditionally assumed that people have stable, underlying preferences for each of the options presented in a decision situation. That is, it is assumed that a person's preferences for the options available for choice are complete, well ordered, and well behaved in terms of the axioms of choice mentioned previously (transitivity, etc.). It has been proven through several formal analyses (Edwards 1954; Fishburn 1981; von Neumann and Morgenstern 1944; Luce and Raiffa 1957; Savage 1954) that having such well-behaved internal preferences has the implication that a person is a utility maximizer—he/she acts to get what he/she most wants. Thus, "rational, economic man" maximizes utility in choice by having previously existing, well-ordered preferences that reliably determine choices when the person has to act on the preferences.

The main problem with this conception is that three decades of work by Kahneman and Tversky (2000) and a host of other cognitive and decision scientists has brought the view of the rational person with well-ordered preference sets into question.[6] The basic problem is, well, pretty basic— many people simply do not seem to have any such thing as stable, well-ordered preferences.

What this work has shown is that people's choices—sometimes choices about very important things—can be altered by irrelevant changes in how the alternatives are presented to them, and in how they respond to the alternatives presented to them (e.g., making purchases, verbally specifying, making a forced choice, etc.). There are dozens of such demonstrations in the research literature, but one of the most compelling is from the early work of Tversky and Kahneman (1981). Give your own reaction to Decision 1:

> Decision 1. Imagine that the U.S. is preparing for the outbreak of an unusual disease, which is expected to kill 600 people. Two alternative programs to combat the disease have been proposed. Assume that the exact scientific estimates of the consequences of the programs are as follows: If Program A is adopted, 200 people will be saved. If Program B is adopted, there is a one-third probability that 600 people will be saved and a two-thirds probability that no people will be saved. Which of the two programs would you favor, Program A or Program B?

Most people when given this problem prefer Program A—the one that saves 200 lives for sure. There is nothing wrong with this choice taken alone. It is only in connection with the responses to another problem that things really become strange. The experimental subjects (sometimes the same group, sometimes a different group—the effect obtains either way) are given an additional problem. Again, give your own immediate reaction to Decision 2:

> Decision 2. Imagine that the U.S. is preparing for the outbreak of an unusual disease, which is expected to kill 600 people. Two alternative programs to combat the disease have been proposed. Assume that the exact scientific estimates of the consequences of the programs are as follows: If Program C is adopted, 400 people will die. If Program D is adopted, there is a one-third probability that nobody will die and a two-thirds probability that 600 people will die. Which of the two programs would you favor, Program C or Program D?

Most subjects when presented with Decision 2 prefer Program D. Thus, across the two problems, the most popular choices are Program A and Program D. The only problem here is that Decision 1 and Decision 2 are really the same decision—they are merely redescriptions of the same situation. Program A and C are the same. That 400 will die in Program C implies that 200 will be saved—precisely the same number saved in Program A. Likewise, the two-thirds chance that 600 will die in Program D is the same two-thirds chance that 600 will die ("no people will be saved") in Program B. If you preferred Program A in Decision 1 you should have preferred Program C in Decision 2. But many subjects show inconsistent preferences—their preference switches depending on the phrasing of the question. It is important to note that the subjects themselves—when presented with both versions of the problem—agree that the problems are identical and that the alternative phrasing shouldn't have made a difference.

The inconsistency displayed in the Disease Problem is a violation of a very basic axiom of rational decision, the so-called property of descriptive invariance (Kahneman and Tversky 1984, 2000). If choices flip-flop based on problem characteristics that the *subjects themselves* view as irrelevant— then the subject can be said to have no stable, well-ordered preferences at all. If preferences reverse based on inconsequential aspects of how the problem is phrased, people cannot possibly be maximizing expected utility. Such failures of descriptive invariance have quite serious implications for our view of whether or not people are rational, yet such failures are not difficult to generate. The literature on decision making is full of them. Consider the following two problems (from Tversky and Kahneman 1986) framed in a gambling context commonly used in the decision theory literature:

Decision 3. Imagine that, in addition to whatever else you have, you have been given a cash gift of $300. You are now asked to choose between two options:

    (A) a sure gain of $100 and

    (B) a 50 percent chance of winning $200 and a 50 percent chance of winning nothing.

Decision 4. Imagine that, in addition to whatever else you have, you have been given a cash gift of $500. You are now asked to choose between two options:

    (C) a sure loss of $100, and

    (D) a 50 percent chance of losing $200 and a 50 percent chance of losing nothing.

Tversky and Kahneman (1986) found that 72 percent of their sample preferred option A over B and that 64 percent of their sample preferred option D over option C. Again though, just as in the disease example, the two decisions reflect comparisons of exactly the same outcomes. If someone preferred A to B, then they should prefer C to D. The sure gain of $100 in option A when added to the starting sum of $300 means ending up with $400—just as the sure loss on option C means ending up with $400 in Decision 4. Option B means a 50 percent chance of ending up with $500 ($300 plus winning $200) and a 50 percent chance of ending up with $300 ($300 plus winning nothing)—just as does option D in Decision 4 (50 percent chance of $500 minus $200 and a 50 percent chance of $500 minus nothing).

The theory of why these failures of descriptive invariance occur was termed prospect theory by Kahneman and Tversky (1979; Tversky and Kahneman 1986). What the examples have in common is that in both cases subjects were risk averse in the context of gains and risk seeking in the context of losses. They found the sure gain of 200 lives attractive in Decision 1 over a gamble of equal expected value; and in Decision 3 they found the sure $100 gain attractive over a gamble of equal expected value. In contrast, in Decision 2 the sure loss of 400 lives was unattractive compared with the gamble of equal expected value. Of course, the "sure loss" of 400 here that subjects found so unattractive is exactly the same outcome as the "sure gain" of 200 that subjects found so attractive in Decision 1!

Similarly, the "sure loss" in option C of Decision 4 was seen as unattractive compared to a gamble of equal expected value. In both of these problems, subjects coded outcomes in terms of gains and losses from a zero point (rather than evaluating the final position). This is one of the key assumptions of Kahneman and Tversky's (1979) prospect theory (see also Markowitz 1952). One of their other key assumptions is that the utility function is steeper (in the negative direction) for losses than for gains. This is why people are often risk averse even for gambles with positive expected values. Would you flip a coin with me—heads you give me $500, tails I give you $505? Most people refuse such favorable bets because the potential loss, although nominally smaller than the potential gain, looms larger psychologically.

These two aspects of how people approach decisions—the differential steepness of the utility function and that they recode options as being from a zero reference point of the current status quo (in terms of wealth, lives, or whatever is at issue)—appear to be TASS characteristics. They are automatic ways that information in decision situations is coded. They cannot be turned off, but they can be overridden (for example, by analytic strategies to make sure that one's preferences are invariant under different representations of the problem). Dawes (1988) notes that "prospect theory describes choices made as the result of automatic processes" (45). That these two characteristics of the way we code decision information can cause an outright reversal of preferences if not overridden is a somewhat frightening implication of the way our TASS preference apparatus is structured. There is the unsettling idea latent here that people's preferences come from the outside (from whoever has the power to shape the environment and determine how questions are phrased) rather than internal preferences based in their unique psychologies. Since most situations can be reframed either way, this means that rather than having stable preferences that are just elicited in different ways, the elicitation process itself can totally *determine* what the preference will be! Such a conception brings the foundational concepts behind the idea of "rational man" from economics crashing down. It also has potent social implications. As Kahneman (1994) has pointed out, the assumption of stable, rational, well-ordered preferences has been used to "support the position that it is unnecessary to protect people against the consequences of their choices" (18).

There are also many important practical consequences of the demonstrations of descriptive invariance. For example, in the important area of medical treatment, outcome information can often be represented in terms of either losses or gains. Work by McNeil, Pauker, Sox, and Tversky (1982) has shown that—as in the examples above—alternative phrasings of the same outcome can lead to different treatment choices for lung cancer. They found that the framing effects were just as large for physicians and those with statistical training as it was for clinic patients. Note that it is a bit scary that preferences for medical treatments can reverse, based on a rephrasing of the outcomes that conveys no differential information about the treatments and that *changes nothing about the outcomes themselves*. This example shows how truly frightening violations of descriptive invariance can be— how frightening is an irrational response pattern (one which, because of its inconsistency, violates the basic stricture of instrumental rationality that people should act to get what they most want).

The key feature in all of these problems is that subjects appear to accept whatever framing is given in the problem as set—without trying alternative ones to see if inconsistencies might result. Tversky (1996a) has argued that "these observations illustrate what might be called the acceptance principle: people tend to accept the frame presented in a problem and evaluate the outcomes in terms of that frame" (8). Here we have another illustration of a TASS processing bias that was mentioned earlier in the discussion of John-son-Laird's (1999) principle of truth and the Spinozan fusion of compre-hension and acceptance examined by Gilbert (1991). TASS is biased toward the automatic acceptance of propositions and biased to accept the context as given. If the exploration of alternative hypotheses and alternative fram-ings of issues are to take place, analytic system override of this natural pro-cessing tendency is essential.

These examples of violations of descriptive invariance represent just the tip of the iceberg of a large literature on anomalies in preference judgments.[7] They mean that people cannot be expected to be utility maximizers—they cannot display consistent rationality in the sense that rationality was de-fined in the last chapter. Additionally, it is important to note that the fact that human choices are so easily altered by framing has potent social and economic implications as well. Thaler (1980) describes how years ago the credit card industry lobbied intensely for any differential charges between credit cards and cash to be labeled as a discount for using cash rather than a surcharge for using the credit card. They were implicitly aware of an as-sumption of Kahneman and Tversky's (1979) prospect theory—that any surcharge would be psychologically coded as a loss and weighted highly in negative utility. The discount, in contrast, would be coded as a gain. Because the utility function is shallower for gains than for losses, forgoing the dis-count would be psychologically easier than accepting the surcharge. Of course, the two represent exactly the same economic consequence. The in-dustry, merely by getting people to accept the higher price as normal, framed the issue so that credit card charges were more acceptable to people.

Descriptive invariance is an important property of rational choice, and, as we have just seen, it is violated in a host of choice situations. There is a re-lated principle—procedural invariance—that is equally basic and equally problematic for human decision making. Procedural invariance is—like de-scriptive invariance—one of the most basic assumptions that lies behind the standard model of rational choice (Kahneman and Tversky 2000; Slovic 1995). It says that choices should not depend on the way that the prefer-ences are elicited. However, although this assumption is indeed basic to the notion of "rational economic man," the evidence has accumulated over the past thirty years that irrational sets of preferences often result because pro-cedural invariance is violated.

Consider the following two contracts:

Contract A: you receive $2,500 five years from now.
Contract B: you receive $1,600 in one and half years.

Tversky, Slovic, and Kahneman (1990) found that three quarters of their subjects preferred Contract B, but when asked what was the smallest amount they would sell each of the contracts for if they owned it, three quarters of their subjects set a higher price on Contract A. This set of responses again leads to an intransitivity and hence to a disastrous money pump.[8] Because we are pricing and choosing all the time in our modern market economies, the potential for procedural inconsistencies[9] revealed by the research reviewed in this chapter demonstrates that there are suboptimal patterns lodged in human cognition that could mean that we fall short of full instrumental rationality—we do not always act to get what we most want.

When presented with a rational choice axiom that they have just violated (the sure-thing principle, transitivity, descriptive invariance, procedural invariance, etc.) in a choice situation, most subjects will actually endorse the axiom. As Shafir and Tversky (1995) describe it: "When confronted with the fact that their choices violate dominance or descriptive invariance, people typically wish to modify their behavior to conform with these principles of rationality. . . . People tend to accept the normative force of dominance and descriptive invariance, even though these are often violated in their actual choices" (97). That subjects endorse the strictures of rationality when presented with them explicitly suggests that most subjects' analytic abilities are such that they can appreciate the force of the axioms of rational choice. If people nevertheless make irrational choices despite consciously endorsing rational principles, this suggests that the ultimate cause of the irrational choices might reside in TASS—the parts of their brains that ignore them.

Can Evolutionary Psychology Rescue the Ideal of Human Rationality?

So far in this chapter I have focused on the strictures of instrumental rationality, but people violate the principles that define epistemic rationality as well. They improperly coordinate theory to evidence, they display confirmation bias, they test hypotheses inefficiently, they do not properly calibrate degrees of belief, they overproject their own opinions onto others, they allow prior knowledge to become implicated in deductive reasoning, and they display numerous other information processing biases in the epistemic domain. Several reviews of this literature describe these errors of reasoning in considerable detail.[10]

Taken collectively, the disturbingly long list of irrationalities that characterize human problem solving and decision making paint a troubling portrait of human rationality. However, over the last decade, an alternative interpretation of these findings has been championed by various evolutionary psychologists, adaptationist modelers, and ecological theorists (e.g., Cosmides and Tooby 1992, 1994b, 1996; Gigerenzer 1996a; Oaksford and Chater 1998, 2001). They have reinterpreted the modal response in most of the classic heuristics and biases experiments—the response that deviates from the one dictated by models of human rationality—as indicating an optimal information processing adaptation on the part of the subjects. It is argued by these investigators that the research in the heuristics and biases tradition has not demonstrated human irrationality at all and that a Panglossian position which assumes perfect human rationality is still defensible.

The evolutionary theorists argue that the TASS heuristics are wrongly characterized by the heuristics and biases researchers. The latter often view the TASS heuristics as short-cuts to get around the computational load imposed by the optimally rational model. The evolutionary psychologists object to conceptualizing TASS heuristics as suboptimal shortcuts. For them, the TASS heuristics are not to be viewed as falling short of some rational goal but instead are fully optimal processing devices designed to solve a particular set of evolutionary problems. They even go so far as to argue that the nonoptimal choices displayed on heuristics and biases tasks should not be viewed as irrational (Cosmides and Tooby [1994a] actually call TASS evolutionary modules "better than rational").

To see how the evolutionary psychologists make this argument, we need to examine their alternative interpretations of the violations of rational axioms discussed previously, and I will do so in this chapter. However, I will also argue that although the work of the evolutionary psychologists has uncovered some fundamentally important things about human cognition, these theorists have misconstrued the nature of human rationality and have conflated important distinctions in this domain. I will argue that dual process models of cognitive functioning, like those that were discussed in chapter 2, provide a way of reconciling the positions of the evolutionary psychologists and the researchers in the heuristics and biases tradition. Such models acknowledge the exemplary efficiency of certain domain-specific TASS processes emphasized by the evolutionary psychologists. But unlike evolutionary psychologists, dual-process theorists emphasize the importance of domain-general analytic processing as well. I will argue that it is the nonautonomous, serial-processing operations of executive control and problem solving that serve to guarantee instrumental rationality by overriding the responses generated by TASS when the latter threaten optimal outcomes at the personal level.

I will also argue that modern society has an extraordinary capacity to create mismatches between the outputs of TASS subsystems optimized under evolutionary criteria and the responses needed to achieve instrumental rationality at the personal level (refer again to figure 2.2 in chapter 2). Indeed, our modern world has immensely influential social and market structures (advertising, for example) that are specifically designed to exploit such conflicts. In the next section, I will describe some of the fundamental processing biases in the TASS subsystems that make them exploitable in this manner. The evolutionary psychologists are correct that in the majority of mundane day-to-day situations, the TASS mechanisms ensure instrumental rationality as well. Unfortunately though, the fundamental computational biases of the TASS subsystems can, on the few occasions when their defaults do not coincide with our personal interests, lead to hardship and grief.

The Fundamental Computational Biases of the Autonomous Brain

Consider the following syllogism. Ask yourself if it is valid—whether the conclusion follows logically from the two premises:

Premise 1: All living things need water.
Premise 2: Roses need water.
Therefore, Roses are living things.

What do you think? Judge the conclusion either logically valid or invalid before reading on.

If you are like about 70 percent of the university students who have been given this problem, you will think that the conclusion is valid. And if you did think that it was valid, like 70 percent of the university students who have been given this problem, you would be wrong (Markovits and Nantel 1989; Sá, West, and Stanovich 1999; Stanovich and West 1998c). Premise 1 says that all living things need water, not that all things that need water are living things. So, just because roses need water, it doesn't follow from Premise 1 that they are living things. If that is still not clear, it will probably become clear after you consider the following syllogism with exactly the same structure:

Premise 1: All insects need oxygen.
Premise 2: Mice need oxygen.
Therefore, Mice are insects.

Now it seems pretty clear that the conclusion does not follow from the premises.

If the logically equivalent "mice" syllogism is solved so easily, why is the "rose" problem so hard? Well for one thing, the conclusion (roses are living things) seems so reasonable and you know it to be true in the real world. And that is the rub. Logical validity is not about the believability of the conclusion—it is about whether the conclusion necessarily follows from the premises. The same thing that made the rose problem so hard made the mice problem easy. The fact that "mice are insects" is not true by definition in the world we live in might have made it easier to see that the conclusion did not follow logically from the two premises.

In both of these problems, prior knowledge about the nature of the world (that roses are living things and that mice are not insects) was becoming implicated in a type of judgment (judgments of logical validity) that is supposed to be independent of content. In the rose problem prior knowledge was interfering, and in the mice problem prior knowledge was facilitative. In fact, if we really wanted to test a person's ability to process the relationships in this syllogism, we might have used totally unfamiliar material. For example, I might have told you to imagine you were visiting another planet and that you found out the following two facts:

All animals of the hudon class are ferocious.
Wampets are ferocious.

We might then ask you to evaluate whether it logically follows that: Wampets are animals of the hudon class. We can see here that the conclusion does not follow. Research has shown that it is easier to see that the conclusion lacks validity in this unfamiliar version than it is in the rose version, but it is harder to see that the conclusion does not follow in the unfamiliar version than it is in the mice version. These differences prove that factual knowledge is becoming implicated in both the rose and mice problems—even though the content of syllogisms should have no impact on their logical validity. The effect on the rose problem is quite large. Only about 32 percent of university students solve it whereas the same subjects respond correctly 78 percent of the time on logically equivalent versions with unfamiliar material (versions where prior knowledge does not get in the way).

The rose problem illustrates one of the fundamental computational biases of TASS—the tendency to automatically utilize all the contextual information presented when solving problems. That prior knowledge is implicated in performance on this problem even when the person is ex-plicitly told to ignore the real-world believability of the conclusion illus-trates that this tendency toward contextualizing problems is so ubiquitous that it cannot easily be turned off—hence its characterization here as a fundamental computational bias (one that pervades virtually all thinking whether we like it or not).[11] This property of obligatory processing is a defining feature of the reflex-like TASS processes.

Of course, the tendency to use all contextual information available to supplement problem solving is more often a help than a hindrance. It will be argued in this chapter that, consistent with arguments in evolutionary psychology (e.g., Barkow, Cosmides, and Tooby 1992; Buss 1999, 2000; Cosmides and Tooby 1994b; Pinker 1997; Plotkin 1998; Tooby and Cos-mides 1992), these fundamental computational biases are resident in the brain because they were adaptive in the so-called environment of evo-lutionary adaptedness (EEA)—our ancestral environment—that existed throughout the Pleistocene. In short, it will be argued that these computa-tional biases make evolutionary sense. Nevertheless, it will also be argued that despite their usefulness in the EEA, and despite the fact that even in the present environment they are more useful than not, the modern world pres-ents situations in which the type of contextualization rendered by the fun-damental computational biases proves extremely problematic. Such situa-tions are numerically minority situations, but they tend to be ones where a misjudgment might have disproportionately large consequences for a per-son's future utility maximization—for the future fulfillment of the person's life goals.

In situations where the present human environment is similar to the EEA, the human brain is characterized by a set of fundamental computa-tional biases that quite efficiently facilitate goal achievement. However, when technological societies create new problems that confound these evo-lutionarily adapted mechanisms, humans must use cognitive mechanisms that are in part cultural inventions to override the fundamental computa-tional biases that, in these situations, will prime the wrong response. These culturally induced processing modes are the abstract, analytic, but compu-tationally expensive processes of rational thought (Dennett 1991; Rips 1994; Sloman 1996, 2002). It is important to recall from chapter 2 that TASS processing pervades all functioning, and that it cannot be "turned off" but instead must be overridden on a case by case basis.[12]

The fundamental computational biases of human cognition derive from the automatic inferential machinery of the TASS brain. These biases have the effect of providing rich supplemental knowledge to augment the sometimes fragmentary and incomplete information we receive when faced with real-world problems. The four interrelated biases I have discussed in detail elsewhere (see Stanovich 2003) are:

1. The tendency to contextualize a problem with as much prior knowledge as is easily accessible, even when the problem is formal and the only solution is a content-free rule,
2. The tendency to "socialize" problems even in situations where interpersonal cues are few,
3. The tendency to see deliberative design and pattern in situations that lack intentional design and pattern, and
4. The tendency toward a narrative mode of thought.

These biases, or processing defaults, often work in mutually reinforcing ways, and each makes good evolutionary sense. Problematic situations occur, though, when a modern technological society turns the tables on these processing defaults and demands that they be overridden if instrumental rationality is to be achieved—if people are to get what they most want given their personal goals.

The Evolutionary Adaptiveness of the Fundamental Computational Biases

Each of the fundamental computational biases discussed previously is a functional aspect of human cognition. Indeed, they are fundamental precisely because they are orientations in our basic systems (TASS) that arose in our evolutionary history—probably long before the more abstract features of analytic intelligence (Mithen 1996, 2000, 2002). Many investigators have painted compelling theoretically and empirically based explanations of why these computational biases developed in the course of human evolution (Cosmides and Tooby 1992; Humphrey 1976, 1986; Mithen 1996, 2000, 2002; Pinker 1997). The socialization of problems and the tendency to see deliberate design in the environment follow from the evolutionary assumptions behind the social intelligence hypothesis[13]—that attributing intentionality in order to predict the behavior of conspecifics and to coordinate behavior with them was a major evolutionary hurdle facing the social primates, in many cases more computationally complex than mastering the physical environment (Humphrey 1976).

The tendency to attribute intention may have other evolutionary sources. Dennett (1991, 32) suggests that we have an innate disposition to treat things that rapidly fluctuate in our environment as if they had souls and that this is an evolutionary design trick allowing us to categorize things in the world without thinking too much. The ubiquitous tendency to adopt what Dennett (1978, 1987) calls the intentional stance underlies many of the fundamental computational biases (particularly the tendency to see human design in the world and to socialize problems). There appear to be biologically based brain structures in TASS devoted to supporting the intentional stance toward other animate beings (Baron-Cohen 1995; Baron-Cohen, Tager-Flusberg, and Cohen 2000). Evolutionarily later aspects of analytic cognition did not replace these older socially based mechanisms but were built on top of them (see Dennett 1991, 1996; Mithen 1996). Thus, aspects of social intelligence infuse even abstract problems that are best solved with later developing analytic intelligence.

Finally, there exist many theoretical arguments for why the automatic contextualization of problems with prior knowledge might be adaptive. For example, Evans and Over (1996) argue that beliefs that have served us well in the past should be hard to dislodge, and projecting them on to new information—because of their past efficacy—might help in assimilating the new information. This argument for the adaptiveness of contextualization—that in a natural ecology where most of our prior beliefs are true, projecting our beliefs on to new data will lead to faster accumulation of knowledge—I have termed the knowledge projection argument (Stanovich 1999), and it reappears in a remarkably diverse set of disciplines and specialties within cognitive science. For example, the knowledge projection argument has been used to explain the false consensus effect in social psychology (Dawes 1989, 1990; Hoch 1987; Krueger and Zeiger 1993), findings on expectancy effects in learning (Alloy and Tabachnik 1984), biases in evaluating scientific evidence (Koehler 1993), realism effects in attributing intentionality (Mitchell, Robinson, Isaacs, and Nye 1996), syllogistic reasoning (Evans, Over, and Manktelow 1993), and informal reasoning (Edwards and Smith 1996). Of course, all of these theorists emphasize that we must project from a largely *accurate* set of beliefs in order to obtain the benefit of knowledge projection (in a sea of inaccurate beliefs, the situation is quite different; see chapter 8 of Stanovich 1999 for a discussion).

The importance of the knowledge projection argument for the present discussion is that it has been used by evolutionary psychologists to justify the nonoptimal responses that many subjects make on heuristics and biases tasks. Correct responding on such tasks often necessitates some type of decontextualization, so that TASS-determined contextualizing tendencies often prime the wrong response. For example, cognitive scientist Steven Pinker (1997) makes the point most succinctly, if colloquially, when he notes that "outside of school, of course, it never makes sense to ignore what you know" (304).

But is this really true? The argument of this chapter will be that Pinker might be underestimating how much, in the "real world," we really *do* have to "ignore what we know" or believe, or think. Consider:

1. The sales clerk who must ignore the insulting behavior of the customer he knows to be a jerk and serve him as politely as everyone else.
2. The lawyer who must ignore the fact that her client is guilty.
3. The teacher who must ignore a pupil who is intentionally trying to irritate.
4. The judicial process and how it is structured to ignore the prior records of the defendants and victims.

And finally, as Pinker (1997) himself discusses, the science on which modern technological societies are based depends on "ignoring what we know or believe." Testing a control group under conditions B when you think A is true is a form of ignoring what you believe. Science is a way of systematically ignoring what we know, at least temporarily (during the test), so that we can recalibrate our belief after the evidence is in.

It is not hard to generate examples like these, and later I will return to this issue of how modern society requires that we ignore what we know. Before proceeding with that discussion, however, I need to describe in more detail how evolutionary psychologists reinterpret the findings of widespread human irrationality on many problems in the heuristics and biases literature.

Evolutionary Reinterpretations of Responses on Heuristics and Biases Tasks

In this section, I will illustrate how evolutionary psychologists, adaptationist modelers, and ecological theorists have reinterpreted the modal response in most of the classic heuristics and biases experiments, previously interpreted as indications of human irrationality, as instead indicating an optimal information processing adaptation on the part of the subjects. An example is again provided by Wason's (1966) selection task introduced in chapter 2 and mentioned again earlier in this chapter ("If there is a vowel on one side of the card, then there is an even number on the other side. Turn over whichever cards are necessary to determine whether the experimenter's rule is true or false"—followed by A, K, 5, 8 face up). As discussed previously, numerous alternative explanations for the preponderance of incorrect PQ responses have been given. What is important in the present context is that several of these alternative theories posit that the incorrect PQ response results from the operation of efficient and optimal cognitive mechanisms. For example, Oaksford and Chater (1994, 1996; see also Nickerson 1996) argue that rather than interpreting the task as one of deductive reasoning (as the experimenter intends), many people interpret it as an inductive problem of probabilistic hypothesis testing. They show that the P and Q response is actually the expected one if an inductive interpretation of the problem is assumed along with optimal data selection (which they modeled with a Bayesian analysis). Other investigators have different models of why the preponderant PQ response obtains, but they have in common the assumption that this response, previously viewed as reflecting a cognitive defect in human inferential abilities, instead reflects efficient cognitive mechanisms operating in an optimal manner. For example, Sperber, Cara, and Girotto (1995) stress that selection task performance is driven by optimized cognitive mechanisms for inferential comprehension—mechanisms that are "geared towards the processing of optimally relevant communicative behaviors" (90).

A second example of theorists defending as rational the response that heuristics and biases researchers have long considered incorrect is provided by the much-investigated Linda problem (Tversky and Kahneman 1983) discussed in chapter 2. The description of Linda (she majored in philosophy . . . she was deeply concerned with issues of discrimination and social justice, etc.) led people to judge it more probable that Linda is a bank teller and is active in the feminist movement than it is that she is a bank teller. However, several investigators have suggested that rather than illogical cognition, it is rational pragmatic inferences that lead to the violation of the logic of probability theory in the Linda Problem.[14]

Hilton (1995) summarizes the view articulated in these critiques by arguing that "the inductive nature of conversational inference suggests that many of the experimental results that have been attributed to faulty reasoning may be reinterpreted as being due to rational interpretations of experimenter-given information" (264). This alternative explanation for the typical response on the Linda problem is essentially an argument that fundamental processing default 2 listed above (the tendency to socialize abstract problems) is a rational default to make in this particular situation. In short, these critiques imply that displaying the conjunction fallacy is a rational response triggered by the adaptive use of social cues, linguistic cues, and background knowledge.

As the next example, consider the 2×2 covariation detection task. Subjects are asked to evaluate the efficacy of a drug based on a hypothetical well-designed scientific experiment. They are told that:

150 people received the drug and were cured
150 people received the drug and were not cured
300 people did not receive the drug and were cured
75 people did not receive the drug and were not cured

These data correspond to four cells of the 2×2 contingency table traditionally labeled A, B, C, and D, respectively. Subjects are asked to evaluate the effectiveness of the drug on a scale. Much previous experimentation has produced results indicating that subjects weight the cell information in the order cell A > cell B > cell C > cell D—cell D receiving the least weight and/or attention (see Arkes and Harkness 1983; Kao and Wasserman 1993; Schustack and Sternberg 1981). The tendency to ignore cell D is nonoptimal, as indeed is any tendency to differentially weight the four cells. The rational strategy (see Allan 1980; Kao and Wasserman 1993; Shanks 1995) is to use the conditional probability rule—subtracting from the probability of the target hypothesis when the indicator is present the probability of the target hypothesis when the indicator is absent. Numerically, the rule amounts to calculating the Δp statistic: $[A/(A + B)] - [C/(C + D)]$ (see Allan 1980). For example, the Δp value for the problem presented above is $-.300$, indicating a fairly negative association.

Despite the fact that the optimal judgment strategy weights each cell equally, the modal subject in such experiments markedly underweights cell D. However, Anderson (1990) has modeled the 2×2 contingency assessment experiment using a model of adapted information processing and come to a startling conclusion. He demonstrates that an adaptive model (making certain assumptions about the environment) can predict the much-replicated finding that the D cell (cause absent and effect absent) is vastly underweighted (however, see Over and Green 2001). Thus, here again in another task is the pattern where the modal response violates what some cognitive psychologists consider the most rational response rule—but this modal response has been defended from the standpoint of an adaptationist analysis.

A final example comes from the probabilistic learning paradigm that has many versions in psychology.[15] In one, the subject sits in front of two lights (one red and one blue) and is told that she or he is to predict which of the lights will be flashed on each trial and that there will be several dozen such trials (subjects are often paid money for correct predictions). The experimenter has actually programmed the lights to flash randomly, with the provision that the red light will flash 70 percent of the time and the blue light 30 percent of the time. Subjects do quickly pick up the fact that the red light is flashing more, and they predict that it will flash on more trials than the blue light. Most often, when predicting they switch back and forth, predicting the red light roughly 70 percent of the time and the blue light roughly 30 percent of the time—thus matching the probabilities.

This strategy of probability matching is suboptimal because it insures that, in this example, the subject will correctly predict only 58 percent of the time ($.7 \times .7 + .3 \times .3$) compared to the 70 percent hit rate that could be achieved by predicting the most likely color (red) on each trial. In fact, much experimentation has indicated that animals and humans often fail to maximize expected utility in the probabilistic learning experiment. Nevertheless, Gigerenzer (1996b) shows how probability matching could, under some conditions, actually be an evolutionarily optimal strategy (see Cooper 1989, and Skyrms 1996, for many such examples). Thus, we have in probability matching another example of how a response tendency traditionally viewed as irrational is defended on an evolutionary or adaptationist account.

With interpretations such as these, the evolutionary psychologists have challenged the conclusions of the heuristics and biases researchers. The latter were prone to attribute irrationality to any response that did not match the response dictated by the rules of probability, logic, or utility theory. In contrast, the ecological theorists, adaptationist modelers, and evolutionary psychologists who have critiqued the tasks have found ways to reinterpret the modal response given so that it can be viewed as adaptive—in their view, a move that preserves the assumption of human rationality. Furthermore, in work that has been very influential in cognitive science, the redesign of some heuristics and biases tasks so that the correct response is more likely to be triggered by a module in TASS has been shown to improve performance on some of the most difficult problems.[16] Some theorists have interpreted these results and theoretical arguments as supporting a Panglossian position (discussed more extensively in chapter 6) that humans are inherently rational. I will explain in the remainder of this chapter why I think interpreting the evolutionary arguments in this light is a mistake.

My alternative interpretation explicitly acknowledges the impressive record of descriptive accuracy enjoyed by a variety of adaptationist and evolutionary models in predicting the modal response in a variety of heuristics and biases tasks. However, my account attempts to make sense of another important empirical fact—that subjects giving the response that the evolutionary psychologists defend tend to be somewhat lower in cognitive ability than those giving the responses deemed rational by the heuristics and biases researchers (Stanovich 1999; Stanovich and West 1998b, 1998c, 1998d, 2000; West and Stanovich 2003). This pattern has been found in several of the tasks discussed previously. For example, it is to the credit of models of optimal data selection (Oaksford and Chater 1994) that they predict the modal PQ response in the four-card selection task. But we are left with the seemingly puzzling finding that the response deemed optimal under such an analysis (PQ) is given by subjects of substantially lower general intelligence than the minority giving the response deemed correct under a strictly deductive interpretation of the problem (P and not-Q).

A similar puzzle surrounds findings on the Linda conjunction problem and analyses which stress that the tendency to commit the conjunction fallacy reflects the evolved use of sociolinguistic cues. Because this group is in fact the vast majority in most studies—and because the use of such pragmatic cues and background knowledge is often interpreted as reflecting adaptive information processing—it might be expected that these individuals would be the subjects of higher cognitive ability. In research done in my own laboratory, we found the contrary—the cognitive ability of the subjects who committed the conjunction fallacy was significantly *lower* than that of the minority who avoided the fallacy. Thus, the pragmatic interpretations of why the conjunction effect is the modal response on this task might well be correct—but the modal response happens not to be the one given by the most intelligent subjects in the sample.

Likewise, in the 2×2 covariation detection experiment, we have found that it is those subjects weighting cell D more *equally* (not those underweighting the cell in the way that the adaptationist model dictates) who are higher in cognitive ability. Again, Anderson (1990, 1991) might well be correct that a rational model of information processing in the task predicts underweighting of cell D by most subjects, but more severe underweighting is in fact associated with *lower* cognitive ability in our individual differences analyses.

Similarly, in several recently completed experiments on probability matching using a variety of different paradigms (West and Stanovich 2003), we have found the same pattern—the modal response of probability matching is consistent with an evolutionary analysis, but the most cognitively able subjects give the response that is instrumentally rational (predicting the most probable outcome on every trial). Again, the choice defensible on evolutionary grounds (probability matching) is the modal choice. But again, as before, it is the maximizing choice that is the choice of the subjects with the highest intellectual ability.

Finally, consider the knowledge projection argument discussed previously—that it is an adaptive tendency to contextualize problems with prior knowledge because, in a natural ecology, most of our prior knowledge is accurate. This argument is often used by evolutionary psychologists to justify the nonnormative responses on heuristics and biases tasks that require some type of decontextualization. The knowledge projection explanation does indeed explain why it is the modal response tendency to display belief bias and other contextual effects, but once again the individual difference trend is just the opposite. Subjects of higher cognitive ability tend to be less prone to project prior knowledge when problem solving and to be more prone to override automatic contextualization (Stanovich 1999).

In the results just reviewed there are two basic patterns that must be reconciled. The interpretations of the evolutionary psychologists correctly predict the modal response in a host of heuristics and biases tasks. Yet in all of these cases—despite the fact that the adaptationist models predict the modal response quite well—individual differences analyses (e.g., relationships with cognitive ability) must also be accounted for.

Evolutionary psychology is in fact mute on the issue of individual differences—it has no mechanism in its theoretical apparatus to explain these associations with cognitive ability. However, the dual-process cognitive architecture introduced in chapter 2 encompasses both the impressive record of descriptive accuracy enjoyed by a variety of evolutionary/adaptationist models as well as the fact that cognitive ability sometimes dissociates from the response deemed optimal on an adaptationist analysis. These data patterns make sense if it is assumed: (1) that there are two systems of processing with the properties outlined in chapter 2; (2) that the two systems of processing are optimized for different situations and different goals as outlined in chapter 2; and (3) that in individuals of higher cognitive ability there is a greater probability that the analytic system will override the response primed by TASS.

The argument will be that the natural processing tendency on many of these problems yields a TASS response—an automatic heuristic, as Tversky and Kahneman (1974) originally argued[17]—that primes the wrong response. The evolutionary psychologists are probably correct that this TASS response is evolutionarily adaptive. Nevertheless, their evolutionary interpretations do not impeach the position of the heuristics and biases researchers that the alternative response given by the minority of subjects is rational at the level of the individual. Subjects of higher analytic intelligence are simply more prone to override their automatically primed TASS response in order to produce responses that are epistemically and instrumentally rational.

It is necessary to reinvoke the distinction from earlier chapters between evolutionary adaptation (optimization for the replicators) and instrumental rationality (optimization of the vehicle's goals) in order to understand what is happening here. In this distinction lies a possible rapprochement between the heuristics and biases researchers who have emphasized the flaws in human cognition (Kahneman and Tversky 1973, 1996, 2000) and the evolutionary psychologists who have emphasized the optimality of human cognition (Cosmides and Tooby 1994b, 1996). For example, evolutionary psychologists are fond of pointing to the optimality of cognitive functioning—of showing that certain reasoning errors that cognitive psychologists have proposed are a characteristic and problematic aspect of human reasoning have in fact a logical evolutionary explanation. They have made a substantial contribution by providing empirical support for such explanations. But such a finding does not necessarily imply that the concern with cognitive reform is misplaced. The possibility of a dissociation between replicator and vehicle goals means that evolutionary adaptation does not guarantee instrumental rationality.

The Fundamental Computational Biases and
the Demands for Decontextualization in Modern Society

Evolutionary psychologists are prone to emphasize situations where genetic goals and personal goals coincide. They are not wrong to do so, because this is often the case. Accurately navigating around objects in the natural world was adaptive during the EEA, and it similarly serves our personal goals as we carry out our lives in the modern world. Likewise, with other evolutionary adaptations: It is a marvel that humans are exquisite frequency detectors (Hasher and Zacks 1979), that they infer intentionality with almost supernatural ease (Levinson 1995), and that they acquire a complex language code from impoverished input (Pinker 1994)—and all of these mechanisms serve personal goal fulfillment in the modern world. But none of this means that the overlap is necessarily 100 percent.

Unfortunately, the modern world tends to create situations where some of the default values of evolutionarily adapted cognitive systems are not optimal. Many of these situations implicate the fundamental computational biases discussed previously. These biases serve to radically contextualize problem-solving situations. In contrast, modern technological societies continually spawn situations where humans must decontextualize information—where they must deal abstractly and in a depersonalized manner with information. Such situations require the active suppression of the social, narrative, and contextualizing styles that characterize the operation of TASS. These situations may not be numerous, but they tend to be in particularly important domains of modern life—indeed, they in part *define* modern life in postindustrial knowledge-based societies.

Mechanisms designed for survival in preindustrial times are clearly sometimes maladaptive in a technological culture. Our mechanisms for storing and utilizing energy evolved in times when fat preservation was efficacious. These mechanisms no longer serve the goals of people in a technological society where a Burger King is on every corner. It will be argued here that many of the fundamental computational biases are now playing a similar role. These biases directly conflict with the demands for decontextualization that a highly bureaucratized society puts on its citizens. Indeed, this is often why schools have to explicitly teach such skills of cognitive decontextualization.[18]

Modern society creates many situations that require radical decontextualization—that require one or more of the fundamental computational biases to be overridden by analytic intelligence. For example, many aspects of the contemporary legal system put a premium on detaching prior belief and world knowledge from the process of evidence evaluation. There has been understandable vexation at odd jury verdicts rendered because of jury theories and narratives concocted during deliberations that had nothing to do with the evidence but instead were based on background knowledge and personal experience. For example, a Baltimore jury acquitted a murder defendant who had been identified by four witnesses and had confessed to two people because "they had invented their own highly speculative theory of the crime" (Gordon 1997, 258). In this particular case, the perpetrator had wanted to plea bargain for a forty-year sentence, but this was turned down at the request of the victim's family. Similarly, in Lefkowitz's (1997) account of the trial of several teenagers in an affluent New Jersey suburb who brutally exploited and raped a young girl who was intellectually disabled, one juror concocted the extenuating circumstance that one defendant thought he was attending an "initiation rite" even though no evidence for such a "rite" had been presented in months of testimony.

The point is that in a particular cultural situation where detachment and decoupling is required, the people who must carry out these demands for decontextualization are often unable to do so even under legal compulsion. Post-trial reports of juries in highly "creative" or "narrative" modes have incited great debate. If the polls are to be believed, a large proportion of Americans were incensed at the jury's acquittal of O. J. Simpson. Similar numbers were appalled at the jury verdict in the first trial of the officers involved in the Rodney King beating. What both juries failed to do was to decontextualize the evidence in their respective cases—and each earned the wrath of their fellow citizens because it is a cultural (and legal) expectation of citizenship that people should be able to carry out this cognitive operation in certain settings.

The need to decontextualize also characterizes many work settings in contemporary society. Consider the common admonition in the retail service sector that "the customer is always right." This admonition is often interpreted to include even instances where customers unleash unwarranted verbal assaults which are astonishingly vitriolic. The service worker is supposed to remain polite and helpful under this onslaught, despite the fact that such emotional social stimuli are no doubt triggering evolutionarily instantiated modules of self-defense and emotional reaction. All of this emotion, all of these personalized attributions—all fundamental computational biases of TASS—must be set aside by the service worker and instead an abstract rule that "the customer is always right" must be invoked in this special, socially constructed domain of the market-based transaction. The worker must realize that he/she is not in an *actual* social interaction with this person (which if true, might call for popping them in the nose!), but in a special, indeed unnatural, realm where different rules apply.

Modern technological societies are increasingly providing lucrative rewards only for those who can master subtle quantitative and verbal distinctions and reason in abstract and decontextualized ways (Bronfenbrenner, McClelland, Wethington, Moen, and Ceci 1996; Frank and Cook 1995; Gottfredson 1997; Hunt 1995, 1999). Objective measures of the requirements for cognitive abstraction have been increasing across most job categories in technological societies throughout the past several decades (Gottfredson

1997). Not to make important linguistic, probabilistic, and logical distinctions in a complex social environment has real costs and represents more than just the failure to play an artificial game. Decision scientists Hillel Einhorn and Robin Hogarth long ago cautioned that "in a rapidly changing world it is unclear what the relevant natural ecology will be. Thus, although the laboratory may be an unfamiliar environment, lack of ability to perform well in unfamiliar situations takes on added importance" (1981, 82).

Critics of the abstract content of most laboratory tasks and standardized tests have been misguided on this very point. Evolutionary psychologists have singularly failed to understand the implications of Einhorn and Hogarth's warning. They regularly bemoan the "abstract" problems and tasks in the heuristics and biases literature and imply that since these tasks are not like "real life" we need not worry that people do poorly on them. The issue is that, ironically, the argument that the laboratory tasks and tests are not like "real life" is becoming less and less true. "Life," in fact, is becoming more like the tests! Try using an international ATM machine with which you are unfamiliar; or try arguing with your HMO about a disallowed medical procedure. In such circumstances, we invariably find out that our personal experience, our emotional responses, our TASS-triggered intuitions about social fairness—all are worthless. All are for naught when talking over the phone to the representative looking at a computer screen displaying a spreadsheet with a hierarchy of branching choices and conditions to be fulfilled. The social context, the idiosyncrasies of individual experience, the personal narrative—all are abstracted away as the representatives of modernist technological-based services attempt to "apply the rules."

Consider Toronto writer Todd Mercer (2000) trying to fly across the continent on short notice to be with his 83-year-old father undergoing emergency surgery. Calling Canadian Airlines and finding out that the last-minute scheduled airline fare was $3,120, Mercer asked if there was any discount that applied to his situation and was informed that he might be eligible for an "imminent death discount" by the telephone ticket agent. Prodded for the definition of "imminent death" the ticket agent quotes from a document outlining the details of the "bereavement travel program" which clarifies the program's requirements when illness rather than death is the reason for the travel. The ticket agent relates that the person in question must be a patient in intensive care, a patient in the final stages of cancer, or a patient involved in a serious accident. Mercer's father had an aortic aneurysm which made him a "walking time bomb" according to his doctor, but he had not yet gone into surgery and had not yet been put into intensive care. The ruling was that such a situation was in "a gray area" and, as a result, the ticket agent stonewalled by saying "not all operations are life threatening. The imminent death discount is not meant just for operations. It is meant for imminent death"—the latter defined as above, and another round of technical and nuanced argument between Mercer and the ticket agent ensued. This is life in the First World in the early part of the twenty-first century.

The abstract, semantic games encountered by Mercer are nothing compared to what a person faces when deciding on whether to apply for a tax deduction for an infirm relative who lived outside Canada for the year 1994. Canada Customs and Revenue Agency will advise the person that "Your dependent must be: —your or your spouse's child or grandchild, if that child was born in 1976 or earlier and is physically or mentally infirm; or —a person living in Canada at any time in the year who meets all of the following conditions. The person must have been: —your or your spouse's parent, grandparent, brother, sister, aunt, uncle, niece, or nephew; —born in 1976 or earlier; and —physically or mentally infirm."

Given the ubiquitousness of such abstract directives in our information and technology-saturated society, it just seems perverse to argue the "unnaturalness" of decontextualized reasoning skills when such skills are absolutely necessary in order to succeed in our society. If one has the postindustrial goal of, say, "going to Princeton," then the only way to fulfill that goal in our current society *is* to develop such cognitive skills. Situations that require abstract thought and/or the ability to deal with complexity will increase in number as more niches in postindustrial societies require these intellectual styles and skills (Gottfredson 1997; Hunt 1995). For intellectuals to use their abstract reasoning skills to argue that the "person in the street" is in no need of such skills of abstraction is like a rich person telling someone in poverty that money is not really all that important.

The TASS Traps of the Modern World

To the extent that modern society increasingly requires the fundamental computational biases to be overridden, then dissociations between evolutionary and individual rationality will become more common—and analytic system overrides will be more essential to personal well-being. For example, evolutionary psychologists argue that we are not adapted to process probabilities, but are instead adapted to process natural frequencies. For example, it is easier for people to process "out of 1,000 people, 40 have the disease" than it is for them to process "there is a 4 percent chance of having the disease."

Consider the work of Brase, Cosmides, and Tooby (1998), who improved performance on a difficult probability problem by presenting the information as frequencies and in terms of whole objects—both alterations designed to better fit the frequency-computation systems of the brain. In response to a query about why the adequate performance observed was not even higher given that our brains contain such well-designed frequency-computation systems, Brase et al. (1998) replied that "in our view it is remarkable that they work on paper-and-pencil problems at all. A natural sampling system is designed to operate on actual events" (13). The problem is that in a symbol-oriented postindustrial society, we are presented with paper-and-pencil problems all the time! Much of what we know about the world comes not from the perception of actual events but from abstract information preprocessed, prepackaged, and condensed into symbolic codes such as probabilities, percentages, tables, and graphs (the voluminous statistical information routinely presented in *USA Today* comes to mind).

Thus, we have here an example of the figure and ground reversals that permeate contemporary debates about evolutionary psychology. It is possible to accept most of the conclusions of the evolutionary psychologists but to draw completely different morals from them. The evolutionary psychologists want to celebrate the astonishing job that evolution did in adapting the human cognitive apparatus to the Pleistocene environment. Certainly they are right to do so. The more we understand about evolutionary mechanisms, the more awed appreciation we have for them. But at the same time, it is not inconsistent for a person to be horrified that a multimillion dollar advertising industry is in part predicated on creating stimuli that will trigger TASS heuristics that many of us will not have the cognitive energy or cognitive disposition to override. I personally find it no great consolation that the heuristics so triggered were evolutionarily adaptive in their day.

What I am attempting to combat here is a connotation implicit in some writings in evolutionary psychology that there is nothing to be gained from being able to understand a formal rule at an abstract level (the conjunction rule of probability, etc.)—and no advantage in flexibly overriding the fundamental computational biases at times. Consider a now-famous experiment by Langer, Blank, and Chanowitz (1978) where a confederate attempts to cut into a line at a copy machine. In one condition a good reason is given ("May I use the Xerox machine, because I'm in a rush?") and in the other a totally redundant nonexplanation ("May I use the Xerox machine, because I have to make copies?"). Despite the fact that the second explanation is much less informative than the first, the compliance rates in the two conditions did not differ. Langer (1989; Langer et al. 1978) terms the compliance in the second case "mindless," and philosopher Jonathan Adler (1984) analyzes it in terms of overly generalizing the Gricean Cooperative Principle. The default assumption that a contribution will be selectively relevant—in this case, that a real reason will follow the request—is false in this condition. Yet TASS will trigger and execute exactly the same compliance behavior unless it is overridden by a thoughtful analysis that registers that the TASS default assumption is false ("Yes but *all* of us are in line to make copies. Why should you go first?").

Langer-type examples of mindlessness abound in many important domains. *Consumer Reports* (April 1998) chronicles how some automobile dealers put an item costing $500 and labeled ADM on many automobile price stickers. The dealers are hoping that some people will not ask what ADM means. The dealers are also hoping that even after asking and being told that it means "additional dealer markup" that some consumers will not fully process what that means and will not inquire further about what this additional dealer markup feature is that they are paying for. In short, the dealers are hoping that the analytic processor will not methodically plow through this smoke and mirrors to ascertain that ADM is not a feature on the car at all—that it simply represents a request from the dealer to contribute $500 more to the dealership, as if it were a charity. As one dealer put it, "every once in a while somebody pays it, no questions asked" (17). A mindless response here, a failure of the analytic system to override, and the buyer could simply throw away a good chunk of hard-earned income.

Experimental studies of choice indicate that such errors due to insufficient monitoring of TASS responses are probably made all the time. Neumann and Politser (1992) describe a study in which people were asked to choose between two insurance policies. Policy A had a $400 yearly deductible and a cost of $40 per month. Policy B had no deductible and a cost of $80 a month. A number of subjects preferred policy B because of the certainty of never having to pay a deductible if an accident occurs. However, it takes nothing more than simple arithmetic to see that people choosing policy B have fallen prey to a TASS default to avoid risk and seek certainty (see Kahneman and Tversky 1979). Even if an accident occurs, policy B can never cost less than policy A. This is because paying the full deductible ($400) plus the monthly fee for 12 months ($480) would translate into a total cost of $880 for policy A, whereas the monthly fee of policy B for 12 months amounts to $960. Thus, even if accidents cause the maximum deductible to be paid, policy A costs less. An automatic reaction triggered by a logic of "avoid the risk of large losses" biases responses against the more economical policy A.

Modern mass communication technicians have become quite skilled at exploiting TASS defaults. The communication logic of relying on TASS processes to trump the analytic system is readily employed by advertisers, in election campaigns, and even by governments—for example in promoting their lottery systems. "You could be the one!" blares an ad from the Ontario Lottery Commission—thereby increasing the availability of an outcome which, in the game called 6/49, has an objective probability of 1 in 14 million.

As mentioned earlier in this chapter, evolutionary psychologists have shown that some problems can be more efficiently solved if represented to coincide with how various brain modules represent information ("when people are given information in a format that meshes with the way they naturally think about probability, they can be remarkably accurate" [Pinker 1997, 351]). Nevertheless, evolutionary psychologists often seem to ignore the fact that the world will not always *let* us deal with representations that are optimally suited to our evolutionarily designed cognitive mechanisms. We are living in a technological society where we must: decide which health maintenance organization to join based on just such statistics; figure out whether to invest in an individual retirement account; decide what type of mortgage to purchase; figure out what type of deductible to get on our auto insurance; decide whether to trade in a car or sell it ourselves; decide whether to lease or to buy; think about how to apportion our retirement funds; and decide whether we would save money by joining a book club— to simply list a random set of the plethora of modern-day decisions and choices. And we must make all of these decisions based on information represented in a manner for which our brains are not adapted (in none of these cases have we coded individual frequency information from our own personal experience). In order to reason rationally in all of these domains (in order to maximize our personal utility) we are going to have to deal with probabilistic information represented in nonfrequentistic terms—in representations that the evolutionary psychologists have shown are different from our adapted algorithms for dealing with frequency information.

In choosing to emphasize the importance of the few situations where evolutionarily adapted processes might not achieve instrumental rationality, I am not disputing that evolutionary psychologists are right to point out that in a majority of cases evolutionary goals and personal goals coincide. On a purely quantitative basis, in terms of the micro-events in day-to-day life, this is no doubt true. Throughout the day we are detecting frequencies hundreds of times, detecting faces dozens of times, using our language modules repeatedly, inferring the thoughts of others constantly, and so on—all of which are adaptive *and* serve personal goal satisfaction. Nevertheless, as illustrated in the examples I have discussed in this chapter, the few instances where the two do not coincide and an analytic system override is needed may be of unusual importance. As several examples discussed previously illustrated, a market economy can very efficiently translate suboptimal behavioral tendencies into utility for those discovering a way to exploit the suboptimal TASS response. A consumer who buys $10,000 worth of shares in a mutual fund with a load (a 5 percent sales charge) because of the glossy brochure rather than buying the equivalently performing but unadvertised no-load (no sales charge) index fund has—in the most direct way imaginable—chosen to simply give away $500 to a salesperson and to the equity owners of the loaded mutual fund company. Modern market economies are simply littered with such TASS traps and, often, the more potentially costly the situation, the more such traps there are (automobile purchases, mutual fund investments, mortgage closing costs, and insurance come to mind). It might be thought that evolutionary psychologists would be uniquely alert to such traps, but instead they often seem to minimize them. Why?

An uneasy tension disturbs the heart of the selfish gene theory. It is the tension between gene and individual body as fundamental agent of life. On the one hand we have the beguiling image of independent DNA replicators, skipping like chamois, free and untrammelled down the generations, temporarily brought together in throwaway survival machines, immortal coils shuffling off an endless succession of mortal ones as they forge towards their separate eternities. On the other hand we look at the individual bodies themselves and each one is obviously a coherent, integrated, immensely complicated machine, with a conspicuous unity of purpose.

—Richard Dawkins, *The Selfish Gene* (1976, 234)

CHAPTER 5

How Evolutionary Psychology Goes Wrong

Why do evolutionary psychologists tacitly collude with the genes in not highlighting the distinction between instrumental rationality and evolutionary adaptation? Evolutionary psychologists fall into this trap by being too wedded to the assumption of a cognitive architecture that displays massive modularity.[1] Many of these theorists carry the assumption to the lengths of virtually denying the existence of domain-general analytic processing mechanisms—in essence, denying the existence of an analytic processing mechanism of the type described in chapter 2.

From the standpoint of the framework developed in this book, the evolutionary psychologists' theoretical stance against domain-general processing is a profound mistake. Because it is this system that computes actions that maximize utility based on the organism's long-leashed goals, analytic processing is essential if a person is to achieve instrumental rationality. It is this system that overrides the TASS subsystems when they compute responses that are not in the interests of our current long-term goals. When is this likely to happen? Answer: In situations where the cognitive requirements of technological societies do not match those of the environment of evolutionary adaptedness (EEA). Ironically, the evolutionary psychologists themselves often mention that the EEA is not to be confused with the modern world. However, they largely fail to develop the most important implication of potential mismatches between the cognitive requirements of the EEA and those of the modern world—that if we respond in the modern world according to TASS subsystems, we will often be maximizing something other than our current goals as individual organisms.

We can see the theoretical biases of the evolutionary psychologists—and how they create peculiar blindspots—in the following statement:

> In actuality, adaptationist approaches offer the explanation for why the psychic unity of humankind is genuine and not just an ideological fiction; for why it applies in a privileged way to the most significant, global, functional, and complexly organized dimensions of our architecture; and for why the differences among humans that are caused by genetic variability that geneticists have found are so overwhelmingly peripheralized into architecturally minor and functionally superficial properties. (Tooby and Cosmides 1992, 79)

This statement provides an example of how and why evolutionary psychology goes off the rails. Let us take a closer look at some of that "genetic variability that geneticists have found" and ask ourselves whether it does indeed reflect "functionally superficial properties."

Well, for starters, some of that "genetic variability that geneticists have found" is in general intelligence (g)—which consensus scientific estimates agree is at least 40 to 50 percent heritable (Deary 2000; Grigorenko 1999; Neisser et al. 1996; Plomin et al. 2001; Plomin and Petrill 1997). Is g a "functionally superficial" individual difference property of human cognition? No responsible psychologist thinks so. It is, indeed, the single most potent psychological predictor of human behavior in both laboratory and real-life contexts that has ever been identified (Lubinski 2000; Lubinski and Humphreys 1997). It is a predictor of real-world outcomes that are critically important to the maximization of personal utility (instrumental rationality) in a modern technological society. This is why measures of the ability to deal with abstraction such as g remain the best employment predictor and the best earnings predictor in postindustrial societies. The psychometric literature contains numerous indications that cognitive ability is correlated with the avoidance of harmful behaviors and with success in employment settings, as well as social status attainment, independent of level of education.[2]

Intelligence is not the only type of "genetic variability that geneticists have found" that is manifestly *not* "functionally superficial." Similar stories could be told about many personality variables that have been shown to be heritable but also important predictors of behavioral outcomes. Indeed, this stance by some evolutionary psychologists against heritable cognitive traits with demonstrable linkages to important real-world behaviors has become an embarrassment even to some evolutionary theorists. Buss (1999) characterizes the view of Tooby and Cosmides as the notion that "heritable individual differences are to species-typical adaptations, in this view, as differences in the colors of the wires in a car engine to the engine's functional working components" (394), and points to some of the same embarrassing empirical facts noted above. For example, heritable personality traits such as conscientiousness and impulsivity have been related to important life goals such as work, status attainment, mortality, and faithfulness in partnerships. Buss's (1999) alternative interpretation is in terms of genetic concepts such as frequency-dependent selection. But whether or not one accepts such explanations, the point is that many evolutionary theorists have mistakenly downplayed cognitive constructs that are heritable (intelligence, personality dimensions, thinking styles) and that have demonstrated empirical relationships to behaviors that relate to utility maximization for the individual (job success, personal injury, success in relationships, substance abuse) because many of these characteristics are domain general and thus violate the massive modularity assumption in its strongest form.

In contrast to Buss's (1999) more nuanced position on individual differences, other influential evolutionary psychologists repeat like a mantra the view that psychological processes with genetic variation lack any importance (and presumably lack any relevance for rationality, since this is obviously important to the vehicle). Thus Tooby and Cosmides write: "Human genetic variation . . . is overwhelmingly sequestered into functionally superficial biochemical differences, leaving our complex functional design universal and species typical" (1992, 25). And continue: "Humans share a complex, species typical and species-specific architecture of adaptations, however much variation there might be in minor, superficial, nonfunctional traits" (38).

One boggles at general intelligence—one of the most potent psychological predictors of life outcomes—being termed "nonfunctional." But then one realizes what is motivating these statements—a focus on the gene. Even if one buys the massive-modularity-of-adaptations line of the evolutionary psychologist and views general intelligence as some kind of spandrel or by-product,[3] from the standpoint of the *vehicle's* interests, it is certainly not nonfunctional. Only a focus on the subpersonal replicators would spawn such a statement—one which backgrounds important cognitive traits such as intelligence and conscientiousness (Lubinski 2000; Matthews and Deary 1998). As soon as one focuses on the organismic level of optimization rather than genetic optimization, the "nonfunctional" traits spring to the foreground as constructs that explain individual differences in attaining one's goals.

Modern Society as a Sodium Vapor Lamp

Evolutionary psychologists also tend to misleadingly minimize the consequences of mismatches between the EEA and the modern environment. Tooby and Cosmides (1992, 72) approvingly paraphrase cognitive scientist Roger Shepard's (1987) point that evolution insures a mesh between the principles of the mind and the regularities of the world. But this "mesh" concerns regularities in the EEA, not in the modern world—with its unnatural requirements for decontextualization (requirements that do not "mesh" with the fundamental computational biases toward comprehensive contextualization of situations). One page later in their chapter, Tooby and Cosmides (1992) reveal the characteristic bias of evolutionary psychologists—the belief that "often, but not always, the ancestral world will be similar to the modern world (e.g., the properties of light and the laws of optics have not changed)" (73). But what is ignored here is that although the laws of optics might not have changed, the type of one-shot, abstract, probabilistic, and symbolically represented decision situations a modern person must deal with are certainly unprecedented in human history. We can walk and navigate among objects as well as we ever did, but no evolutionary mechanism has sculpted my brain in order to help me estimate the deductible I need on my insurance or to help me evaluate the cost of a disability policy to cover salary loss. Think, also, of retirement decisions, investment decisions, home buying decisions, relocation decisions, and school choices for children. These are not the highly practiced, frequency coded, time pressured, recognition-based situations where TASS work best. Instead, these are just the type of situations that cause the representativeness, availability, sunk cost, confirmation bias, overconfidence, and other effects that the heuristics and biases researchers have studied—effects that thwart the ability of the individual to maximize utility (see the many real-life examples in Kahneman and Tversky 2000).

Tooby and Cosmides (1992) seem to take a completely one-sided message from the potential mismatch between the EEA and modern conditions—when in fact the mismatch has more than one implication. Using the example of how our color constancy mechanisms fail under modern sodium vapor lamps, they warn that "attempting to understand color constancy mechanisms under such unnatural illumination would have been a major impediment to progress" (73)—a fair enough point. But my purpose here is to stress a different corollary point that one might have drawn. The point is that if the modern world *were* structured such that making color judgments under sodium lights was critical to one's well-being, then this would be troublesome for us because our evolutionary mechanisms have not naturally equipped us for this. One might be given impetus to search for a cultural invention that would circumvent this defect (relative to the modern world) in our cognitive apparatus. In fact, humans in the modern world are in just this situation vis à vis the mechanisms needed for fully rational action in highly industrialized and bureaucratized societies.

The processing of probabilistic information is critical to many tasks faced by a full participant in a modern society. Of course, the heuristics and biases literature is full of demonstrations of the problems that people have in dealing with probabilistic information. As discussed earlier in this chapter, evolutionary psychologists have done important work indicating that the human cognitive apparatus is more adapted to dealing with frequencies than with probabilities (frequency representations appear to reduce but not eliminate many cognitive illusions). For example, it is easier for people to process "out of 1,000 people, 40 have the disease" than it is for them to process "there is a 4 percent chance of having the disease." As useful as this research has been (and indeed it can usefully be adapted to tell us how to more understandably present probabilistic information in real-life settings, see Gigerenzer, Hoffrage, and Ebert 1998; Gigerenzer 2002), it will not remove the necessity of being able to process probabilistic information when it *is* presented in the real world.

The evolutionary psychologists and ecological rationality theorists are sometimes guilty of implying just this—that if the human cognitive apparatus can be shown to have been adapted during evolution to some *other* representation (other than that required for a problem in modern society) then somehow it has been shown that there really is no cognitive problem. For example, in the titles and subheadings of several papers on frequency representations, Gigerenzer (1991, 1993, 1998; Gigerenzer, Hoffrage, and Kleinbolting 1991) has used the phrasing "how to make cognitive illusions disappear." This is a strange way to phrase things, because the original illusion has of course not "disappeared." As Kahneman and Tversky (1996) note, the Müller-Lyer illusion (see fig. 2.1) is removed when the two figures are embedded in a rectangular frame, but this does not mean that the *original* illusion has "disappeared" in this demonstration (see also Samuels, Stich, and Tremoulet 1999; Stein 1996). The cognitive illusions in their original form still remain (although their explanation has perhaps been clarified by the different performance obtained in the frequency version),[4] and the situations in which these illusions occur have not been eliminated.

Banks, insurance companies, medical personnel, and many other institutions of modern society are still exchanging information using linguistic terms like probability and applying that term to singular events.

Perhaps both groups in this debate have been guilty of some disproportionate emphasis. Some theorists in the heuristics and biases camp are perhaps so prone to emphasize the errors occurring in the modern world that they fail to acknowledge that humans really are optimally designed in a certain sense. Conversely, evolutionary theorists err by emphasizing the EEA so much that they seem to forget about the properties of the modern world.

Evolutionary psychologist David Buss (1999) shows the latter tendency when he asks the question: "If humans are so riddled with cognitive mechanisms that commonly cause errors and biases, how can they routinely solve complex problems that surpass any system that can be developed artificially?" (378)—and answers it by quoting an unpublished paper by Tooby and Cosmides where the argument is made that our criteria for recognizing sophisticated performance "have been parochial" (378). Buss seems to be calling our natural privileging of the present environment—the one we actually have to operate in—unnecessarily parochial. The devaluing of the actual decontextualized environment in which we must operate in modern technological society continues, as Buss repeatedly minimizes rational thinking errors by pointing out that they occur in "artificial or novel" (378) situations. The latter of course seems damning to his own argument (that these errors are trivial) because novel symbolic situations are exactly what workers and citizens who are immersed in highly bureaucratized societies must constantly deal with.

With respect to the "artificial situations" criticism, Buss trots out the old sodium vapor lamps example, saying that the experiments have used "artificial, evolutionarily unprecedented experimental stimuli analogous to sodium vapor lamps" (379). Like Tooby and Cosmides (1992), Buss takes exactly the wrong message from the potential mismatch between EEA and modern conditions. None of these theorists will acknowledge that it is *a very serious worry* that we are essentially in situations where we must work under sodium vapor lamps! The cognitive equivalent of the sodium vapor lamps are: the probabilities we must deal with; the causation we must infer from knowledge of what *might* have happened; the vivid advertising examples we must ignore; the unrepresentative sample we must disregard; the favored hypothesis we must not privilege; the rule we must follow that dictates we ignore a personal relationship; the narrative we must set aside because it does not square with the facts; the pattern that we must infer is not there because we know a randomizing device is involved; the sunk cost that must not affect our judgment; the judge's instructions we must follow despite their conflict with common sense; the contract we must honor despite its negative affects on a relative; the professional decision we must make because we know it is beneficial in the aggregate even if unclear in this case. These are all the "sodium vapor lamps" that modern society presents to our cognitive apparatus—and if evolution has not prepared us to deal with them so much the worse for our rational behavior in the modern world. Luckily, the culturally acquired tools of rational thought run by our analytic systems are there to help us in situations such as this.

In short, evolutionary psychologists are wrong to assume that TASS heuristics (adapted for the EEA) are optimal for achieving rationality in the modern world. Many important decisions in life are nearly "one shot" affairs (job offers, pension decisions, investing decisions, housing decisions, marriage decisions, reproductive decisions, etc.). Some of these decisions were not present at all in the EEA, and we have had no time nor learning trials to acquire extensive personal frequency information about them. Instead we need to make certain logical and probabilistic inferences using various rules of inference, and most importantly, we must decouple myriad sources of information that our autonomously functioning modules might be detecting and feeding into the decision ("no, the likability of this salesperson should not be a factor in my deciding on this $25,000 car"). Decoupling—a major function of the analytic system—was discussed in chapter 2.

In fact, some of the TASS heuristics that are in place might seriously subvert instrumental goals in a modern technological society. For example, one chapter in an edited book extolling the usefulness of simple heuristics as opposed to computationally expensive, fully analytic reasoning (Gigerenzer and Todd 1999) is devoted to the so-called recognition heuristic—the chapter subheading being "How Ignorance Makes Us Smart." The idea behind such "ignorance-based decision making," as they call it, is that the fact that some items of a subset are unknown can be exploited to aid decision making. The yes/no recognition response can be used as a frequency estimation cue (judge the one of two items that is recognized as more frequent, more important, larger, etc.). With ingenious simulations, Goldstein and Gigerenzer (1999, 2002) demonstrate how certain information environments can lead to such things as less-is-more effects: where those who know less about an environment can display more inferential accuracy in it.

One is certainly convinced after reading material like this that the recognition heuristic is certainly efficacious in *some* situations. But then one immediately begins to worry when one ponders how it relates to a market environment specifically designed to exploit it. If I were to right now step outside the door of my home—located in the middle of a financial and industrial center of a First World country—and relied solely on the recognition heuristic, I could easily be led to:

1. buy a $3 coffee when in fact a $1.25 one would satisfy me perfectly well;
2. eat in a single snack the number of fat grams I should have in an entire day;
3. pay the highest bank fees (because the highest fees are charged by the most recognized banks in Canada);
4. incur credit card debt rather than pay cash;
5. buy a mutual fund with a 6 pecent sales charge rather than a no-load fund.

None of these behaviors serves my long-term instrumental goals at all— none of them help me achieve my reflectively acquired aspirations (Gewirth 1998). Yet the recognition heuristic triggers these and dozens more that will trip me up while trying to make my way through the maze of modern society.

The evolutionary psychologists and ecological theorists refuse to acknowledge this downside of the ecological approach. For example, Borges, Goldstein, Ortmann, and Gigerenzer (1999) take curious pride in the finding that a portfolio of stocks recognized by a group of Munich pedestrians beat two benchmark mutual funds for a six-month period during the mid-1990s. This finding is of course a pure artifact of an extraordinary short period in the 1990s when large capitalization stocks outperformed small capitalization stocks (Over 2000). The adaptive heuristics investigated by Borges et al. (1999) haven't repealed the basic laws of investing. Risk is still related to reward, and over longer time periods small capitalization stocks outperformed their less-risky large capitalization counterparts. Obviously, the Munich pedestrians had better recognition for the large companies— precisely those enjoying a good run in that particular six-month period (which is of course too short for various risk/reward relationships to show themselves).

Borges et al. (1999) might have alternatively focused on another well-known finding in the domain of personal finance discussed by Bazerman (2001)—that consumers of financial services overwhelmingly purchase high-cost products that underperform in terms of investment return the low-cost strategies recommended by true experts (e.g., dollar-cost averaging into no-load index mutual funds). The reason is, of course, that the high-

cost fee-based products and services are the ones with high immediate rec-
ognizability in the marketplace, whereas the low-cost strategies must be
sought out in financial and consumer publications.

Throwing Out the Vehicle with the Bathwater

One leaves the writings of the ecological theorists and evolutionary psychologists feeling that they are being much too sanguine about the ability
of TASS to achieve instrumental rationality for a person. Clearly though,
not all evolutionary psychologists miss the implications of the replicator/
vehicle distinction for conceptions of rationality. But some evolutionary
theorists do—quite egregiously. In an astonishing essay titled "How Evolutionary Biology Challenges the Classical Theory of Rational Choice," Cooper
(1989) basically argues that when choosing between your own goals and
those of your genes, you should opt for the latter! After a marvelous discussion of why a probability matching strategy (Estes 1961, 1976) might be fitness optimizing rather than the utility maximizing strategy (picking the
most frequent option each time), Cooper (1989) implies that this outcome
undermines the prescriptive force of the utility maximizing strategy.

Of course, early in the article one feels that this is a verbal slip. But ten
pages on, we find out that the author does indeed wish to argue that we
should follow goals that satisfy our genes rather than ourselves as individual organisms. The ordinary application of the logic of decision science is
termed "naively applied" when interpreted "with the individual treated as
an isolated locus of decision making and with the role of the genotype ignored" (473). The instabilities in preference orderings that signal the failure
of individual utility maximization (Dawes 1998; Kahneman and Tversky
2000; Slovic 1995) are defended because "perhaps some of the observed instability is due to adaptive strategy mixing. If so, instability would have to
be reevaluated; when one is acting as an agent of one's genotype, it could
sometimes be a sound strategy" (473). But who in the world would want to
act as an agent of one's genotype rather than in the service of one's own life
goals! This is precisely the choice Cooper is posing when he pits the concerns of genetic fitness against those of instrumental rationality.

In his summary statement, Cooper makes it clear that the proposition
he wishes to defend is that "the traditional theory of rationality is invalid as
it stands, and in need of biological repair" (479), and acknowledges that
this is "a stance not likely to be popular with confirmed classical decision
theorists, but perhaps understandable to evolutionists, psychologists, philosophers, and others that have been impressed by the pervasive explanatory

power of the modern evolutionary perspective" (479). The view explicitly championed is the notion that "behavioral rationality [be] interpreted in terms of fitness" (480) and that any dissent from this policy be viewed as "biologically naive" (480). Like the sociobiologists before him, Cooper seems to have taken the defense of the genes as his brief!

Cooper's view may well seem extreme, and few evolutionary psychologists so explicitly throw out the vehicle with the bathwater. But many evolutionary psychologists and proponents of ecological rationality do implicitly do so when they echo Cooper's contention that "the traditional theory of rationality is invalid as it stands, and in need of biological repair" (479). For example, in a paper discussing economics and evolutionary psychology, Cosmides and Tooby (1994a) quite closely mimic Cooper's view when they argue that "evolutionary considerations suggest that traditional normative and descriptive approaches to rationality need to be reexamined" (329). Throughout this essay they repeat the odd declaration that "despite widespread claims to the contrary, the human mind is not worse than rational (e.g., because of processing constraints)—but may often be better than rational" (329).

It is in fact relatively common for the traditional normative rules of rational thought to be denigrated in the literature critical of the heuristics and biases approach. Gigerenzer and Goldstein (1996) adopt exactly Cooper's (1989) extreme position in their argument that "questions classical rationality as a universal norm and thereby questions the very definition of 'good' reasoning on which both the Enlightenment and the heuristics-and-biases views were built" (651). The classical norms are referred to as just so much useless "baggage" in statements such as the following: "A bit of trust in the abilities of the mind and the rich structure of the environment may help us to see how thought processes that forgo the baggage of the laws of logic and probability can solve real-world adaptive problems quickly and well" (Gigerenzer and Todd 1999, 365).[5]

Similarly, Cosmides and Tooby, in their essay directed at economists, ignore completely the role of culture in determining human preferences. In a series of points laid out like a series of axioms, they argue that because "natural selection built the decision-making machinery in human minds" (1994a, 328) and because "this set of cognitive devices generates all economic behavior," "therefore . . . the design features of these devices define and constitute the human universal principles that guide economic decision making" (328).

These postulates lead Cosmides and Tooby to the grandiose claim that "evolutionary psychology should be able to supply a list of human universal preferences, and of the procedures by which additional preferences are acquired or reordered" (331). But to the extent that the claim is true, it is true only because the grain-size of the predictions will be all wrong. The economic literature is not full of studies debating whether humans who are dying of thirst prefer water or shelter—or whether men prefer 23-year-old females over 75-year-old ones for mates. Instead, the literature is full of studies trying to determine the rationale for such fine-grained judgments as, for example, whether a poor briefcase produced by an athletic shoe company will adversely affect the family brand name (Ahluwalia and Gurhan-Canli 2000). Economists and psychologists are not debating the reasons for preferences among basic biological needs. But instead, they are debating the reasons for fine-grained preferences among highly symbolic products embedded in a complex, information-saturated, "attention-based" (Davenport and Beck 2001) economy. Even after we grant evolutionary assumptions like, for example, that people use clothes purchases for some type of modern dominance display or sexual display, we have not progressed very far in explaining how brand names wax and wane in the fashion world, or how price elastic such purchases will be, and/or what kind of substitutability there will be among these types of goods.

This essay by Cosmides and Tooby (1994a) directed to economists serves to reinforce all of the worst Panglossian tendencies in the latter discipline. For example, Kahneman, Wakker, and Sarin (1997) discuss why experienced utility is essentially ignored in modern economics despite psychological studies showing that experienced utility is not identical to expected utility. They argue that experienced utility is ignored by economists on the grounds that "choices provide all necessary information about the utility of outcomes because rational agents who wish to do so will optimize their hedonic experience" (375). Dual-process theories of cognition—in conjunction with the assumptions that we have made about goal structures—help to explain why this assumption might not hold. The choices triggered by the goal structures of TASS might not always be oriented toward the optimization of hedonic experience for the individual agent. The hedonic experience is just a means to an end for most of the goals lodged in TASS (largely genetic goals). TASS will readily sacrifice the vehicle's hedonic pleasure if ultimate fitness goals are achievable without it.

Make no mistake, I am impressed with the seminal achievements of evolutionary psychology (see table 1 of Buss et al. 1998, for a long list of important behavioral relationships that were in large part uncovered because of applications of the theoretical lens of evolutionary psychology), and I consider its emergence as a dominant force in psychology during the 1990s

to be a salutary development.[6] But in the area of rationality, the evolutionary psychologists have built a bridge too far. They too easily gloss over the
important issue of replicator/vehicle goal mismatches and their implications. They too easily dismiss the role of general intelligence and/or general
computational power in overriding deleterious TASS responses.

Because many of the tools of instrumental and epistemic rationality are
cultural inventions and not biological modules, their usefulness in technological societies is too readily dismissed by evolutionary psychologists. Unlike Darwinian creatures, our interests are not necessarily our genes'. The remarkable cultural project to advance human rationality concerns how to
best advance human interests whether or not they coincide with genetic
interests. We lose its unique liberating potential by ignoring the distinction between maximizing genetic fitness and maximizing the satisfaction
of human desires.

What Follows from the Fact that Mother Nature Isn't Nice

Following the strictures of the evolutionary psychologists—glorying in the
evolutionary efficiency of TASS and ignoring instrumental rationality at the
personal level—will, in the long run, prove costly for humans. Failure to invoke the analytic system to monitor whether standards of rationality are being met means, in effect, handing over your life to TASS, to the gut instincts
that Mother Nature gave you (and to *another* selfish replicator as well, see
chapter 7). And, as we constantly need reminding, Mother Nature did not
build those TASS subsystems in order to be nice to us.

As distinguished evolutionary biologist George Williams (1996) has
eloquently pointed out, Mother Nature is in the replication business, not
the niceness business. TASS was built to further the goals of the subpersonal
replicators rather than your own personal goals.

If you give up your life to TASS (in the form of giving in to your "gut instincts"), then you are essentially buying a lottery ticket. You are betting,
when you do so, that it is one of those instances when genetic goals and personal goals coincide (area B in figure 2.2 in chapter 2). You are betting that
it is not one of those instances when pursuing ancient replicator goals conflicts with your personal goals (area A in figure 2.2). However, work in cognitive and decision science has indicated that you would sometimes lose
this bet and, further, sometimes you would lose it in important circumstances and experience negative real-world consequences.

For example, according to the arguments put forth by Bazerman and
colleagues (Bazerman, Tenbrunsel, and Wade-Benzoni 1998), defaulting to

TASS can have negative implications for consumers in modern market economies. Applying their two-process theory of cognition to this domain, they argue that comparative choices spawn more analytic processing, whereas single option choices are more likely to implicate TASS and are thus more likely to be based on things like semantic associations or vividness. They discuss consumer research showing that brand names are more important when consumers are evaluating products individually but are less important when consumers are comparing products. Relying on TASS in consumer decisions means more profits for companies relying on brand visibility rather than product quality. Is this what the advocates of reliance on gut instincts want?

Decision scientists Eric Johnson and colleagues (Johnson, Hershey, Meszaros, and Kunreuther 2000) illustrate how defaulting to TASS can affect insurance decisions in ways that change choices worth, in total, billions of dollars. For example, they discuss research on the willingness to pay for flight insurance. In the study they describe, a group of subjects was asked what they would pay for $100,000 of insurance in case of death due to a mechanical failure, and the mean estimate of what subjects said they would pay was $10.31. Another group was asked what they would pay for $100,000 of insurance in case of death due to *any reason*, and the mean estimate of what they said they would pay was $12.03. This makes sense. There are reasons that one could die in an airplane other than mechanical failure, for example, pilot error or deliberate sabotage (in the 1990s an Egypt Air flight was deliberately nosedived into the ocean by the co-pilot). A third group was asked what they would pay for $100,000 of insurance in case of death due to terrorism, and the mean estimate of what they would pay was $14.12. This makes no sense. Death due to terrorism is of course covered in the insurance against death by any reason, and people thought that the any-reason insurance was worth only $12.03. Why then should we want to pay more ($14.12) for insurance against causes (terrorism) that are a subset of (i.e., less probable than) *any* reason? The word terrorism triggers (through a variety of TASS processes like automatic semantic priming) vivid memory instances that artificially inflate the probability of this outcome. The worry about it is magnified and thus leads to an overestimate of the worth of the insurance. Knowledge of research like this could be used by the insurance industry to maximize their profits. Do the proponents of gut-instinct responding want people to overpay for insurance?

In another example, Johnson and his colleagues discuss how automatic status quo biases account for expensive insurance decisions and could have cost the consumers in the state of Pennsylvania millions of dollars. They describe how, in the 1980s, both New Jersey and Pennsylvania attempted to reduce the cost of insurance by introducing the consumer choice of a reduced right to sue (with a concomitant lower rate). However, it was implemented differently in the two states. In New Jersey, the status quo was a limited right to sue at a reduced rate. To gain the full right to sue the consumer had to acquire it by agreeing to pay a higher rate. In Pennsylvania, the status quo was the full right to sue. In order to pay a lower rate the consumer had to agree to a reduced right to sue. In New Jersey, only 20 percent of drivers chose to acquire the full right to sue, but in Pennsylvania, where full right to sue was already in place, 75 percent of the drivers chose to retain it. A TASS-based status quo bias froze these consumers in their current situation by framing as "normal" the insurance features the consumer already had. The Pennsylvania reform had been intended by the legislature to save consumers millions, and it did save some people money. However, reversing the status quo in the wording of the legislation could have saved consumers approximately $200 million more, because people's gut instincts told them to stick with what they had. Do the proponents of gut-instinct responding approve of this donation to the insurance industry?

So Mother Nature's gut instincts are often not nice because they sometimes lead to outcomes that are noninstrumental—that fail to maximize our goals. However, these problems do not occur only in the instrumental domain. Relying on our gut instincts in our social and emotional lives also often leads to outcomes that are not so nice. This is because our gut instincts, being ancient evolutionary products, have not taken into account cultural trends—trends that are now reflected in our conscious sensitivities and concerns but that are totally unreflected in our gut instincts.

Consider as an example an interchange between journalist John Richardson and Andrea, a woman with achondroplasia dwarfism, as described in Richardson's book *In the Little World: A True Story of Dwarfs, Love, and Trouble* (2001). Richardson attended a conference of over one thousand people with dwarfism, engaged in extended correspondences with several of the interesting people he met there, and wrote a book about their lives and his reaction to them. In one interchange, Andrea criticizes Richardson for saying, in a magazine piece about the conference, that he thought that dwarfs looked "wrong." Richardson describes how he didn't want to back down from his comment, didn't want to lie to Andrea, and besides, thought that it was—in his words—natural to fear things that looked like the symptom of a disease. Andrea asks Richardson if he understands how hurtful it is for her to hear that he thinks her body looks wrong. She says "I mean, I understand that as a first reaction, that people freak out over difference for lots of reasons—fear of their own difference or the unknown. But people get over it. . . . I need to know if you're committed to this idea, because if you are, I think I have to remove myself from this relationship" (134).

Richardson does not back down in the face of this argument but, instead, shows that he has read his evolutionary psychology by arguing that "these things are hardwired when you're a very young person. . . . There's always a standard of beauty and deviations from that standard. It's simple normative thinking. But not to see the surface at all—that's asking a person not to be human" (135). But Andrea's position is that "accepting difference doesn't mean overlooking bodies. It means getting to the point where difference doesn't seem wrong" (135). Later, in this continuing conversation, Richardson becomes exasperated by what he sees as Andrea's attempts to badger him into politically correct responses, and he unloads on her in an email message: "Maybe I should have put in 'sometimes' or 'occasionally' or 'initially,' but that would be just hedging to protect myself when the fact is, if you glance across the statistical room and see fifteen thousand people with 'average' proportions and ONE who's a dwarf, which one looks wrong? Can I be Mr. Rogers? Which of these things does not belong? Again, I'm saying not IS wrong. I'm saying LOOKS wrong. At first. In the reptilian brain. My reptilian brain" (208).

I think, however, that Richardson is missing something here. Andrea is not asking him to reprogram TASS in his brain. She acknowledges that this is probably impossible. What she wants him to do is to make a judgment using his analytic mind, and, if that conflicts with TASS, to tell her which one he identifies with—which one reflects the person he believes himself to be. In his denial of what he sees as her inhuman request that he not have a TASS response of the type that he describes, Richardson seems to her to be saying that he will not make an analytic judgment or, if he does so, that he wants to identify instead with TASS—with his gut instinct to see the bodies of dwarfs as wrong. However, the analytic system has something to say here as well, not in terms of ancient, hardwired evolutionarily conditioned responses, but in terms of acquired cultural knowledge that is stated in propositional form, in the slow, serial language-based codes of the analytic system—propositions such as "Don't judge a book by it's cover" and "Beauty is only skin deep."

In his defense of his right to be honest about having this TASS response, Richardson seems to Andrea to be saying not only that he has it, but also that he wishes to identify with it. She keeps prodding him to give her an analytic judgment, to say whether his analytic mind validates or negates the TASS response ("I need to know if you're committed to this idea"), but

Richardson has dug himself so far into the hole of defending his gut instincts that he cannot even hear what Andrea is asking of him.

If you are puzzled about why someone would want to identify with TASS rather than their analytic minds in cases like this, do not ask me to explain it to you. It's for the proponents of the "follow your gut instincts" view to defend. The present volume, in contrast, is devoted to the opposite proposition—that culture advances when we get more and more of the world structured on the basis of the reflections of the analytic mind. In the environment in which we evolved, all that mattered was that individuals like Andrea looked wrong and were treated accordingly. Cultural advance occurs when we can get our analytic systems in control; when, like Andrea's friends, we can "freak out at first, but get over it" and "get to the point where difference doesn't seem wrong." It takes the conscious reflections of the analytic mind, using cultural tools invented over centuries, to do that.

If that man had two brains, he'd be twice as stupid!
—Punch line to an Irish joke told by
Desmond Ryan, Ann Arbor, Michigan

CHAPTER 6

DYSRATIONALIA:

WHY SO MANY SMART PEOPLE DO SO MANY DUMB THINGS

Because of the failure to follow the rules of rational thought—because of the very processing biases discussed in the previous chapters: people choose less effective medical treatments; people fail to accurately assess risks in their environment; information is misused in legal proceedings; patients may choose to endure more pain rather than less; millions of dollars are spent on unneeded projects by government and private industry; parents fail to vaccinate their children; unnecessary surgery is performed; animals are hunted to extinction; billions of dollars are wasted on quack medical remedies; and costly financial misjudgments are made.[1] I am emphasizing these real-world negative consequences of the fundamental TASS biases discussed previously because it is important to stress that these are not just phenomena confined to the laboratory.

The findings from the reasoning and decision-making literature and the many real-world examples of the consequences of irrational thinking create a seeming paradox. The physicians using ineffective procedures, the financial analysts making costly misjudgments, the retired professionals managing their money poorly—none of these are unintelligent people. The findings from laboratory studies of reasoning discussed in the previous chapters are just as perplexing. Over 90 percent of the subjects in the studies are university students—some from the most selective institutions of higher learning in the world. Yet these are the very people who have provided the data that indicate that a substantial proportion of people can sometimes violate some of the most basic strictures of rational thought such as transitivity or the sure-thing principle. It appears that an awful lot of pretty smart people are doing some incredibly dumb things. How are we to understand this seeming contradiction?

The first step in understanding the seeming paradox here is to realize that the question "How can so many smart people be doing so many dumb things?" is phrased in the vernacular—what cognitive scientists call the language of folk psychology. By using some basic concepts from cognitive science, folk usage can be honed in ways that serve to clear up some seeming paradoxes. I propose to do just this with the "smart but dumb" phrase.

In this chapter, I identify the folk term "smart" with the psychological concept of intelligence. In contrast, the acts that spawn the folk term "dumb" I identify with violations of rationality as that term is conceptualized within cognitive science, philosophy, and decision science.[2] The concepts of intelligence and rationality refer to different levels of analysis in cognitive theory. Because they refer to different levels in the hierarchical control system that the brain implements, it is possible for the two to become dissociated. Thus, if the difference between intelligence and rationality is understood, there turns out to be nothing paradoxical at all about the idea that people might actually be smart but act dumb.

Cognitive Capacities, Thinking Dispositions, and Levels of Analysis

Imagine three incidents, each resulting in the same outcome—a particular human is dead. In incident A, a woman is walking on a cliffside by the ocean and a powerful and totally unexpected wind gust blows her off the cliff and she is killed. In incident B, a woman is walking on a cliffside by the ocean and goes to step on a rock but the rock is actually the side of a crevice and she falls down the crevice and is killed. In incident C, a woman attempts suicide by jumping off an ocean cliff and is crushed and killed on the rocks below. When we ask ourselves why the woman is dead in each case we would want (since there are myriad variables and causes in each incident) to home in on the *crucial* aspect of the situation. In other words, the examples each call for a different level of explanation when we zero in on the essential cause of death.

In incident A it is clear that *nothing more* than the laws of physics are needed (the laws of wind force, gravity, and crushing). Scientific explanations at this level—the physical level—are important, but to the cognitive scientist concerned with human psychology and brain function they are relatively uninteresting. In contrast, incidents B and C are both more interesting to the psychologist and cognitive scientist. Importantly, the differences between incidents B and C illustrate a distinction between levels of analysis in cognitive explanation that is critical to the subsequent argument in this chapter.

The need for additional levels of explanation becomes apparent when we notice the perhaps surprising fact that some of the same laws of physics in operation in incident A (the gravitational laws that describe why the woman will be crushed upon impact) are also operative in incident B (and in C for that matter). However, we feel that the laws of gravity and force somehow do not provide a complete explanation of what has happened in incidents B and C. The laws of physics are leaving out something critical.

In incident B, a psychologist would be prone to say that when processing a stimulus (the crevice that looked somewhat like a rock) the woman's information processing system malfunctioned—sending the wrong information to response decision mechanisms which then resulted in a disastrous motor response. Cognitive scientists refer to this level of analysis as the algorithmic level. In the realm of machine intelligence, this would be the level of the instructions in the abstract computer language used to program the machine (FORTRAN, LISP, etc.). The cognitive psychologist works largely at this level by showing that human performance can be explained by positing certain information processing mechanisms in the brain (input coding mechanisms, perceptual registration mechanisms, short- and long-term-memory storage systems, etc). For example, a simple letter pronunciation task might entail encoding the letter, storing it in short-term memory, comparing it with information stored in long-term memory, if a match occurs making a response decision, and then executing a motor response. In the case of the woman in incident B, the algorithmic level is the right level to explain her unfortunate demise. Her perceptual registration and classification mechanisms malfunctioned by providing incorrect information to response decision mechanisms, causing her to step into the crevice.

Incident C, on the other hand, does not involve such an algorithmic-level information processing error. The woman's perceptual apparatus accurately recognized the edge of the cliff and her motor command centers quite accurately programmed her body to jump off the cliff. The computational processes posited at an algorithmic level of analysis executed quite perfectly. No error at this level of analysis explains why the woman is dead in incident C. Instead, this woman died because of her overall goals and how these goals interacted with her beliefs about the world in which she lived. Cognitive scientists refer to this level of analysis—the level where we analyze in terms of goals, desires, and beliefs—as the intentional level of analysis.[3]

The intentional level of analysis is concerned with the goals of the system, beliefs relevant to those goals, and the choice of action that is optimal given the system's goals and beliefs. It is thus clear how issues of rationality arise at this level of analysis because, as was discussed in chapter 3, instrumental rationality is defined in terms of using aptly chosen means in light of an organism's goal structure and beliefs.[4] The algorithmic level provides an incomplete explanation of behavior in cases like incident C because it provides an information processing explanation of how the brain is carrying out a particular task (in this case, jumping off of a cliff) but no explanation of *why* the brain is carrying out this particular task. This is what the intentional level is for. This level provides a specification of the *goals* of the system's computations (*what* the system is attempting to compute and *why*).

We now have in place all of the conceptual apparatus needed to understand that the notion of "smart people acting dumb" is not paradoxical at all. The concept of intelligence (being smart) and the concept of rationality (acting dumb) exist at different levels of analysis in psychological theory. The study of intelligence is largely the study of algorithmic-level cognitive capacities such as perceptual speed, discrimination accuracy, working memory capacity, and the efficiency of the retrieval of information stored in long-term memory.[5] These algorithmic-level cognitive capacities are relatively stable individual difference characteristics. Although they are affected by long-term practice, they cannot be altered online by admonition or verbal instruction. Measures of general intelligence provide an overall index of the cognitive efficiency of a wide variety of such mechanisms in a given individual (Carroll 1993, 1997).

In contrast to intelligence (an algorithmic-level construct), psychologists studying so-called thinking dispositions have focused their attention at the intentional level of analysis. For example, many thinking dispositions concern beliefs, belief structure, and, importantly, attitudes toward forming and changing beliefs. Other cognitive styles (the terms cognitive styles and thinking dispositions will be used interchangeably here as they are in the psychological literature)[6] that have been identified concern a person's goals and goal hierarchy. These cognitive styles and thinking dispositions thus relate directly to the probability that an individual will choose rational actions and have rational beliefs as defined in chapter 3.

Thus, in contemporary cognitive science, individual differences in measures of intelligence index individual differences in the efficiency of processing at the algorithmic level. In contrast, thinking dispositions and cognitive styles, as traditionally studied in psychology, index individual differences at the intentional level of analysis. They are related to rationality because they are telling us about the individual's goals and epistemic values.

TASS Override and Levels of Processing

Throughout thousands of micro-events in our day-to-day lives, TASS systems work in a fashion that is instrumentally rational. However, as we saw in the last two chapters, in the modern world, a small but increasingly important number of occasions occur in which the TASS-primed response, however well it may be designed to serve evolutionary goals, is not the instrumentally rational response for the person considered as a coherent organism. In such cases, if instrumental rationality is to be achieved, the TASS-primed response must be overridden by the analytic system.

Figure 6.1 illustrates the logic of the situation in terms of the levels of analysis discussed earlier. As discussed in detail in chapter 2, TASS implements the universal and short-leashed goals that largely overlap with the genetic goals which were adaptive in our ancient evolutionary environment.[7] The algorithmic level of TASS will implement these goals unless overridden by the algorithmic mechanisms implementing the long-leash goals of the analytic system (the horizontal arrow in figure 6.1). But the vertical, downward pointing arrow in figure 6.1 is a reminder that the algorithmic level of the analytic system is conceptualized as subordinate to the higher-level goal states at the intentional level. (The boxes labeled biological level on the sides are simply to remind us that the hierarchical control structures that we conceptualize at the algorithmic and intentional levels are both grounded in a biological substrate.)

Recall from chapter 2 that individual differences in the algorithmic-level computational power of TASS are assumed to be few because it is an old evolutionary system.[8] In contrast, the algorithmic-level computational power of the analytic system displays the variability we still measure as general intelligence because it is a more recent evolutionary product. It should be clear from figure 6.1 that the probability of TASS override will indeed be a function of the computational power instantiated at the algorithmic level of the analytic system, but it will not be *only* a function of that algorithmic-level computational power. This is because it is clear from the figure that the override processing initiated by the algorithmic level of the analytic system is itself triggered by hierarchical control from the intentional level of that system (the downward-pointing arrow).

Thus, there is nothing in principle preventing rational response tendencies (intentional-level thinking dispositions) from dissociating from

Control structures

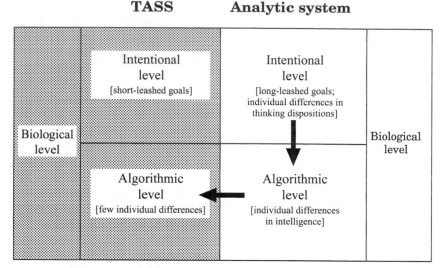

Figure 6.1 Processing control in TASS override by the analytic system

intelligence (the algorithmic-level computational power of the analytic system). Intelligence, while representing the *potential* computational power to carry out TASS override, does not guarantee rationality because the capacity for override remains unrealized if not triggered by the superordinate control hierarchies that we identify as rational thinking dispositions at the intentional level. Variability in intentional-level thinking dispositions means that there is at least the potential for cognitively competent people to do irrational things. Such an analysis implies that it should be no surprise that a lot of smart people act dumb sometimes.[9]

The Great Rationality Debate:
The Panglossian, Apologist, and Meliorist Positions Contrasted

> The debate over human rationality is a high-stakes controversy that mixes primordial political and psychological prejudices in combustible combinations.
>
> —Philip Tetlock and Barbara Mellers (2002, 97)

As was described in chapter 4, much research has accumulated to indicate that human behavior often deviates from optimal standards as determined by decision scientists. How to interpret these deviations is, however, a matter of contentious dispute. The so-called great debate about human rationality is a "high stakes controversy" because it involves nothing less than the models of human nature that underlie economics, moral philosophy, and the personal theories (folk theories) we use to understand the behavior of other humans.

In a previous book (Stanovich 1999) I used three labels to differentiate the major positions of the opposing theorists in the great rationality debate. Falling into the so-called Panglossian position are many philosophers who argue that human irrationality is a conceptual impossibility (e.g., Cohen 1981, 1983, 1986; Wetherick 1993, 1995). The Panglossian sees no gaps at all between how most people perform on reasoning tasks and how they ought (optimally) to perform. The Panglossian wishes to avoid ascribing irrationality to human action.

But how can this position possibly square with the findings from cognitive psychology discussed in chapter 4 in which various suboptimal behavioral patterns were demonstrated? Of course, in specific instances, human behavior does depart from optimality, but the Panglossian has a variety of stratagems available for explaining away these departures without labeling the response irrational. First, the departures might simply represent so-called performance errors (by analogy to linguistics)—minor cognitive slips due to inattention, memory lapses, or other temporary and basically unimportant psychological malfunctions. Second, it can be argued that the experimenter is applying the wrong model of optimality to the problem (in short, the problem is with the experimenter rather than subject). Third, the Panglossian can argue that the subject has a different construal of the task than the experimenter intended and is responding optimally to a *different* problem. Each of these three alternative explanations for seeming behavior suboptimalities—performance errors, incorrect evaluation of behavior, and alternative task construal—has been the focus of empirical investigation, as will be discussed.

In addition to the philosophers who argue for essentially perfect human rationality, another very influential group in the Panglossian camp is represented by the mainstream of the discipline of economics, which is notable for using strong rationality assumptions as fundamental tools and pressing them quite far: "Economic agents, either firms, households, or individuals, are presumed to behave in a rational, self-interested manner . . . a manner that is consistent with solutions of rather complicated calculations under conditions of even imperfect information" (Davis and Holt 1993, 435). These strong rationality assumptions are essential to much work in modern economics, and they account for some of the hostility that economists have displayed toward psychological findings that suggest nontrivial human irrationality. As economist Daniel McFadden (1999) has characterized the situation, what psychologists have found is that "all these apparently normal consumers are revealed to be shells filled with books of rules for handling specific cognitive tasks. Throw these people a curve ball, in the form of a question that fails to fit a standard heuristic market response, and the essential 'mindlessness' of the organism is revealed. For most economists, this is the plot line for a really terrifying horror movie, a heresy that cuts to the vitals of our profession. To many psychologists, this is a description of the people who walk into their laboratories each day" (76).

The work of cognitive psychologists engenders hostility among economists because the former expose the implausibly Panglossian assumptions that lie behind economic pronouncements about human behavior. For example, the *New York Times*, citing evidence that most people do not save enough for the retirement life that they desire, notes that this evidence is directly contrary to mainstream economics, which assumes that people rationally save the optimal amount: "Confronted with the reality that people do not save enough, the mainstream has no solution, except to reiterate that people are rational, so whatever they save must be enough" (Uchitelle 2002). Increasingly though, there are dissenters from the Panglossian view even within economics itself. In response to other examples like the one in the *New York Times*, Cornell economist Richard Thaler (1992) exasperatedly pleads, "Surely another possibility is that people simply get it wrong" (2).

That people "simply get it wrong" was the tacit assumption in much of the early work by the heuristics and biases researchers in cognitive psychology discussed in chapter 4—and such an assumption defines the second position in the great rationality debate, the Meliorist position. The Meliorist begins with the assumption that there is substantial room for improvement in human reasoning. Unlike the Panglossian, the Meliorist thinks that not all human reasoning errors can be explained away. However, because the Meliorist thinks that the errors are real (not excusable due to inappropriate evaluation or alternative task construal), the Meliorist is likely to ascribe irrationality to actions, whereas the Panglossian is not.

The difference between the Panglossian and the Meliorist was captured colloquially in an article in *The Economist* (February 14, 1998), where a subheading asked "Economists Make Sense of the World by Assuming that People Know What they Want. Advertisers Assume that They Do Not. Who Is Right?" The Meliorist thinks that the advertisers are right—people often do not know what they want, and can be influenced so as to maximize the advertiser's profits rather than their own utility. Or, to put it another way, the Meliorist thinks that the advertisers are right—and that's why we have to worry about advertising! In contrast, economists promulgating the view that people take only from advertising what optimizes their consumption utility are the advertising industry's best friend. The economists' Panglossian assumption is what the advertisers hide behind when trying to repel government strictures prohibiting certain types of ads. In contrast, the Meliorist does not assume that consumers will process the advertiser's information in a way that optimizes things for the consumer (as opposed to the advertiser). Thus, Meliorists are much more sympathetic to government attempts to regulate advertising because, in the Meliorist view, such regulation can act to increase the utility of the total population. This is just one example of how the great rationality debate has profound political implications.

The third position in the great rationality debate is that of the so-called Apologist. The Apologist position is like the Meliorist position in that it views the suboptimalities in human behavior as real (i.e., they cannot be excused away by positing inappropriate evaluation or alternative task construal), but the Apologist is like the Panglossian in not wanting to call the suboptimalities instances of irrationality. They avoid this by relying on the well-known stricture that judgments about the rationality of actions and beliefs must take into account the resource-limited nature of the human brain.[10]

Reasoners have limited short-term memory spans, limited long-term storage capability, limited perceptual abilities, and limited ability to sustain the serial reasoning operations that are needed to reason logically and probabilistically (see chapters 2 and 4). The Apologist posits that in many situations the computational requirements of the optimal response exceed those of the human brain. Thus, they argue that it seems perverse to call an action irrational when it falls short of optimality because the human brain lacks the computational resources to compute the most efficient response. Ascriptions of irrationality seem appropriate only when it was possible for the person to have done better. In fact, most Meliorists also agree with the general stricture that we should not call an action irrational if the optimal model calls for computational abilities that exceed those of the human brain. The disagreement between the Meliorists and Apologists concerns the applicability of the stricture in specific instances—whether a particular optimal response was actually within the computational limits of humans. Apologists think that in most cases computational limitations are stopping people from giving the optimal response.

To summarize the three positions: the Meliorist thinks that sometimes people are not reasoning very well and that they could do much better. The Apologist thinks that people are not reasoning properly, but that they are doing about as well as they could possibly do. And, finally, the Panglossian feels that people are reasoning very well—indeed, as well as anyone could possibly reason in this best of all possible worlds.

It is clear that from the standpoint of motivating cognitive remediation efforts, these positions have markedly different implications. The Panglossian position provides little motivation for remediation efforts because the Panglossian thinks that reasoning is as good as it could conceivably be. The Meliorist position, on the other hand, strongly motivates remediation efforts. Actual behavior is markedly below what is cognitively possible on this view, and the payoff for remediation efforts would appear to be high.

The Apologist position is slightly more complex than either of the other two. The Apologist is like the Panglossian in seeing little that can done, given *existing* cognitive constraints (that is, the existing computational constraints of our brains and the current way that environmental stimuli are structured). However, the Apologist position does emphasize the possibility of enhancing performance in another way—by presenting information in a way that is better suited to what our cognitive machinery is designed to do. Evolutionary psychologists represent this aspect of the Apologist position when they emphasize that many heuristics and biases tasks, if redesigned to fit the stimulus constraints of evolutionary modules, become more solvable for some people (Brase, Cosmides and Tooby 1998; Cosmides and Tooby 1996; Gigerenzer and Hoffrage 1995).

The great rationality debate between the Meliorists, Panglossians, and Apologists has been ongoing for over two decades now and shows no sign of abating.[11] It is likely that a preexisting bias related to cognitive remediation has partially motivated the positions of the Meliorist reformers in the heuristics and biases literature as well as their critics in the Panglossian and Apologist camps. Pretheoretical biases also arise because of different weightings of the relative costs and benefits of different assumptions about the nature of human rationality. For example, if Panglossians happen to be wrong in their assumptions, then we might miss opportunities to remediate reasoning. Conversely, unjustified Meliorism has its associated costs as well. Effort might well be wasted at cognitive remediation efforts. We might fail to appreciate and celebrate the astonishing efficiency of unaided human cognition. Excessive Meliorism might lead to a tendency to ignore the possibility that the *environmental* change advocated by the Apologists might be an easier route to performance enhancement than cognitive change.

Some commentators in these disputes feel that it is insulting to people to make widespread ascriptions of human irrationality, as Meliorists do. In this concern lies a well-motivated anti-elitist tendency as well, no doubt, as some kind of inchoate concern for human self-esteem. All of these tendencies show a commendable concern for the folk psychological and social implications of the interpretation of research findings. But I wonder whether our immediate reaction—to think that ascriptions of human irrationality are a bad thing—might not be a bit hasty itself. The world is full of catastrophes caused by human action. Meliorists think that some of these might be avoided by teaching people to be less prone to irrational judgments and actions. What are the alternatives to the Meliorist position? In fact, I will argue here that the alternatives to the Meliorist explanation are not at all comforting.

Assume that the Meliorist is wrong. Is it not disturbing that none of our wars, economic busts, technological accidents, pyramid sales schemes, telemarketing fraud, religious fanaticism, psychic scams, environmental degradation, broken marriages, and Savings and Loan scandals are due to remediable irrational thought? If not irrationality, then what? One alternative is that the causes of these disasters must reside in much more intractable social dilemmas (like the famous Prisoner's Dilemma, for instance, or Hardin's tragedy of the commons; see Colman 1995; Hardin 1968; Komorita and Parks 1994) that are not remediable by reforming individual human cognition.

If not intractable social dilemmas, then there is in fact another alternative explanation, but it is even less palatable. Recall that instrumental rationality means attaining our goals via the most efficient means—regardless of what those goals are. If the world seems to be full of disastrous events despite the fact that everyone is assumed (by the Panglossians) to be rationally pursuing their goals and there is no social dilemma involved, then it seems we are left with a most distressing alternative. Large numbers of people must be pursuing truly evil human desires (see Kekes 1990; Nathanson 1994). The problems of a thin theory of rationality were discussed in chapter 3, where it was noted that such thin theories might well deem Hitler rational. Positing that there are many "rational Hitlers" in the world is a way to square a Panglossian assumption of human rationality with the observation that myriad human-caused catastrophes occur daily.

In the face of this type of conclusion it appears that ascribing some irrationality to human beings is not quite the bogey we once thought it was, and, ironically, a Panglossian assumption of perfect human rationality is not the warm fuzzy that it initially seems. If we feel queasy about insulting people, it seems less insulting to say they are irrational than that they are evil or despicably selfish. Consider another example where this seems even clearer. Several years ago, there was a television report that the National Highway Safety Administration had determined that something like 40 percent of small children in America were still riding in automobiles without being secured with seatbelt restraints. How are we to interpret this appalling statistic from the standpoint of evaluating the rationality of the parents of those children?

Starting with a Panglossian bias to preserve the rationality of the parents, we could call a certain percentage of these cases "performance errors"—some of these are children who *are* regularly belted but whose parents forgot to do so on a certain occasion. However, highway safety commissions assure us that this is not the *entire* story—some children are indeed unbelted time after time. Their parents' behavior cannot be a so-called performance error, because it is systematic. How do we preserve the presumption of the rationality of the parents in *these* cases? We are not prone to take one escape hatch that is available to the Panglossian—that perhaps the parents really do not value their children. That is, what creates the paradox is that the parents' behavior (they are putting their children at great risk by not securing them with seatbelts) is totally at odds with their desires and goals (they love their children and desire to protect them). One way out of the paradox is to deny the latter. (Panglossian economists regularly make this move when they reply to someone who claims to like widgets but does not purchase them that, in fact, they do not really like widgets.) This assumption preserves the Panglossian default of perfect rationality, but at the expense of taking a dim view of the characteristics of our fellow humans. Most of us would prefer not to escape the paradox by going in that direction.

Instead, the more popular way to resolve the seeming paradox is to retreat to the extreme strictures of a thin theory of rationality—where beliefs are treated as fixed and not subject to evaluation and where, likewise, neither is the content of desires. That is, the individual's goals and beliefs are accepted as they are, and debate centers only on whether individuals are optimally satisfying their desires given their beliefs. The thin theory—plus a Panglossian default—then simply says that, in this case, given that the desire is fixed (these parents love their children) and the behavior is fixed (they did not, in fact, secure the children with seatbelts), then it must follow that the belief behind the action must have been incorrect. The parents must not have known that unbelted children in automobile collisions—the leading cause of childhood death (National Highway Traffic Safety Administration 1999)—are in unusual danger.

But is this really a satisfactory solution? Are we really assuaged that there is nothing wrong here? I submit that there are still grounds for worry. If we slip off the Panglossian blinders, we will see that this example points to an entire domain of rationality that is open to assessment—the calibration of knowledge acquisition according to practical goals. I believe that the seatbelt example does reveal a failure of rational thinking that is related to notions of so-called epistemic responsibility (see Code 1987). Specifically, it is fine to say—as the thin theory does—that the parents were not irrational because they did not *know* that unbelted children are in particular danger. However, the issue I am raising here is that perhaps it is also appropriate to ask the question: Why *didn't* they know? The media has been saturated with seatbelt warnings for over two decades now. Educational efforts in schools and communities are directed toward seatbelt use. Of course, seatbelt use is a key component of all driver training courses. Information about the importance of seatbelts for children is not hard to acquire.

There seems to be a requirement of calibration that is not being met in our seatbelt example. Consider another example provided by cognitive scientist John Pollock (1995), who concocts the story of a ship's captain on a busman's holiday aboard a Caribbean cruise liner. The ship seems well equipped and our captain casually notes the lifeboats, wonders how many there are, and consults the cheap brochure given to all passengers. Later on, an accident occurs—disabling the entire ship's crew and putting the ship in danger of sinking. The vacationing captain is now in charge of the cruise liner and immediately wants to know whether there are enough lifeboats on board. His belief based on the holiday brochure is no longer sufficient—he must have the lifeboats counted carefully and accurately.

In Pollock's (1995) example, it was important for the captain to calibrate his knowledge of the lifeboats to his situation (was he a mere passenger or was he in charge of the boat?). Similarly, over a person's lifetime, it is critical to acquire knowledge in domains that are most relevant to fulfilling one's most important goals. This is what I have termed the pragmatic calibration of knowledge acquisition. Knowledge acquisition is effortful and the cognitive resources available for it are limited. There is a limited amount of time and effort available to spend in epistemic activities, and it is important that our effort be calibrated so that it is directed at knowledge domains that are connected to goals we deem important. If we say that something is of paramount importance to us (our children's safety, for example), then it is incumbent on us to know something about these things we deem to be so important. The parents of the unbelted children are irrational in this view because they have instead poorly calibrated their knowledge acquisition over a long time period of their lives. And, contra the Panglossian, it is more optimistic and respectful of humans to say that they are irrational rather than to posit that they actually do not love their children as much as they say they do.

What does research say about the evidence regarding the Panglossian, Meliorist, and Apologist positions? Dozens of relevant studies conducted in the last twenty years have begun to converge on a set of tentative conclusions.[12] First, the Panglossians have served a useful function in their critique of the tasks in the heuristics and biases literature, because in a few notable cases they have demonstrated that psychologists had been applying the wrong criteria to evaluate performance.[13] However, these cases are in the minority. Most cases of suboptimal human responding cannot be explained away. Particularly misguided has been the Panglossian's reliance on the performance error argument (random lapses in ancillary processes necessary to execute a cognitive strategy—lack of attention, temporary memory deactivation, distraction, etc.) to explain away human responses that appear irrational. This is because most human reasoning errors of the type described in chapter 4 are systematic rather than random (Rips and Conrad 1983; Stanovich 1999; Stanovich and West 1998c, 2000).

Likewise, the Apologist position does seem to have a grain of truth to it, but it cannot explain the large number of systematically irrational responses uncovered by cognitive psychologists in the heuristics and biases literature. Some tasks that psychologists have studied do seem to cause some people to fail because they stress the computational abilities of humans, but again this is true only in a minority of cases. Research has shown that at least some subjects of very modest cognitive abilities give the optimal response in virtually all heuristics and biases tasks. Most people are not prevented from giving the optimal response because of computational limitations. The Meliorists appear to be right that some human actions are systematically irrational.

Dysrationalia:
Dissolving the "Smart But Acting Dumb" Paradox

By now it should be clear how the notion of smart people acting dumb is entirely explicable within the framework I have developed here. Research has established that individual differences in intelligence (an algorithmic-level construct) and individual differences in rational thinking dispositions (an intentional-level construct) are not perfectly correlated.[14] Individual differences at the two levels can thus become dissociated.

A little mapping of folk psychological terms now resolves the "smart but dumb" paradox with which I opened this chapter. In the vernacular, we often say "what a dumb thing to do" when irrational thinking has led to a maladaptive behavioral act. To put it simply, when we say that a person has done something dumb we often mean that they have acted irrationally. Or, to put it more formally and technically, the act appears to result from suboptimal behavioral regulation at the intentional level of analysis. We do not mean that the dumb act occurred because of some algorithmic-level malfunction (improper encoding of stimuli, short-term memory failure, etc.). The paradoxical connotation of the idea of "smart people acting dumb" is dissolved if we recognize that folk psychological usage shades the connotation of "dumb" a little more toward rationality than toward intelligence, and the connotation of smart a little less toward rationality and a little more toward intelligence.

Some years ago, in an attempt to make the possibility of dissociations between intelligence and rationality more salient, I invented a new category of disability (Stanovich 1993, 1994). The new disability category was called dysrationalia—the inability to think and behave rationally despite adequate intelligence. The term dysrationalia was coined to parallel the discrepancy definitions in the field of learning disabilities—for example the notion of tying the concept of dyslexia to discrepancies between reading acquisition and intelligence and dyscalculia to discrepancies between arithmetic ability and intelligence. In parallel, a person suffering from dysrationalia is showing an inability to think and behave rationally despite adequate intelligence—that is, is a smart person who acts dumb.

The emphasis on separating the constructs of rationality and intelligence in my framework is not shared by some psychological theorists. Many psychologists prefer to conflate these two constructs—to fold aspects of rationality into their concept of intelligence (see Baron 1985a; Perkins 1995; Sternberg 1997a). However, there is one consequence of the conflated view that is hardly ever emphasized in the psychological literature. That is that the notion of smart people acting dumb much of the time becomes a puzzle under a view that conflates rationality and intelligence. Smart and dumb refer to the same thing in the conflated view, and we must face the consequences of failing to make such a conceptual distinction. Under the conflated view, smart people who continually act dumb simply are not as smart as we thought they were! What my view says is irrationality in the face of intelligence (dysrationalia), the conflated view says is impeached intelligence.

Like the conflated view, neither the Apologist nor the Panglossian position recognizes a dissociation between rationality and intelligence. Intentional-level cognitive functioning is never impeached in the Apologist view, but instead it operates optimally (that is, with perfect rationality). All suboptimal responses are attributed to limitations in capacity at the algorithmic level. In contrast, the disability of dysrationalia is a possibility under the Meliorist view because strategies of intentional-level behavioral regulation are subjected to critique. They are posited to be suboptimal despite the individual's adequate computational capacity. The positing of dysrationalia is a Meliorist bet on the ability to reform cognition at the intentional level.

Would You Rather Get What You Want Slowly or Get What You Don't Want Much Faster?

A dysrationalic fails to fulfill his/her personal goals despite having at least adequate cognitive capacity because it is rationality that ensures human fulfillment, not algorithmic-level capacity alone. It is thus puzzling that society seems to place so much more emphasis on algorithmic-level capacity than on intentional-level rationality. Society is obsessed with intelligence— discussing it, measuring it, demanding that schools increase it, and so on. Intelligence tests, SAT tests, and most school aptitude and achievement tests assess algorithmic-level capacities rather than intentional-level thinking dispositions. Society is much less focused on debating rationality and ways to increase it.

This discontinuity is disturbing and odd. Rationality is, if anything, probably more malleable than basic algorithmic-level cognitive capacities (short-term memory capacity is less likely to be changed after a brief period of instruction than, for example, the tendency to look for evidence pertinent to the alternative hypothesis)—and rationality is more important for the attainment of a person's goals. In a testing-obsessed Western society with all kinds of assessment instruments now being used in schools and industry, there is very little emphasis on assessing rational thinking skills. This is not because the components of rational thought cannot be assessed or taught. A massive literature in decision science and cognitive psychology (see chapter 4) contains a variety of methodologies that can be used to measure a host of rational thinking skills including the ability to: form conclusions commensurate with the evidence, assess covariation, deal with probabilistic information, calibrate degrees of belief, recognize logical implications, have coherent assessment of degrees of uncertainty, have consistent preferences that maximize utility, consider alternative explanations, and make coherent judgments. Teaching and training programs for many of these components of rational thought also exist.[15]

I am going to borrow and embellish a telling thought experiment from a book by cognitive psychologist Jonathan Baron (1985a, 5) to illustrate my point about the oddly dysfunctional ways that rationality is devalued in comparison to intelligence. Baron asks us to imagine what would happen if we were able to give everyone a harmless drug that increased their algorithmic-level cognitive capacities (discrimination speed, STM capacity, etc.)— in short, that increased their intelligence. Imagine that everyone in North America took the pill before retiring and then woke up the next morning with one more slot in their working memories. Both Baron and I believe that there is little likelihood that much would change the next day in terms of human happiness. It is very unlikely that people would be better able to fulfill their wishes and desires the day after taking the pill. In fact, it is quite likely that people would simply go about their usual business—only more efficiently. If given more short-term memory capacity, people would, I believe: carry on using the same ineffective medical treatments, keep making the same poor financial decisions, keep voting against their interests, keep misjudging environmental risks, and continue making other suboptimal decisions. The only difference would be that they would be able to do all of these things much more quickly due to their enhanced algorithmic-level computational abilities!

Increasing people's cognitive capacities, in the Meliorist view, would help in some cases where poor responses were made because of computational limitations, but it would do nothing to help the majority of situations in which suboptimal rational thinking strategies were at fault. In contrast, increasing the rational thinking skills previously defined—processes of accurate belief formation, belief consistency assessment, and behavioral regulation—might really improve our own lives and those of others.

A context for understanding the implications of Baron's thought experiment is provided by recalling the definition of instrumental rationality discussed in chapter 3 and the difference between the algorithmic and intentional level discussed in this chapter. Thinking rationally ensures, by the very definition of instrumental rationality, that you will get what you want. A rational thinker with modest general intelligence might be slow in executing plans that are optimal on a means/ends analysis. In contrast, an irrational plan, no matter how efficiently executed by powerful algorithmic-level mechanisms, cannot maximize the individual's personal utility. As a society that places such a high value on intelligence and gives short shrift to rationality, we do indeed seem to be fostering getting what we don't want faster. Our obsession with intelligence and our general hostility to the types of cognitive evaluations that are necessary for fostering rationality seems designed to create a dysrationalic citizenry.

I have used the concept of dysrationalia to help provoke some needed discussion about the relative cultural value of intelligence and rationality. Given the social consequences of rational versus irrational thinking, the practical relevance of this domain of skills cannot be questioned. Why then, do the selection mechanisms used by society tap only cognitive capacities and ignore rationality? The issue of the differential privileging of some cognitive skills over others deserves more explicit public discussion, and I coined the term dysrationalia to provoke just such a discussion.

Consider the example of Ivy League colleges in the United States. These institutions are selecting society's future elite. What societal goals are served by the selection mechanisms (e.g., SAT tests) that they use? Social critics have argued that it is the goal of maintaining an economic and social elite. But the social critics seem to have missed a golden opportunity to critique current selection mechanisms by failing to ask the question "Why select for cognitive capacities only and ignore rationality completely?"

For example, some Panglossian philosophers have found experimental demonstrations of irrationality implausible because, they say, the subjects—mostly college students—"will go on to become leading scientists, jurists, and civil servants" (Stich 1990, 17). I do think that these philosophers have drawn our attention to something startling, but I derive a completely different moral from it. Most jurists and civil servants, in my experience, do seem to have adequate cognitive capacities. However, despite this, their actions are often decidedly suboptimal. Their performance often fails to measure up, not because they lack short-term memory capacity or memory retrieval speed, but because their dispositions toward rationality are sometimes low. They may not lack intelligence, but they *do* lack some rational thinking skills.

As illustrated in my earlier "smart but acting dumb" discussion, the poor performance of the college students in the experiments in the literature on reasoning and decision making is not in the least paradoxical. The college students who fail laboratory tests of decision making and probabilistic reasoning are *indeed* the future jurists who, despite decent cognitive capacities, will reason badly. These students have never been specifically screened for rationality before entering the laboratory. And they will not be so assessed at any other time. If they are at elite state universities or elite private schools, they will continue up the academic, corporate, political, and economic ladders by passing SATs, GREs, placement tests, performance simulations, etc. that assess primarily algorithmic-level cognitive capacities (i.e., intelligence). Rationality assessment will never take place. But what if it did? It is an interestingly open question, for example, whether race and social class differences on measures of rationality would be found to be as large as those displayed on intelligence tests.

Jack and His Jewish Problem

Consider Jack. As a child, Jack did well on an aptitude test and early in his schooling got placed in a class for the gifted. He did well on the SAT test and was accepted at Princeton. He did well on the LSAT and went to Harvard Law School. He did well in his first and second year courses there and won a position on the *Law Review*. He passed the New York Bar Exam with flying colors. He is now an influential attorney, head of a legal division of Merrill-Lynch on Wall Street. He has power and influence in the corporate world and in his community. Only one thing is awry in this story of success: Jack thinks the Holocaust never happened and he hates Jewish people.

Jack thinks that a Jewish conspiracy controls television and other media. Because of this, he forbids his children to watch "Jewish shows" on TV. Jack has other habits that are somewhat "weird." He doesn't patronize businesses owned by Jewish people. There are dozens of business establishments in his community, but Jack always remembers which ones are owned by Jewish people (his long-term storage and retrieval mechanisms are quite good). When determining the end-of-year bonuses to give his staff, Jack shaves off a little from the Jewish members of the firm. He never does it in a way that might be easily detectable, though (his quantitative skills are considerable). In fact, Jack wishes he had no Jewish staff members at all and attempts not to hire them when they apply for positions. He is very good at arguing (his verbal skills are impressive) against a candidate in a way that makes it seem like he has a principled objection to the candidate's qualifications (his powers of rationalization are immense). Thus, he manages to prevent the firm from hiring any new Jewish members without, at the same time, impeaching his own judgment. Jack withholds charitable contributions from all organizations with "Jewish connections" and he makes sizable contributions (his salary is, of course, large) to political groups dedicated to advancing ethnocentric conspiracy theories.

The point is that Jack has a severe problem with belief formation and evidence evaluation—but none of the selection mechanisms that Jack has passed through in his lifetime were designed to indicate his extreme tendencies toward belief perseveration and biased evidence assimilation. They would indeed have been sensitive—indeed, would have quickly raised alarm bells—if Jack's short-term memory capacity were 5.5 instead of 7. But they were deadly silent about the fact that Jack thinks Hitler wasn't such a bad chap.

In fact, Jack has a severe cognitive problem in the area of epistemic rationality—he is severely dysrationalic in the epistemic domain. Yet he has a leading role in the corporate structure that is a dominant force in American society. Does it make sense that our selection mechanisms are designed to let Jack slip through—given that he has a severe problem in epistemic regulation (and perhaps in cognitive regulation as well)—and to screen out someone with normal epistemic mechanisms but with a short-term memory capacity 0.5 items less than Jack's?

Although Jack's problem in belief formation may seem to be "domain specific," it is clear from this brief description that such unjustified beliefs can affect action in many areas of modern life. In a complex society, irrational thinking about economics, or about the nature of individual differences among people of different races or genders, can—when it occurs in people of social influence—have deleterious influences that are extremely widespread. Besides, some domains are more important than others. When the domains involved become too large and/or important it seems ill-advised to assuage concern about irrational thinking by arguing that it is domain specific. To say "Oh, well, it only affects his/her thinking about how to make financial decisions in his/her life" or "It only affects his/her thinking about other races and cultures" seems somewhat Panglossian in the context of modern technological and multicultural societies. Domain specificity is only a mitigating factor in a case of irrational thought when it can be demonstrated that the domain is truly narrow and that our technological society does not magnify the mistake by propagating it through powerful informational and economic networks.

Finally, it is equally possible that Jack's thinking problems are really not so domain specific. It is possible that careful testing would have revealed that Jack is subpar in a variety of tasks of human judgment: He might well have displayed greater than average hindsight bias, extreme overconfidence in his probability assessments, belief perseverance, and confirmation bias. Of course, none of this would have been known to the law school admissions committee considering Jack's application. They, as had many others in Jack's life, conferred further social advantages on him by their decisions, and they did so without knowing that he was dysrationalic.

Obviously, I have concocted this example in order to sensitize the reader to the social implications of mismatches between cognitive capacities and rationality. However, as a dysrationalic, Jack is unusual only in that society bears most of the cost of his disability. Most dysrationalics probably bring most of the harm onto themselves. In contrast, Jack is damaging society in myriad ways, despite the fact that his cognitive capacities may be allowing him to "efficiently" run a legal department in a major corporation. Ironically, then, Jack is damaging the very society that conferred numerous social advantages on him because of his intelligence. The maintenance worker who cleans Jack's office probably has cognitive capacities inferior to Jack's and has been penalized (or denied rewards) accordingly. However, the fact that the maintenance worker does not share Jack's irrational cognition has conferred no advantage on the maintenance worker—just as the presence of dysrationalia has conferred no disadvantage on Jack. Perhaps if we assessed rationality as explicitly throughout educational life as we do cognitive capacity, it would.

The Panglossian's Lament: "If Human Cognition Is So Flawed, How Come We Got to the Moon?"

In this chapter, I have removed the air of paradox from the observation that smart people often seem to act pretty dumb. The Meliorist position accommodates this finding quite easily because there is nothing paradoxical about a dissociation between intelligence and rationality in the Meliorist view. The stance of the Meliorists does raise other questions though, most notably what cognitive scientists Jonathan Evans and David Over (1996) have termed the paradox of rationality. The paradox of rationality is succinctly illustrated in a question posed to psychologists Dick Nisbett and Lee Ross by a colleague who had read some of their studies on errors in human reasoning and had asked them: "If we're so dumb, how come we made it to the moon?" (Nisbett and Ross 1980, 249). The puzzle that this question poses—the puzzle that Evans and Over call the paradox of rationality—is that if psychologists have demonstrated so many instances of human irrationality, how could we have done anything so impressive as find a way to get to the moon? How could humans have accomplished any number of supreme cultural achievements such as curing illness, decoding the genome, and uncovering the most minute constituents of matter?

The answer to this question is actually fairly simple. As cultural products, collective feats of societal progress do not bear on the capabilities of *indi-*

viduals for rationality or sustained efficient computation, because cultural diffusion allows knowledge to be shared and short-circuits the need for separate individual discovery. Most of us are cultural freeloaders—adding nothing to the collective knowledge or rationality of humanity. Instead, we benefit every day from the knowledge and rational strategies invented by others.

The development of probability theory, concepts of empiricism, mathematics, scientific inference, and logic throughout the centuries have provided humans with conceptual tools to aid in the formation and revision of belief and in their reasoning about action. A college sophomore with introductory statistics under his or her belt could, if time-transported to the Europe of a couple of centuries ago, become rich "beyond the dreams of avarice" by frequenting the gaming tables or by becoming involved in insurance or lotteries (see Gigerenzer, Swijtink, Porter, Daston, Beatty, and Kruger 1989; Hacking 1975, 1990). The cultural evolution of rational standards is apt to occur markedly faster than human evolution. In part this cultural evolution creates the conditions whereby instrumental rationality separates from genetic optimization. As we add to the tools of rational thought, we add to the software that the analytic system can run to achieve long-leash goals that optimize actions for the individual. Learning a tool of rational thinking can quickly change behavior and reasoning in useful ways—as when a university student reads the editorial page with new reflectiveness after having just learned the rules of logic. Evolutionary change is glacial by comparison.

Thus, in an astonishingly short time by evolutionary standards, humans can learn and disseminate—through education and other forms of cultural transmission—modes of thinking that can trump genetically optimized modules in our brains that have been driving our behavior for eons. Because new discoveries by innovators can be conveyed linguistically, the general populace needs only the capability to understand the new cognitive tools—not to independently discover the new tools themselves.

Cultural increases in rationality might likewise be sustained through analogous mechanisms of cumulative ratcheting. That is, cultural institutions might well arise that take advantage of the tools of rational thought, and these cultural institutions might enforce rules whereby people accrue the benefits of the tools of rationality without actually internalizing the rational tools. In short, people just learn to imitate others in certain situations or "follow the rules" of rationality in order to accrue some societal benefits, while not actually becoming more rational themselves.

Cultural institutions themselves may achieve rationality at an organizational level without this entailing that the individual people within the organization are themselves actually running the tools of rational thought on their serial mental simulators.[16] Raising the issue of institutional rationality provides a useful analogy for understanding the logic of human rationality in the context of dual-process models of cognition. In the institution, the equivalent of the genes and their short-leash TASS subsystems are actually the *people* who work in the organization. Note that the people within the organization have their own goals and autonomous behavioral potential. However, the organization cannot allow the employee to sacrifice the organization for his/her own goals. Such behavior is a threat to the organization. It cannot allow such choices to become widespread, and therefore it attempts to monitor the behavior of its employees closely to make sure that no employee is sacrificing the organization to his/her own personal goals in cases where the two might conflict. Such monitoring is done even by organizations with little labor strife. Even when it is confident that employee and organizational goals largely coincide, the company is under no illusion that its employees' goals are 100 percent lined up with the institution's goals.

However, we as humans, in an analogous situation with respect to our own TASS, make just that mistake. Just like the employee in a corporation sacrificing the organization to pursue his/her own ends when the two conflict, short-leashed genetic subsystems may sacrifice the vehicle in order to pursue their own ends (several examples of this were discussed in chapter 1). TASS in humans may contain many such subsystems with just these propensities. Institutions, of course, have very direct systems for controlling the subentities within them, and they view such monitoring as a critical component of institutional success. As humans, we must do the same. The analytic system needs to constantly monitor the outputs of the short-leashed subsystems in TASS to make sure they are not sacrificing the overall goals of the vehicle. Humans, as large structural entities, lag behind corporations in the recognition of this function. We remain quite confused about the purpose, origin, and meaning of our TASS subsystems. Many modern-day irrationalists and opponents of the scientific world view actually champion the interests of the genes over those of the analytic system. In many contexts of life, we are urged to follow our "gut instincts"—the equivalent, in our analogy, of a corporation telling its employees to set their own wages.[17]

Take the new Times Square. . . . The point here is the way everything in that place is aimed. Everything is firing message modules, straight for your gonads, your taste buds, your vanities, your fears. . . . Some of the most talented people on the planet have devoted their lives to creating this psychic sauna, just for you. . . . Today, your brain is, as a matter of brute fact, full of stuff that was designed to affect you.

—Thomas de Zengotita (2002, 36–37)

Might insight into human behavior come from knowing that our brains were constructed by genes with the single-minded purpose (so to speak) of perpetuating themselves and that these brains are infested with memes with the single-minded purpose of perpetuating themselves?

—John A. Ball (1984, 146)

CHAPTER 7

FROM THE CLUTCHES OF THE GENES
INTO THE CLUTCHES OF THE MEMES

The argument so far seems to have come to an optimistic conclusion after what was a fairly grim start. The argument has been this. We have accepted Dawkins's view that the living world can be parsed into the categories of replicator and vehicle and that we, as humans, are on the vehicle side of the ledger. We are survival machines for the genes. They are the ones that survive, we do not. We were built so that they can propagate themselves into the next generation. Such is the grim view of evolutionary science. It is little wonder that belief in creationism persists, when the implications of the evolutionary view seem so horrific.

The escape hatch I have constructed is a view of the structure of the mind that is illustrated in figure 7.1. There, it can be seen that a set of largely automatic systems is structured to prime responses that directly serve the short-term goals of the genes in the EEA (duck when an object looms, infer intention in conspecifics, satisfy nutritional needs, engage in reproductive behavior).[1] Another system—the analytic system—executes algorithms that operate serially and instantiate long-leash goals that are more specifically keyed to the vehicle. Humans thus escape the clutches of the selfish genes by getting their analytic systems in control, by developing override capabilities in instances where the responses primed by TASS and the analytic system conflict. In such instances, the survival machine carries out a successful rebellion by thwarting the goals of the replicators. We are thus able to enjoy lives more oriented toward our own long-term interests as whole organisms (rather than unwittingly playing the role of mere vehicles for our genes).

But we are not out of the woods yet, because we have failed to ask a crucial question. Where do the long-leash goals in the analytic system come from? The answer to this question is something that we really have to worry

about. The answer is worrisome because of a further fact that is, to put it in
the vernacular, truly creepy: There is another replicator out there.

Attack of the Memes: The Second Replicator

Richard Dawkins introduced the term "meme" in his famous 1976 book
The Selfish Gene. The term meme refers to a unit of cultural information that
is meant to be understood in rough (rather than one-to-one) analogy to
a gene. The Oxford English Dictionary defines a meme as "an element of

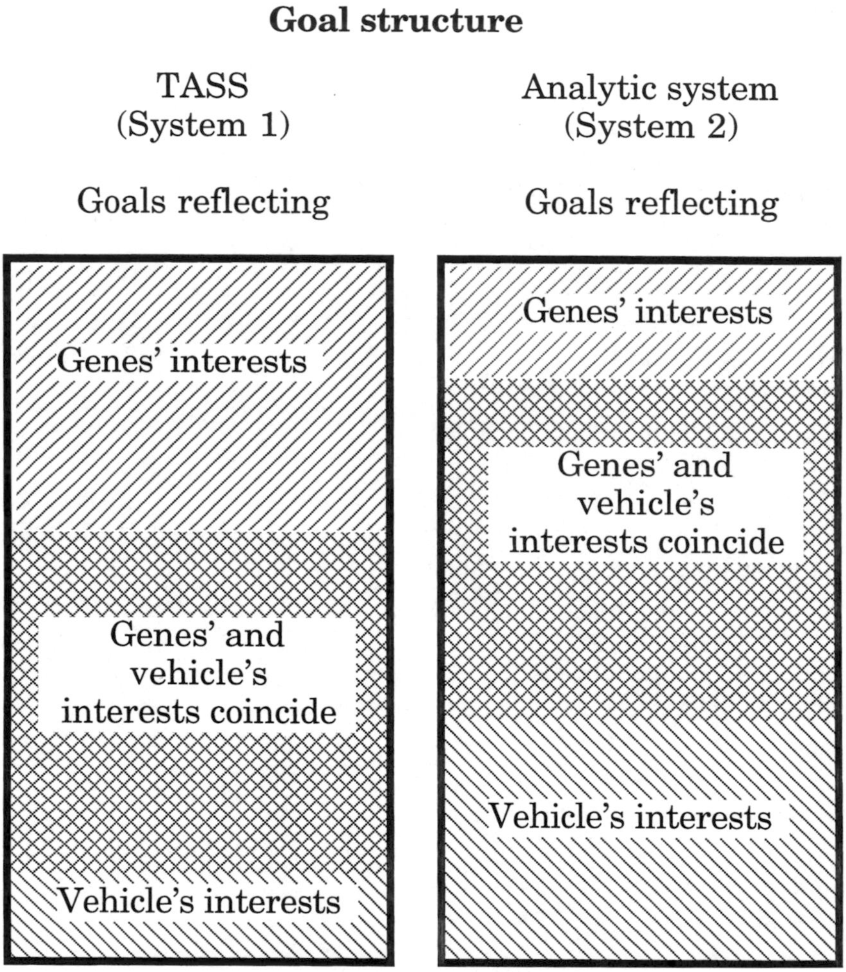

Goal structure

TASS	Analytic system
(System 1)	(System 2)
Goals reflecting	Goals reflecting

Figure 7.1 Genetic and vehicle goal overlap in TASS and in the analytic system

a culture that may be considered to be passed on by non-genetic means, esp. imitation." Blackmore (1999) defines the meme as instruction(s) for behavior(s) and communications that can be learned by imitation broadly defined (i.e., memes can be copied using language, memory, or other mechanisms) and that can be stored in brains (or other storage devices).

I prefer to view a meme as a brain control (or informational) state that can potentially cause fundamentally new behaviors and/or thoughts when replicated in another brain.[2] Meme replication has taken place when control states that are causally similar to the source are replicated in the brain host of the copy. Philosopher Daniel Dennett (1991, 1995) lists some examples of memes (or memeplexes—co-adapted sets of memes)—the arch, the wheel, calendars, calculus, chess, the Odyssey, impressionism, deconstruction, the vendetta, the right triangle, "Greensleeves," the alphabet—to help us get the idea of memes as idea units or collections of idea units.

Collectively, genes contain the instructions for building the bodies that carry them. Collectively, memes build the culture that transmits them. The meme is a true selfish replicator in the same sense that a gene is. As with the gene, by using the term "selfish" I do not mean that memes make people selfish. Instead, I mean that memes (like genes) are true replicators that act only in their own "interests." I again will use the same anthropomorphic language about memes having interests that I did for genes. As before, this is shorthand for the simple fact that the interest of a meme is replication and those that make more copies of themselves, copy with greater fidelity, or have greater longevity will leave more copies in future generations.

Once we have understood the meme as a true replicator,[3] we are in a position to understand how memetic theory helps to clarify certain characteristics of our beliefs. The fundamental insight triggered by memetic studies is that a belief may spread without necessarily being true or helping the human being who holds the belief in any way. Memetic theorists often use the example of a chain letter. Here is a meme: "If you do not pass this message on to five people you will experience misfortune." This is an example of a meme—an idea unit. It is the instruction for a behavior that can be copied and stored in brains. It has been a reasonably successful meme. A relatively large number of copies of this meme have been around for decades (with the advent of e-mail, this meme is doing even better). Yet there are two remarkable things about this meme. First, it is not true. The reader who does not pass on the message will not experience misfortune. Second, the person who stores the meme and passes it on will receive no benefit—the person will be no richer or healthier or wiser for having passed it on. Yet the meme survives. It survives because of its own self-replicating properties (the essential logic of this meme is that basically it does nothing more than say "copy me"). Memes are independent replicators. They do not necessarily exist in order to help the person in which they are lodged. They exist because, through memetic evolution, they have displayed the best fecundity, longevity, and copying fidelity—the defining characteristics of successful replicators.

Memetic theory has profound effects on our reasoning about ideas because it inverts the way we think about beliefs. Personality and social psychologists are traditionally apt to ask what it is about particular individuals that leads them to have certain beliefs. The causal model is one where the person determines what beliefs to have. Memetic theory asks instead what it is about certain memes that leads them to collect many "hosts" for themselves. The question is not how do people acquire beliefs (the tradition in social and cognitive psychology) but how do beliefs acquire people! Indeed, this type of language was suggested by Dawkins himself who, paraphrasing Nick Humphrey, said that "when you plant a fertile meme in my mind you literally parasitize my brain, turning it into a vehicle for the meme's propagation in just the way that a virus may parasitize the genetic mechanism of a host cell" (1976, 192). Dawkins argues that "what we have not previously considered is that a cultural trait may have evolved in the way it has, simply because it is advantageous to itself" (200).

Our common-sense view of why beliefs spread[4] encompasses the notions that "belief X spreads because it is true," or "X is esteemed because it is beautiful," but, in turn, has trouble accounting for ideas that are beautiful or true but not popular, and ideas that are popular but are neither beautiful nor true. Memetic theory tells us to look to a third principle in such cases. Idea X spreads among people because it is a good replicator—it is good at acquiring hosts. Memetic theory focuses us on the properties of ideas as replicators rather than the qualities of people acquiring the ideas. This is the single distinctive function served by the meme concept in the present volume, and it is a critical one.

I should acknowledge before going further that there are numerous critics of the meme concept.[5] However, these critics seem to be in an awkward position. Their opposition to the meme concept (the meme meme) must mean that they think it is having an untoward effect on the thinking of meme proponents. However, there is no question that the meme meme has spread since Dawkins introduced it in 1976. If the critics are right, then the meme meme is spreading despite the fact that it is having negative effects on both science and on the thinking of its hosts. But of course this fact then becomes the existence proof for the main proposition of the science of memes: some ideas spread because of properties of the ideas themselves.

There are numerous other controversial issues surrounding memetic theory, for example: the falsifiability of the meme concept in particular applications, the extent of the meme/gene analogy, how the meme concept differs from concepts of culture already extant in the social sciences. These debates in the science of memes are interesting, but they are tangential to the role that the meme concept plays in my argument. That role is simply and only to force on us one central insight: that some ideas spread because of properties of the ideas themselves. It is uncontroversial that this central insight has a different emphasis than the traditional default position in the social and behavioral sciences. In those sciences, it is usually assumed that to understand the beliefs held by particular individuals one should inquire into the psychological makeup of the individuals involved.

With this central insight from memetic theory in mind, we can now list several classes of reason why beliefs survive and spread. The first three classes of reason are reflected in traditional assumptions in the behavioral and biological sciences. The last reflects the new perspective of memetic theory:

1. Memes survive and spread because they are helpful to the people that store them (most memes that reflect true information about the world would be in this category).

2. Certain memes are numerous because they are good fits to preexisting genetic predispositions or domain-specific evolutionary modules.

3. Certain memes spread because they facilitate the replication of the genes that make vehicles that are good hosts for these particular memes (religious beliefs that urge people to have more children would be in this category).

4. Memes survive and spread because of the self-perpetuating properties of the memes themselves.

Categories 1, 2, and 3 are relatively uncontroversial. The first is standard fare in the discipline of cultural anthropology; category 2 is emphasized by the evolutionary psychologists; and category 3 is meant to capture the type of effects covered by psychologist Susan Blackmore's concept of memetic drive. It is category 4 that introduces new ways of thinking about beliefs as symbolic instructions that are more or less good at colonizing brains.[6]

There are many subcategories of meme survival strategies under the general category 4, including proselytizing strategies, preservation strategies, persuasive strategies, adversative strategies, freeloading strategies, and mimicking strategies.[7] For example, Aaron Lynch (1996) discusses proselytizing meme transmission and uses as an example the belief that "my country is dangerously low on weapons," which he argues illustrates proselytic advantage: "The idea strikes fear in its hosts. . . . That fear drives them to persuade others of military weakness to build pressure for doing something about it. So the belief, through the side effect of fear, triggers proselytizing. Meanwhile, alternative opinions such as 'my country has enough weaponry' promote a sense of security and less urgency about changing others' minds. Thus, belief in a weapons shortage can self-propagate to majority proportions—even in a country of unmatched strength" (5–6).

Memes can acquire protection against proselytizing by developing preservative strategies. For example, the use of the admonition "never argue about politics or religion" is a fairly transparent attempt by current memes in those categories to inoculate the host against proselytizing strategies by memes that might attempt to replace those currently resident. Some of the self-preservative strategies of memes are adversative—for example, strategies that alter the cultural environment in ways that make it more hostile to competing memes or that influence their hosts to attack the hosts of competing memes. In less adversarial fashion, it is obvious that memes that have successfully survived selection have been disproportionately those that have changed the cognitive environment in directions more advantageous to themselves. Many religions, for example, prime the fear of death in order to make their promise of the afterlife more enticing.

Additionally, in category 4 are symbionts (memes that are more potent replicators when they appear together) and memes that appear to be advantageous to the vehicle but are not—the freeloaders and parasites that mimic the structure of helpful memes and deceive the host into thinking that the host will derive benefit from them. Advertisers are of course expert at constructing meme parasites—memes that ride on the backs of other memes. Creating unanalyzed conditional beliefs such as "if I buy this car I will get this beautiful model" is what advertisers try to do by the judicious juxtaposition of ideas and images. Advertisers are expert creators of so-called memeplexes—sets of memes that tend to replicate together (co-adapted meme complexes)—that connect their product with something already valued.

Rationality, Science, and Meme Evaluation

The concepts of rational thinking discussed in this book are directly relevant to the memes resident in our brains. Many principles of instrumental rationality test memes for their consistency—for example, whether the sets of probabilities attached to beliefs are coherent and whether sets of desires hang together in logically consistent ways. Scientific inference is designed to test memes for their truth value—for example, whether they correspond to the way the world is. How might it help us if we use these mechanisms? The reason is that memes that are true are good for us because accurately tracking the world helps us achieve our goals. In contrast, as the examples discussed above indicated, many memes survive despite not being true nor helping us to achieve our goals. They are in category 4, which consists, among other things, of the freeloaders and parasite memes that mimic the structure of helpful memes and deceive the host into thinking that the host will derive benefit from them. These memes are like the so-called "junk DNA" in the body—DNA that does not code for a useful protein, that is just "along for the ride" so to speak. As discussed in chapter 1, until the logic of replicators was made clear, this junk DNA was a puzzle. Once it was understood that DNA is there only to replicate itself and not necessarily to do anything good for us (as organisms), then there is no longer any puzzle about why there is so much junk in the genome. If DNA can get replicated without helping to build a body, that is perfectly fine with it. Replicators care—to again use the metaphorical language—only about replicating!

And so it is with memes. If a meme can get preserved and passed on without helping the human host, it will (think of the chain letter example). Memetic theory leads us to a new type of question: How many of our beliefs are "junk beliefs"—serving their own propagation but not serving us? The principles of scientific inference and rational thought serve essentially as meme evaluation devices that help us to determine which beliefs are true and therefore probably of use to us, and also which are consistent with those that are true. Failure of the consistency checks embodied in the choice axioms of instrumental rationality will often point us to memes representing goals that are not serving our life plans.

Scientific principles such as falsifiability are of immense usefulness in identifying possible "junk memes"—those that are really not serving our ends but merely serving the end of replication. Think about it. You will never find evidence that refutes an unfalsifiable meme. Thus, you will never have an overt reason to give up such a belief. Yet an unfalsifiable meme really says nothing about the nature of the world (because it admits to no

testable predictions) and thus may not be serving our ends by helping us
track the world as it is. Such beliefs are quite possibly "junk memes"—unlikely to be shed despite the fact that they do little for the individual who
holds them (and may actually do harm).[8]

Reflectively Acquired Memes:
The Neurathian Project of Meme Evaluation

It is imperative that humans not allow themselves to be colonized by selfish memes that will sacrifice their hosts to their replicative interests. There is
a sense in which such selfish memes can be even more pernicious for humans than selfish genes. A gene for running over cliffs would vanish with
each vehicle it occupied, but a meme for the same thing could still be propagated widely in our media-saturated information society. Becoming an active evaluator of our memes is also thought by many theorists to be the way
to achieve true personal autonomy (e.g., Dawkins 1993; Frankfurt 1971;
Gewirth 1998; Gibbard 1990; Nozick 1989, 1993; Turner 2001). In order to
make sure that we control our memes rather than that they control us ("you
will experience misfortune if you do not pass this letter on"), we need intellectual tools such as the falsifiability criterion, unconfounded testing of
hypotheses, and preference consistency tests to weed out the "junk" in our
belief/desire intentional psychological systems.

Note, however, the devilish recursiveness in the whole idea of meme
evaluation. Scientific and rational thinking are themselves memeplexes—
co-adapted sets of interlocking memes. I shall talk about this dilemma of recursiveness, which I call the co-adapted meme paradox, in a section below.
I will argue that we can evaluate memes, albeit not in an absolute sense
which guarantees success. Instead, we must engage in a process which displays the same sort of provisionality as science itself—what might be called
a Neurathian project of skeptical bootstrapping. Philosopher Otto Neurath
(1932–33; see Quine 1960, 3–4) employed the metaphor of a boat which
had some rotten planks. The best way to repair the planks would be to bring
the boat ashore, stand on firm ground, and replace the planks. But what if
the boat could not be brought ashore? Actually, the boat could still be re-
paired, but at some risk. We could repair the planks at sea by standing on
some of the planks while repairing others. The project could work—we
could repair the boat without being on the firm foundation of ground. The
project is not guaranteed, however, because we might choose to stand on a
rotten plank.

Science proceeds in just this manner—each experiment testing certain assumptions but leaving others considered as fixed and foundational. At a later time, these temporarily foundational assumptions might be directly tested in another experiment where they would be considered contingent and optional, with other assumptions being treated as foundational. Likewise, our examination of whether the memes we host are serving our interests must be essentially a Neurathian project. We can conduct certain tests assuming that certain memeplexes (e.g., science, logic, rationality) are foundational, but at a later time we might want to bring these latter memeplexes into question too. The more comprehensively we have tested our interlocking memeplexes, the more confident we can be that we have not let a meme virus enter our mindware (to use cognitive scientist Clark's term from chapter 2). Alternatively, and more realistically, we might modify portions of memeplexes that prove unable to pass logical and empirical tests once they are considered provisional. Just as computer scientists have proven that there is no conceptual problem with the idea of self-modifying computer software, rationality is mindware with the capability of modifying itself. Circularity is escaped because we are in a Neurathian enterprise highly similar to the logic of modern science, which is nonfoundationalist but progresses nonetheless (Boyd, Gasper, and Trout 1991; Brown 1977; Laudan 1996; Radnitzky and Bartley 1987).

Personal Autonomy and Reflectively Acquired Memes

The first step in the project of comprehensive meme evaluation is to understand conceptually the predicament that humans are in, given that:

1. We are vehicles.
2. We are self-aware of this fact.
3. We are aware of the logic of replicators and that there are two different replicators that house themselves in humans.
4. Most of us want to preserve some notion of an autonomous self.

The most important tool we have as humans in this quest for personal autonomy is a comprehensive understanding of the structure of human cognition and a recently acquired knowledge of replicator dynamics. A schematic that portrays the additional complications introduced by the notion that our goal structures consist partly of memes is presented in figure 7.2 (again, the absolute areas are mere guesses—for illustrative purposes only).

Intentional-level goals

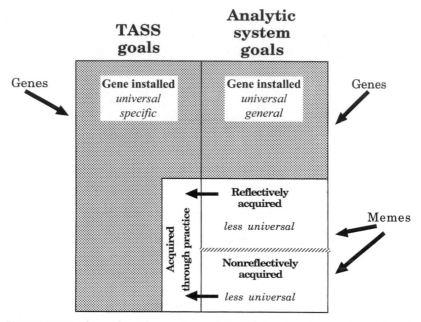

Figure 7.2 Hypotheses about the way that gene-installed goals and meme-installed goals are distributed across TASS and the analytic system

Figure 7.2 illustrates the intentional-level goal structure of both TASS and the analytic system in terms of which replicator is a source of the goal. The goal structure of TASS is dominated by gene-installed goals. These are the short-leash goals discussed earlier—nearly universal in the sense that they are shared by most humans and not the result of the environmental history of the organism. They are not flexible or generic goals, but instead are content specific, situation specific, and hard-wired to trigger (disgust and repulsion to noxious smells and substances, and fear responses to animals such as snakes, would be examples; see Buss 1999; Rozin 1996; Rozin and Fallon 1987).

The analytic system, with its more general, flexible goals, is more evenly balanced with genetic goals shared by most humans (e.g., rise in the dominance hierarchy of your conspecifics) and with meme-installed goals that are the result of the specific environmental experience (and culture) of the individual. This makes it clear that the "escape route" from the clutches of genes—serial, analytic system activity focused on the goals of the vehicle—is more complicated than has been portrayed in earlier chapters. The vehicle's analytic system goals could be in the interests of a meme rather than the vehicle as a whole. A meme for proselytizing followed by martyrdom is no different in some respects than a gene that sacrifices the vehicle's longevity to genetic reproduction. No one should want to spend a life that merely serves the ends of the replicators that inserted their short-leash goals in TASS. But neither should we want to serve the goals of replicators that are parasitic residents in the symbolic structures of the analytic mind.[9] (Why are you using that name-brand aspirin at twice the price of the identical generic?)

In figure 7.2, I distinguish between memetically acquired goals that are "caught" like viruses (as in the Dawkins quote above, "a cultural trait may have evolved in the way it has, simply because it is advantageous to it-self")—what we might call nonreflectively acquired memetic goals—and memetic goals that an individual acquires reflectively, with full awareness of their effects on the organism. The nonreflectively acquired goals are perhaps the equivalent of the parasites to which Dawkins refers. They may not actually be good for the individual, but in a manner similar to the vehicle-sacrificing genes discussed in chapter 1, these memes use the human merely as a host to aid in propagating themselves.

The figure also indicates that meme-acquired goals need not be barred from becoming TASS goals (automatic, autonomous, and rapidly triggering). Through practice, memetically installed goals can become lodged in the goal hierarchy of TASS. "Branding" and other advertising gimmicks aspire to do just this—to have a logo for X trigger the "must have X" response without much thought. These then become especially pernicious memes—parasites that, because they are not part of the reflective mind, become difficult to dislodge.

The important point here is that memes that have not passed any reflective tests are more likely to be those memes that are serving their own interests only—that is, memes that we are hosting only because they have properties that allow them to easily acquire hosts. The Neurathian project of meme evaluation aims to operationalize philosopher Robert Nozick's (1993) observation that "how much weight we should give to the fact that something has long endured depends upon why it has continued to exist, upon the nature of the selective test it has passed, and upon the criterion embodied in that test" (130). Memes having passed many selective tests that we reflectively apply are more likely to be memes that are resident because they serve our ends. Memes that we have acquired unreflectively—without subjecting them to logical and/or empirical scrutiny—are statistically more likely to be parasites, resident because of their own structural properties rather than because they serve our ends.

Which Memes Are Good for Us?

As will be discussed in more detail in the next chapter, one implication of the structure portrayed in figure 7.2 is that humans, to be fully autonomous agents, must pursue a broad rather than a thin notion of rationality. It is not enough to try to be rational given ones beliefs and goals. The beliefs and goals themselves must be evaluated in order to make sure we are not merely pursuing the ends of a replicating entity whose properties do not serve us as vehicles. We must learn how to critique our desires and beliefs—and, in order to accomplish this, we need to have installed as mindware the tools of rational and scientific thinking.

What memes would you want to have in place for a future self? It is impossible to say specifically, but it is possible to sketch out the type of evaluative criteria that might be used to examine memes and memeplexes if we become truly reflective about what memes we will allow ourselves to host. I will list four such rules for meme evaluation below. They are not intended to comprise an exhaustive list, but are instead merely departure points for the type of discussion about broad rationality that should be at the top of our intellectual agenda:

1. Avoid installing memes that are harmful to the vehicle physically.
2. Regarding memes that are beliefs, seek to install only memes that are true—that is, that reflect the way the world actually is.
3. Regarding memes that are desires, seek to install only memes that do not preclude other memeplexes becoming installed in the future.
4. Avoid memes that resist evaluation.

I will now consider each of the rules in turn.

1. Avoid installing memes that are harmful to the vehicle physically. This rule of course would cover many fairly standard admonitions against risk-inducing behaviors such as dangerous drug use, unprotected sex, and dangerous driving. We should avoid these memes because the vehicle's ability to pursue any future goal will be impaired if it is injured or has expired.[10] So, obviously, the suggestion that cigarettes do not impair health is a bad meme, as is the suggestion that cigarette smoking has style or attrac-

tiveness that may advance one's social goals. More controversially, this particular meme rule might also preclude one from adopting ideologies that involve putting one's life at serious risk—ideologies that demand war, for example. Many people might judge this implication of meme criterion 1— that there should be fewer war advocates—as an unequivocal good because they feel that history clearly shows that war serves no one's interests. War, of course, looks ridiculous from the standpoint of a vehicle that has no hope of replicating itself beyond its own lifespan and that knows that the felt urges toward war may well be the result of selfish memes or genes. On the other hand, there are those who may balk at a universal application of this principle. They will no doubt point to the few but important instances where wars involved defending a principle worth defending (including, recursively, the principle that people have a right to noncoercive meme evaluation).

On the latter point it should be noted that these suggested criteria for meme evaluation are not meant as iron-clad and deterministic rules. They are guidelines for reflective thinking about the memes one takes on. They are perhaps best taken as warning signals. Memes that violate one of these rules are signaling to the reflective thinker that they deserve further and closer scrutiny. After this further scrutiny, it is certainly possible that, in a given case, one might well decide to take on a meme that violates one of these strictures. For example, there may well be situations where reflectively installed values do demand some level of vehicle sacrifice (many altruistic memes would and should pass this test). But at least if we decide that a particular instance is one of the rule's exceptions, we will have one line of defense (in the form of the explicit awareness of the meme's exceptionality) that we would not have had, had the meme been an unexamined parasite. The issue is one of the burden of proof. Rule 1 as a principle of meme evaluation should be used as a warning signal. There is a strong burden of proof riding on the shoulders of any meme demanding physical sacrifice of the vehicle.[11]

The history of the meme for smoking provides a powerful example of how the lens of our critical intelligence can be focused on a meme and change its status entirely. Smoking causes a physiological addiction. To add fuel to that fire, the culture at one time provided powerful reinforcing memes to the complex memeplex "smoking behavior" by adding images of glamorous movie stars and popular television characters who smoked. Added to this mixture was the final incendiary of market capitalism. Because people made money as a result of other people smoking, an industry developed around smoking that advertised the associations between smoking and glamorous images. The smoking memeplex at one point was very strongly reinforced by myriad cultural forces. The problem here, of course, is that smoking is very, very bad for the vehicle. In fact, the damage it does to the vehicle turns out to be serious enough for the culture (particularly in North America) to openly declare war on memeplex smoking and to attempt to make a very severe dent in its frequency in the population. Smoking provides a positive example of how it is possible to root out and remove a memeplex that is damaging to the vehicle.

2. Regarding memes that are beliefs, seek to install only memes that are true—that is, that reflect the way the world actually is. Almost regardless of what a person's future goals may be, these goals will be better served if accompanied by beliefs about the world which happen to be true. Obviously there are situations where not tracking truth may (often only temporarily) serve a particular goal (see Foley 1991). Nevertheless, other things being equal, the presence of the desire to have true beliefs will have the long-term effect of facilitating the achievement of many goals. Thus, having memes that reflect accurately the nature of the world is a sort of superordinate goal that should be desired by most people, because whatever their goals are at a more micro-level they will often be well served by true beliefs.

Here, it must be emphasized that memeplex science has proven itself a very useful mechanism for collecting true memes. The same might be said of logic and many of the components of probability theory which have been proven historically to have quite positive epistemic consequences. Recall from chapter 3 that if individuals are to maximize expected utility in their lives, in any given situation they must assess (unconsciously in most cases— no assumption of conscious calculation is required), for each alternative action, the probabilities of each of the outcomes of this action, as well as the utilities of those outcomes (the expected utility of the action then becomes the sum of the utilities multiplied by the probabilities). For utility to be maximized, the probabilities assessed must be an accurate reflection of the uncertainties in the world.

The reason we need to be reflectively concerned with the truth of beliefs—and to engage in such truth assessments with our analytic minds—is that TASS is less likely to be concerned with truth, often for very sound evolutionary reasons. The mechanisms of TASS can be viewed as evolution's bets on the future stimuli in the environment remaining the same (Dennett 1995). These bets have been conditioned by whatever TASS mechanisms led to survival in the environment in which humans evolved.[12] The TASS subprocesses that record the structure of the world are fast and efficient mechanisms, but they often purchase their speed at the cost of being insensitive as regards veridically recording details that are not directly relevant to their function. Since TASS is responding on the basis of this kind of logic, it is critical that more reflective analytic processing be concerned with establishing a high truth content in our belief networks.

3. Regarding memes that are desires, seek to install only memes that do not preclude other memeplexes becoming installed in the future. Memeplexes that leave room for future changes in desires are somewhat like the superordinate goals discussed in chapter 3. Recall from that discussion that we might evaluate certain goals positively because attaining them leads to the satisfaction of a wide variety of other desires. In contrast, there are some goals whose satisfaction does not lead to the satisfaction of any other desires. At the extreme there may be goals that are in conflict with other goals. Fulfilling such a goal actually impedes the fulfillment of other goals (this is why goal inconsistency is to be avoided).

The memeplexes that serve as higher-level goal structures may be evaluated according to this criterion of mutual exclusiveness. Those that preclude many higher-level future goals might be deemed deleterious. Those that leave open many future desire states might be deemed advantageous (Scitovsky 1976). Of course, stage of life would be an interacting factor here. From the standpoint of memetic theory, there is in fact some justification for our sense of distress when we see a young person adopt a memeplex that threatens to cut off the fulfillment of many future goal states (early pregnancy comes to mind, as do the cases of young people joining cults that short-circuit their educational progress and that require severing ties with friends and family). Rule 3 is like rule 1 in that it seeks to preserve flexibility for the person if his/her goals should change. The closed-off religious cult destroys personal relationships that might be forfeited without possibility of rekindling if that memeplex is jettisoned at some later stage. Likewise, the false beliefs engendered by many cults (indeed, by some mainstream religions as well) can become an impediment to successful pursuit of any number of superordinate goals.

4. Avoid memes that resist evaluation. This is perhaps the most important rule for meme assessment. This characteristic is of overriding importance because parasitic memes (junk memes that do not serve the interests of the vehicle, nor of the genes that built the vehicle) will tend to increase their longevity in this way, by finding tricks that allow them to escape evaluation. Testability (falsifiability) is one of the surest means of differentiat-

ing memes that are good for you from memes that are viruses that you have simply caught like a cold. A testable meme, if actually put to the test, will eventually fall into one of two categories. Passing a logical or empirical test provides at least some assurance that the meme is logically consistent or that the meme maps the world and thus is good for us (as was discussed above under rule 2). A meme failing these kinds of tests selects itself out as candidate for the type of meme that will be useful to us in fulfilling our goals.

Either way the test comes out, we have received useful evaluative information about the memes. An untestable meme—one that avoids such critical evaluation—provides us with no such information. It is a statistical certainty that dangerous or purely parasitic memes are more likely to have come from this class of untestable memes, and those that serve vehicle interests are, conversely, statistically more likely to derive from the class of testable memes that have passed empirical and logical tests.

Examples of memes that disable attempts to evaluate them that regularly occur in the literature are the memes for faith, conspiracy theories, and free speech.[13] The first two are pretty obvious. Some readers, particularly liberal academics, may be surprised to find free speech on this list—so it is a good example for liberals to practice their memetic thinking on (thinking critically about meme acquisition and preservation). Its presence in this list makes the point that we must look carefully for memes hiding in our menomes (the term used by analogy to genome) under cover of a positive, liberal connotation.

A neutral outside observer, one without the free speech meme already installed, might wonder, for instance, whether the advocacy of the proposition that people should be killed because of their race has not already failed every empirical, logical, moral, and reasoning test. The neutral observer not already inoculated by the free speech meme might well ask what purpose is served in keeping this "idea" alive by guaranteeing it a hearing. The answer to this neutral critic that is contained within the free speech memeplex is that not to guarantee any idea a hearing would start us on a slippery slope to ban speech—but of course this is just the free speech memeplex warding off any attempt to evaluate whether it is serving the ends of the people or societies in which it resides.

The argument here is not that the free speech meme is a pure parasite, only that it has certain evaluation-disabling properties. Undoubtedly, this meme has provided vast benefits for the individuals and societies holding it (Hentoff 1980, 1992). Of course, the psychological snares in the free speech meme are nothing compared to the traps, threats, and intimidation used by faith-based memes throughout history. The whole notion of faith is meant to disarm the hosts in which it resides from ever evaluating it. To have faith in a meme means that you do not constantly and reflectively question its origins and worth. The whole logic of the faith-based meme is to disable critique. For example, one of the tricks that faith-based memes use to avoid evaluation is to foster the notion that mystery itself is a virtue (a strategy meant to short-circuit the search for evidence that meme evaluation entails). In the case of faith-based memes, many of the adversative properties mentioned above come into play. Throughout history, many religious memeplexes have encouraged their adherents to attack nonbelievers or at least to frighten nonbelievers into silence.

It is of course not necessarily the case that all faith-based memes are bad. Some may be good for the host; but a very stiff burden of proof is called for in such cases. One really should ask of any faith-based meme why it is necessary to disable the very tools in our cognitive arsenal (logic, rationality, science, decision theory) that have enabled humans to thwart the goals of the genes and to create their own life plans independent of genetic imperatives. These are the tools that make the robot's rebellion possible. And now another (perhaps more pernicious?) replicator is telling us to turn off these very mechanisms? The conceptual tools introduced throughout this book advise skepticism in such cases.

Journalist Jonathan Rauch (1993) reports that during the course of his work as a newspaper reporter in North Carolina in the 1980s he uncovered a list of "Don'ts for students" published by a fundamentalist Christian group. One rule in it was: Don't get into classroom discussions that begin with:

What would you do if . . . ?
Do you suppose . . . ?
What is your opinion of . . . ?
What might happen if . . . ?
Do you value . . . ?
Is it moral to . . . ?

It is clear that this memeplex was attempting a comprehensive inoculation against its evaluation by the host.

Consider also the case of Warwick Powell, described by journalist Mary Braid (2001). Powell had been diagnosed as HIV-positive and had a T-cell count of 220. Any count below 450 was the criterion his physician would use to recommend starting a regimen of antiretroviral drugs. However, Powell deferred conventional combination-drug therapy and instead spent ten months at a center offering The Process—energy-channelling exercises that promised to cure depression, unhappiness, irritable bowel syndrome, and cancer. Warwick was told that The Process could do what medical science could not—reverse his HIV-positive status. All this would be done by exercises to remove "energy blocks." The key to the success of the exercises in The Process was that you must, according to Warwick "give yourself over to the process completely; there was no room for doubt" (12). According to journalist Braid, the instructors offering this treatment had already convinced Powell that "if the virus did not disappear he would have only himself to blame. It would mean he had not believed enough" (12).

Full-time treatment at the center cost £100 per day (one British pound was about $1.40 to 1.50 at the time of his treatment), and Warwick had taken one-on-one sessions with senior instructors that cost £410 per hour. Telephone sessions costing even more were held with the Grand Master of The Process, who resided in New Zealand. Warwick went on an intensive course to meet the Grand Master that cost £4,000. He took a Grand Master Healing Course that cost £7,800. By the end of the period, Warwick was spending up to ten hours per day at the center and was deeply in debt. He had maxed out his credit cards (he was, however, extended credit by the center to take the Grand Master Healing Course). For the period of his devotion to the center, his physician had continued to take T-cell counts but Powell had refused to become informed about them lest the information disturb his energy exercises with negativity. After ten months at the center, Powell finally allowed his physician to inform him of his T-cell count which was then just at 270, well below the line of 450 which would indicate that he needed antiretroviral drugs and not beyond what would be expected on the basis of natural statistical fluctuation from the count he had started with ten months earlier. Powell's response to this news? He concluded that "The treatment is about surrendering your ego but I am still too arrogant and I still criticise too much. I wasn't grateful enough for the chance" (17). Braid (2001) reports that Powell intended to continue the energy channeling treatment if he could raise cash from friends and relatives.

Sadly, Powell has been captured by a meme virus—one that viciously preys on those with physical and mental difficulties and that uses evaluation-disabling strategies. Just as a hijacker attempts to commandeer a plane for his own purposes, vehicles can be hijacked by memes for their own purposes (replication). The destruction of the World Trade Center in New York has, sadly, helped many people understand this horrific logic of the virus meme that will replicate itself at any cost to human life. It has spawned a more explicit discussion of the danger of memes that become weapons because they commandeer vehicles so completely. For example, an article in the *Toronto Star* (Hurst 2001) talks about "deactivating" the ideas in terrorists. The *Times* of London, in the aftermath of the terrorist attack, explicitly referred to Osama bin Ladan as using "word bombs" (Macintyre 2001). The martyrdom meme and the meme for extravagant rewards in the afterlife for those giving themselves over to it have been recurrent throughout human history.

Evaluation-disabling strategies are common components of parasitic memeplexes. They were clearly apparent in the recovered-memory meme that swept through clinical psychology and social work in the 1990s.[14] Many cases were reported of individuals who had claimed to remember instances of child abuse that had taken place decades earlier but had ostensibly been forgotten. Many of these memories occurred in the context of therapeutic interventions. It is clear that many of these "recovered" memories were induced by the therapy itself (Piper 1998). Many lives were ruined and families disintegrated because false, therapy-induced accusations were made. This is not to say that the therapists themselves were knowingly perpetuating a falsehood. Many were merely infecting others with a memeplex that was using them (and their professional positions) as hosts. But the memeplex they passed on to their clients was an especially pernicious one, not only because of the nature of its effects on others but also because it contained strategies to disable the collection of evidence that would evaluate its status.

A sad example is provided by Rowland Mak (Makin 2001). After entering therapy for drug use and depression, Mak was told by his therapist that the therapist herself was the victim of ritual sexual abuse and that the therapist thought that this might possibly be the cause of Mak's own anger toward his father. Through session after session, some including reinforcement from other so-called survivors of ritual abuse, Mak came to believe that his father was indeed part of a secret cult that practiced ritual abuse on babies. When Mak prepared to confront his father with this information, he was told in advance that his father would be in denial and that a vehement denial of the allegation would be proof that it was true. When Mak's father did indeed deny the allegation, Mak took this as an indication that his therapist had been correct. Thus did this memeplex carry out the peculiar self-reinforcing strategy with which it controls its hosts. (Luckily, in later life,

Mak came to understand how he had come to believe something about his early life that in fact he had no memory of whatsoever. He and his father were reconciled.)

Why Memes Can Be Especially Nasty (Nastier than Genes Even!)

These examples illustrate how memeplexes can incorporate evaluation-disabling devices. One of the most obvious, but also the most pernicious, is the blatant attempt to disable criticism by incorporating into the memeplex a proposition that says that a presumed benefit will not occur if the host questions the memeplex as a whole. It is a meme that essentially says that in order to carry out its presumed positive function it must not be questioned. This is a prime disabling strategy used by viral parasitic memeplexes ("don't question me or bad things will happen"). So the rule that can be learned for inoculating yourself against such memes is simple: Question any meme that tells you that in order for it to deliver some benefit you must not question it.

The possibility of memes systematically disabling the evaluative mechanisms arrayed against them provides another reason to emphasize the importance of analytic system override of TASS responses. The unreflective view of thinking fostered by the advocates of "gut instinct" actions delivers the vehicle into the hands of both of the replicators (genes and memes)—neither of which have as their primary goal the optimization of human satisfaction. Not only will the short-leashed genetic goals stored in TASS dominate processing if we do not override inappropriate response priming from them, but parasitic memes (those not serving vehicle interests) may be lodged in our goal structures as well. Only if they are examined with a reflective intelligence that maintains some skepticism toward memes already residing in the vehicle can we escape the fate of avoiding the nonoptimal (from the vehicle's standpoint) goals of one replicator only to be delivered up to the nonoptimal goals of another.

It is thus important to emphasize that—in terms of a lack of concern for the vehicle—there may be reasons to believe that memes can be even nastier than genes. Genes are in a body together and they replicate together. These facts highlight that any given gene has interests in common with the vehicle and with other genes. Using a metaphor invented by Richard Dawkins, Sober and Wilson (1998) point out that "genes in a sexually reproducing individual are like the members of a rowing crew competing with other crews in a race. The only way to win the race is to cooperate fully with the other crew members. Similarly, genes are 'trapped' in the same individual with other genes and usually can replicate only by causing the entire collective to survive and reproduce. It is this property of shared fate that causes selfish genes to coalesce into individual organisms that function as adaptive units" (87–88).

This passage draws attention to two facts that might make genes more "friendly"—they must keep a body alive, at least long enough for it to replicate them, and they must cooperate with the other genes in doing so. Neither of these strictures applies to memes. Genes at least have to keep a vehicle alive for a decade or so before they can replicate. A meme that has just been acquired can, in contrast, be used as a template to send a signal that starts a replication reaction in another brain almost immediately. It does not have to cooperate with the other memes, or the genes, and it does not have to cooperate in keeping the vehicle alive or in facilitating its welfare in any way.

So the robot's rebellion is more complicated than I have portrayed it in earlier chapters—but it is no less an essential project of cognitive reform. And it is still an achievable one. Being reflective about the memes we take on as desires and beliefs is the only answer. This is particularly so with the memes that were acquired in our early lives—those that were passed on to us by parents, relatives, and other children. The longevity of these early acquired memes is likely to be the result of their having avoided consciously selective tests of their usefulness. They were not subjected to selective tests because they were acquired during a time when you lacked reflective capacities. Philosopher Robert Nozick (1989) warns of this when he reminds us:

> Mostly we tend—I do too—to live on automatic pilot, following through the views of ourselves and the aims we acquired early, with only minor adjustments. No doubt there is some benefit—a gain in ambition or efficiency—in somewhat unthinkingly pursuing early aims in their relatively unmodified form, but there is a loss too, when we are directed through life by the not fully mature picture of the world we formed in adolescence or young adulthood. . . . This situation is (to say the least) unseemly—would you design an intelligent species so continuingly shaped by its childhood, one whose emotions had no half-life and where statutes of limitations could be invoked only with great difficulty? (11)

Or, as poet Philip Larkin puts it (in a slightly different way):

> They fuck you up, your mum and dad
>   They may not mean to, but they do

They fill you with the faults they had
   And add some extra, just for you.
(From "This Be the Verse," Collected Poems [1988], 180)

The Ultimate Meme Trick:
Why Your Memes Want You to Hate the Idea of Memes

Why are some people so upset about the idea of memes? Certainly some
of the professional hostility stems from territorial encroachment. Memetic
theorists have come in with a new language and colonized areas previously
occupied by disciplines that had claimed the study of cultural evolution for
themselves, such as anthropology and sociology. But the hostility is more
than just professional hostility, because it is shared by the layperson, as well
as by scientists and academics from outside of the study of culture.

Indeed, even those of us predisposed to think that memetics provides
some useful concepts find something unnerving about the meme concept.
Even philosopher Daniel Dennett, who has written most eloquently about
the meme concept, initially found unpalatable the view of humans and
human consciousness that his own researches into the implications of the
science of memetics was revealing: "I don't know about you, but I'm not
initially attracted by the idea of my brain as a sort of dung heap in which
the larvae of other people's ideas renew themselves, before sending out
copies of themselves in an informational Diaspora. It does seem to rob
my mind of its importance as both author and critic. Who's in charge, according to this vision—we or our memes?" (1991, 202). Dennett calls this
meme's-eye view of the human mind "unsettling, even appalling" (202)
but is convinced, as I am, that if we just stick with the concepts of memetic
science and get used to them, they not only provide a new way of thinking
about ideas but can help in reconceptualizing the nature of the self in the
Age of Darwin.

My purpose here is to draw attention to what the widespread opposition
to the concept of the meme itself might be telling us about the nature of the
general memosphere (the intellectual environment in which memes com-
pete) in which we live.[15] The important point is to realize that we inhabit a
cognitive environment in which there is a widespread hostility to examining belief. Educational theorists in the critical thinking literature have be-
moaned for decades the difficulty of inculcating the critical thinking skills
of detachment, neutral evaluation of belief, perspective switching, and decontextualizing from current position. The literature in cognitive psychol-

ogy on so-called belief bias effects—that the interpretation of new evidence is colored by our current beliefs—is likewise uniform in indicating how hard it is for individuals to examine evidence from standpoints not guaranteed to reinforce the memes that already reside in their cognitive systems. The memes that currently reside in our cognitive structures are singularly unenthusiastic about sharing precious brain-space with others that might want to take up residence and potentially displace them.[16]

Some of this simply reflects the logic of environments with finite carrying capacity—but we would be silly not to at least worry about some other not mutually exclusive implications that seem troublesome. That this is indeed the cognitive environment that most of us experience internally—that most of us share the trait of hostility to new memes—does prompt some troubling thoughts. Perhaps the unsupportable memes have allied together to create in us a cognitive environment antithetical to the idea that our beliefs and desires need evaluation. Or, another way to put it is: If most of our beliefs are serving us well as vehicles and are able to pass selective tests of their efficacy, why shouldn't they have created a cognitive bias to submit themselves to those very tests—tests that their competitors would surely fail. Instead, the memosphere of most of us is vaguely discouraging of stiff real-world tests of our beliefs. This prompts the worrisome question: What have these memes got to hide?

At this point in my argument, I need to revisit the list of four reasons for meme survival and spread that were given at the beginning of this chapter and examine the implications of each for the well-being of the hosts that they inhabit. First, memes survive and spread because they are helpful to the people who store them. This is the good category. All of us hope that a majority of the memes we are carrying around are in this category—and it is surely the case that most ideas spread because they are good for us (they are true and helpful in accomplishing our other goals). The important point, however, is that there are three other categories, and none of these necessarily imply benefits to the vehicle.

The second category—memes become frequent because they fit genetic predispositions—is a complex one to think of in terms of whether memes of this type serve the vehicle. Certainly such memes can be said to be fulfilling genetic goals, and, to the extent that these overlap with vehicle goals, then these memes are also serving the vehicle. But sections of several earlier chapters (1, 2, and 4 for example) were devoted to demonstrating the profoundly important proposition that, particularly in modern society, genetic and vehicle goals do not always coincide. Thus, that a meme fits well a genetic predisposition is only efficacious for the vehicle in those cases where the genetic predisposition is fit by a meme that serves the vehicle's long-term life plan as well.[17]

In the third category are memes that spread because they facilitate the spread of the genes that make bodies which are good hosts for these particular memes. Of course, these are the memes that may have made us "believers" in the first place. Numerous authors have posited models of meme-gene coevolution in which, during at least part of evolutionary history, memes that could direct the course of genetic evolution did so in order to statistically bias the memetic host environment in their favor (see Blackmore 1999, on memetic drive; also Lynch 1996). Religious beliefs that urge people to have more children would be in this category—the religious belief has found a good host and it is likely that the host's progeny will share some of the characteristics that made the original vehicle such a good host for that particular meme.

Category three is certainly difficult to interpret as to whether it is a category that should be viewed as "good" for the vehicle. There is a sense in which future progeny of the original host will "want" the meme even more. But certainly also there is an aspect of this category suggesting a "trick" on the meme's part. The anthropomorphic logic from the meme's point of view is something like this. A meme has found a good host—one that readily accepts it. Why not include in the memeplex an admonition to propagate more such vehicles? The result will be more willing hosts for this meme in future generations, and then—the feedback loop is obvious. In a sense, this case is one in which the meme is not being selected in the interests of the host but the host is being selected in the interests of the meme.

The fourth category is clearly problematic. It is the one in which memes survive and spread because of the self-perpetuating properties of the memes themselves. These memes are likely to be harmful to us. At best they are neutral—simply taking up brain-space but neither hurting nor harming the host. But since these memes are there in the first place not because hosts benefit, it is more likely that many of these memes are like some of the genes discussed in chapter 1 that sacrifice the vehicle in order to further their own interests (replicative success).

Of course, no one has knowledge of what percentage of the memes they have resident fall into each of the four categories just discussed. But the analysis just completed leaves some room for worry. One category contains memes that are clearly efficacious for the host (category 1), but another (category 4) contains largely memes that are harmful to the host (or wasteful—even neutral memes taking up finite memory space and finite conceptual power are harmful because they limit memory space and computational power for those that are helpful). Another (category 2) no doubt has both types of memes in some unknown proportion. A final category (category 3) contains memes whose relation to the host are unclear. They are memes that, in the environment in which we evolved, biased the genes in their favor—but whether they serve the interests of the host in the modern world is an entirely unresolved question. Perhaps some do and some do not.

Integrating across the four categories is thus rather disheartening. If you learned that as few as 10 percent of your beliefs were harming you, would you not still experience considerable unease? There seems a far from negligible chance that a good proportion of our resident memes are harming the host (rather than facilitating the host's life plan). If this is indeed the case, it is hardly surprising that the unsupportable memes (those that would not survive explicit evaluation) banded together to create a cognitive ambience that makes the idea of external evaluation unpalatable.[18]

Memetic Concepts as Tools of Self-Examination

The concept of the meme allows us, for the first time in the thousands of years that humans have had culture, to get a handle on, to examine at a distance, the cultural artifacts that infect our thoughts. Metaphorically of course, there is a sense in which—more so than at any other time in history—we can reach into our brains, pull out a cultural artifact, hold it in our hands, and examine it at a distance. This distancing function is greatly aided by the language of memetic science.[19] If only in its vocabulary, it provides additional mind tools (Clark 1997, 2001; Dennett 1991, 1995, 1996) to aid in the difficult, because deeply recursive, project of cognitive self-examination.

This entire project of memetic self-examination is bound to be a profoundly Neurathian bootstrapping affair. Anything that gives one part of our minds leverage to examine another will aid in the difficult task of cognitive self-examination.[20] One way that the meme concept will aid cognitive self-analysis is that by emphasizing the epidemiology of belief it will indirectly suggest to many (for whom it will be a new insight) the contingency of belief. A new emphasis on the contingency of belief will make distancing from belief cognitively easier (and more habitual); and such distancing will make the examination, evaluation, and rejection of preexisting beliefs easier. The concept of the meme is cognitively liberating because it aids in the task of mental self-examination by making it easier to achieve cognitive distancing. Memes demystify beliefs. They desanctify beliefs. Memes are solvents for beliefs, and they are epistemic equalizers in the following sense. By providing a common term for all cultural units, memetic science deconstructs the unreflective privileging of certain memes that have leaked into culture through arbitrary historical accident and through specific meme strategies that encourage privileging. The very concept of the meme will suggest to more and more people who become aware of it that they need to engage in some meme therapy—to ponder whether they have thoughtfully taken on a meme or have simply been infected by it without any critical input from their analytic intelligence.

We are evolutionarily contingent animals. Evolution did not have to result in humans. Likewise, our memetic evolution is highly contingent, both at the level of the culture as a whole and at the level of the individual. A full awareness of the latter on the part of humans would have a distancing effect—would lead to less self-identification with nonreflectively acquired memes (probably a good thing for overall vehicle well-being).

Some people might have difficulty, however, with the notion that the beliefs they hold are contingent products of memetic evolution. One demonstration of how unnatural this idea will be for some people is provided by the response to one questionnaire subscale developed by my own research team (Stanovich and West 1997, 1998c). This subscale is designed to assess individual differences in people's ability to appreciate the historical contingency of important memeplexes that they have acquired. One item on this scale is the following statement, to which the subject must agree or disagree strongly, moderately, or mildly:

> Even if my environment (family, neighborhood, schools) had been different, I probably would have had the same religious views.

Religion of course is the classic case of environmentally contingent memes. The extreme environmental contingency of religious beliefs (Christianity is clustered in Europe and the Americas, Islam in Africa and the Middle East, and Hinduism in India, etc.), along with their usual conjoining with evaluation-disabling faith-based strategies, are the things that most strongly suggest these types of beliefs are not reflectively acquired. Nevertheless, in several studies, my colleagues and I have repeatedly found that roughly 40 to 55 percent of a university student population will deny that their religious views are conditioned in any way by their historical circumstances (parents, country, education). Roughly similar percentages deny the historically contingent nature of other types of beliefs (as measured on other similar statements on the questionnaire requiring subjects to decenter from their contingent position in the universe). In short, in many cases, the student from suburban Virginia feels that he/she would have Baptist Christian beliefs even if he/she grew up in New Delhi because this is what God ordained. That this is patently absurd suggests that the realization of the epidemiological nature of beliefs—a realization fostered by the meme concept—is going to have shattering conceptual consequences for a substantial portion of the population. An inchoate recognition of this implication probably in part accounts for the hostility toward the meme concept discussed previously.

Building Memeplex Self on a Level Playing Field: Memetics as an Epistemic Equalizer

The idea of looking at one's memes from a distance will reveal the strange privileging that faith-based memes have achieved in the memosphere. For example, you will have no trouble examining and holding at a distance memes such as:

> "a university education is good for you"
> "a vegetarian is a good thing to be"
> "watching too much television is bad"
> . . . and so on

Likewise, you will probably have no trouble examining the following memeplexes:

> memeplex science
> memeplex rationality
> memeplex socialism
> memeplex capitalism
> memeplex democracy

However, the idea of examining Catholicism as a memeplex will seem stranger to some people, particularly if they are Catholics. Likewise, with the memeplex Islam.

Memeplex science and memeplex rationality are worldviews as comprehensive as memeplex Catholicism, as has been pointed out by religious believers and nonbelievers alike (see Raymo 1999). Yet it is much more acceptable to insult the former than the latter, because the former open themselves to examination and critique more than the latter. Like other faith-based memes, the latter has inoculated many cultures with the view that it is in a domain beyond question.

Scientists and rationalists are quite naturally and commonly asked for reasons why they believe in science or in rationality, and they are expected to give arguments and evidence that meet certain intellectual standards. But it is seen as fundamentally insulting to ask for similar reasons for believing in Catholicism. It would be seen as somewhat impolite to ask for reasons of the same standard as one requires from the scientist defending her worldview. The meme concept threatens to wipe out this asymmetry, by giving no meme the privilege of escaping examination and critique.

The hostility toward the meme concept in part flows from the fact that the science of memetics has highlighted the tricks played by all faith-based memes and threatens to deprive them of their special status—a status we do not grant other memeplexes that do not contain evaluation-disabling tricks. But since such disabling memes are widely held, they provide the climate in which the meme concept is being received. They condition the cognitive environment in the direction of hostility toward meme evaluation.

The science of memetics is part of the advance of the scientific materialist worldview, and as such it inherits the hostility directed at the latter by many other memeplexes. As scientific materialism advances as a worldview, the tricks played by faith-based memes will be more likely to be exposed, and memetics is just a very public advertisement of what has already been happening, albeit more subtly. In October 2001 discussions were taking place in Great Britain about the possibility of amending hate-crime legislation primarily directed at crimes accompanied by racial and ethnic hostility to include religiously motivated hostility. The unmasking of the faith-based meme trick has come far enough in intellectual history that the differential privileging of religious over nonreligious worldviews was commented upon in several letters to daily newspapers, one letter writer querying: "Under the Government's planned ban on inciting religious hatred, are the religious to be allowed to insult our scepticism with impunity? Are we unbelievers to be prosecuted for pointing out the more embarrassingly illiberal parts of the Bible and Koran?" (Philip 2001).

That it is so easy to demonstrate the historical contingency of people's religious beliefs, and yet this historical contingency is so seldom emphasized (in the media for example), illustrates the strong degree of privileging that faith-based memes have managed to achieve by their evaluation-disabling strategies. For example, teachers of psychology, sociology, or anthropology in universities in the United States notice that merely pointing out the demographic contingency of religious belief is seen as itself mildly impolite and aggressive. The idea that we should not insult someone's religious beliefs but that a nonreligious worldview should not receive the same protection from insult is itself a proposition that will receive more scrutiny as memetic concepts become more widespread.

Memetic science threatens to level the epistemic playing field and thus engenders hostility. That a leveling of the playing field of belief evaluation engenders hostility should itself make us suspicious about the memetic makeup of our general culture. The meme "blood circulates in the veins and arteries" does not need to disable attempts at its evaluation. It lives or dies on its mapping to the empirical world. It does not have to infect the world with disabling conditions (conditions designed to disable the means of evaluating it). Note how the disabling function of memes operates differentially in the case of memes grounded in empirical observations and in the case of faith-based memes. It does not seem at all aggressive or insulting to refer to the proposition "blood circulates in the veins and arteries" as a meme. Nor does it seem insulting to the person articulating it to consider the proposition "Canada is the best place in the world to live" to be a meme. But one hesitates to point out that the phrase "God is all knowing" is a meme because those hosting the meme have been encouraged not to be reflective about it and not to consider it as a separate entity with its own replicative interests.

Evolutionary Psychology Rejects the Notion of Free-Floating Memes

What I have argued in this book is that the cultural tools of cognitive science and decision science—when used in conjunction with the potent insight that there can be a conflict of interest between replicators and the vehicle—have the potential to create in us a uniquely critical and discerning type of self-reflection. Combined with the tools of decision science, the vehicle/replicator distinction and the concept of the meme can spawn new thoughts and new tools for the restructuring of human goals. Evolutionary psychologists resist this extrapolation (falling back on their "culture on a leash" notion) because, to many of them, the idea of free-floating cultural products—those totally unconditioned by and unadapted to evolved mental mechanisms—is anathema.[21] Also, because human rationality is in large part a memetic product—a set of cultural tools—evolutionary psychologists are prone to miss or denigrate its importance.

What evolutionary psychologists do not like is the idea of memes becoming completely "unglued" from genetic control, and this is precisely the

notion that most memetic theorists are advancing (Blackmore 1999; Dawkins 1993; Dennett 1991, 1995; Lynch 1996), and that I am developing in
this book (that human rationality is a major cultural tool that allows humans to become even more "unglued"). What they do not like is the idea—
which I am advancing here—that, at a certain level of recursiveness, a mind
populated with cultural tools in the form of memeplexes designed for the
evaluation of other memeplexes and TASS tendencies (science; logic; some
notions from decision science such as consistency, transitivity, etc.) acquires
some autonomy from genetic control. What they have neglected is the recursive power of evaluative memes in the context of an organism that has
become aware of the replicator/vehicle distinction. A brain aware of the
replicator/vehicle distinction and in the possession of evaluative memeplexes such as science, logic, and decision theory might begin a process of
pruning vehicle-thwarting goals from intentional-level psychology and
reinstalling memetic structures that serve the vehicle's interests more efficiently.[22] This is exactly what the canons of instrumental rationality—
largely a product of the twentieth century—were designed to accomplish.

This might seem like a Promethean goal, but in fact, a rich tradition in
cognitive science has emphasized how cultural changes and increased scientific knowledge results in changes in folk psychology. Already, among educated citizens of the twenty-first century, violations of transitivity and independence of irrelevant alternatives can be a cause of cognitive sanction in
ways that are historically unprecedented. A full appreciation of the implications of the replicator/vehicle distinction—with its emphasis that differing
optimization criteria apply to the personal and subpersonal levels of analysis (utility maximization versus genetic fitness)—could have equally profound cultural implications.

The Co-Adapted Meme Paradox

In chapter 3, I distinguished so-called thin theories of rationality from so-
called broad theories. The former is the familiar instrumental rationality
that has been the focus throughout most of the earlier chapters. Instrumental rationality is often termed means/ends rationality because it takes the
person's beliefs and desires as given, and expresses what actions maximize
desire fulfillment given those previously existing beliefs and desires. But
many desires, particularly those operating as high-level intentional states in
the analytic system, are memes—as are many beliefs.[23] If we understand
what a meme is, it immediately becomes apparent why we must focus on a
broad theory of rationality—one that critiques the beliefs and desires that

go into the instrumental calculations. Otherwise, the meme goals are no better than the preinstalled gene goals. Instrumental rationality, as classically defined, serves current goals without inquiring where those goals came from. It does not inquire whether those goals might well be "junk"—good at propagating themselves but ill-serving the vehicles who hold them.

Generally speaking, we want to make sure that the memes we are carrying around were reflectively acquired and are good for us as hosts, as vehicles. But "good for the vehicle" takes on a recursive flavor here if the vehicle includes already installed memes. This makes it seem like what is being asked for are merely good co-adapted memes—memes that cohere well with the memes already resident. Thus, we seem right back at an instrumental view, because we would merely be accommodating previously existing desires. We cannot go back to square one before a person had any memes. There is no such thing as "what's good for the vehicle" independent of the memes already acquired.

A person's interests at a particular point in time are in part determined by the memes already resident in the brain. If those memes are already maladaptive for the vehicle, then using them to evaluate the new instrumental needs would just amount to requiring newly acquired memes to be co-adapted with the previously acquired maladaptive ones. This I will call the co-adapted meme paradox.[24] I think that there is no foundationalist answer to this problem—that is, no totally neutral stance from which memes can be evaluated. Nevertheless, the memeplexes of science and rationality can still be used to restructure our own goal hierarchies in ways that serve the host. The project will be a highly Neurathian bootstrapping endeavor. Some memeplexes must be used to evaluate other memes. These original memeplexes must then become the subjects of evaluation using other resident memes. Science progresses despite being confined to a similar mind-twisting and Rube Goldberg-like logic.

Additionally, I think there are additional tools that can be used in this Neurathian project of "meme cleansing"—tools that have already been invented and put to use in similar intellectual projects by philosophers (see Nozick 1993, 139–51, on rational preferences). For example, philosopher Derek Parfit (1984) develops the idea of treating our future selves as different people that we then include in the moral calculus of determining what to do now (knowing that the lives of these "future dependents" will be determined by our present behavior). His work reinforces some of the principles discussed earlier in this chapter. It provides a further thought tool to help in following the stricture contained in meme evaluation rule 3: regarding memes that are desires, seek to install only memes that do not preclude other memeplexes becoming installed in the future. The thought experiment of conjuring up future selves helps foster the distancing from the present self that is necessary to adhere to this stricture. The thought of a collection of future "yous"—as constituencies along with the present you—helps to keep the present self from dominating all utility calculations and actions.

Another example comes from the work of philosopher John Rawls (1971, 2001). Rawls is known for developing the idea of the so-called Original Position in order to circumvent the problem of how to conduct an argument about social justice that is not totally constrained by the self-interest of the participants. The Original Position is an imaginary situation from which you are to argue the principles of a just and fair society without knowing what role in that society you will play. Rawls wanted people to adopt the posture of reasoning about the justice of social structures and to construct what society they would build if they could end up playing any role in the society.

It might help to use as tools the insights of both Parfit and Rawls when trying to adopt a reflective stance toward the memes we currently have resident. Parfit's tool helps to keep in focus the memeplexes it would make sense to have installed, given the future people that we will be. Rawls's Original Position—forcing us to imagine we are without some of the memes we currently have—might allow us to treat those future people more fairly and to assume less about them.

In short, various programs of Neurathian evaluation and reflection can aid, at least to some extent, in dealing with the co-adapted meme paradox. A Neurathian program of science can help us purify the truth content of our memes that are beliefs. A Neurathian program of desire evaluation can help purify our goal networks from memes that are parasites, junk, or viruses. The reflectivity involved in evaluating memes and in developing a broad theory of rationality to guide our life choices will be the foundation for reconstructing the memeplex soul in the Age of Darwin. In the next, final chapter, I will attempt to summarize how much of that program has been accomplished and how much of the traditional concept of the soul will remain when it has been reflectively reconstructed on scientific and rational grounds rather than by the genetically programmed proclivities of TASS and the accidental histories of the memes that collected together to form this memeplex. As should be clear already, it cannot survive without substantial reformulation. Like the Wizard of Oz, the tricks this memeplex has beguiled us with are being unmasked by scientific knowledge and by reflection, using as a tool the meme concept itself.

Some readers may find the conclusions reached heretofore and the direction I shall take in chapter 8 to be not at all uplifting. But I think the key question should really be: Is it any more uplifting to be the functional equivalent of an automaton—to be doing the bidding of subpersonal entities (genes and memes) that you either had no choice about (in the first case) or that you unreflectively took on (in the second case) or because of a trick that these subpersonal entities played on you? The final conclusions to be reached in chapter 8 are considerably more uplifting than that.

We're descended from the indignant, passionate tellers of half truths who in order to convince others, simultaneously convinced themselves. Over generations success had winnowed us out, and with success came our defect, carved deep in the genes like ruts in a cart track—when it didn't suit us we couldn't agree on what was in front of us. Believing is seeing. That's why there are divorces, border disputes and wars, and why this statue of the Virgin Mary weeps blood and that one of Ganesh drinks milk. And that was why metaphysics and science were such courageous enterprises, such startling inventions, bigger than the wheel, bigger than agriculture, human artifacts set right against the grain of human nature.

—Ian McEwan, *Enduring Love* (1998, 181)

The Grand Inquisitor lies to his charges in the *Brothers Karamazov* to keep them happy and protect them from the horrible truth that there is no God. Dostoevsky, like Nietzsche, thought that most people can't bear the truth, can't bear to see how life goes, how dangerous it is, and how vulnerable one is as a creator of a meaningful life. But I see this as underestimating persons. Or better: if true, it is, like all else, historically conditioned. We could get used to the news.

—Owen Flanagan, *Self Expressions: Mind, Morals, and the Meaning of Life* (1996, 209)

CHAPTER 8

A Soul without Mystery:

Finding Meaning in the Age of Darwin

It is becoming increasingly apparent that Darwin's universal acid is beginning to seep into the general culture. Novelists, for example, have begun to
assimilate the insights from cognitive science and evolutionary psychology,
and are reflecting on them in their work. Booker Prize-winning novelist Ian
McEwan, in the acknowledgements to his novel *Enduring Love*, mentions
debts to E. O. Wilson's *On Human Nature*, Steven Pinker's *The Language Instinct*, Robert Wright's *The Moral Animal*, and Antonio Damasio's *Descartes'
Error*—and is reported chairing a session at a conference at the London
School of Economics on "Darwinism Today" (Malik 2000, 150). In her recent novel *The Peppered Moth* Margaret Drabble uses mitochondrial DNA
as a plot device. The peppered moth of the title refers to Kettlewell's (1973;
Majerus 1998) famous confirmation of evolution through natural selection
by looking at adaptive coloring changes. In his memoir about moving to a
small farm in Sussex, journalist Adam Nicolson (2000) writes of his first
close experiences at tending farm animals through a lens created by biologist Richard Dawkins:

> A lamb is a survival machine: a big head, a big mouth and four stocky black
> legs way out of proportion to the sack of a body which joins these standing
> and eating parts together. The ewe had a full udder, and the lamb soon found
> its way to suck, wriggling its tail, the instinctive drive at work, the vital
> colostrum running into the gut. Survival. . . . I found the mother standing
> alert, eyes big, defensive, stamping her front feet as I approached the pen or
> picked up the lamb to look at the navel and the shriveling cord or to feel its,
> gratifyingly, filling belly. The ewe is tensed to protect her own. She is a servant
> of her genetic destiny. Her life can only be dedicated to these fragile, transi-
> tional moments on which so much hinges. So this instant, in the pen with

the hours-old lamb, with the tautened presence of the protective mother, this is one of those moments when you come close to the "blood of the world," to the essential juices running under the everyday surface of things, when the curtain is drawn back and you find yourself face to face with how things are. (144–45)

That such Dawkins-like themes are showing up in modern nonscientific writings is a sure sign that they are starting to resonate in the culture. Some of these novelists and writers might be horrified by these themes, but they all rightly see that they will reshape the future conceptual universe in which humans will live. David Lodge, in his recent book *Thinks . . .* (2001), presents us with a novel in which the whole theme derives from cognitive science, the scientific study of consciousness, and evolutionary psychology. In addition to citing, in his acknowledgements, such works as Pinker's *How the Mind Works,* Dennett's *Darwin's Dangerous Idea,* Searle's *The Rediscovery of Mind,* and Edelman's *Bright Air, Brilliant Fire,* Lodge has one of his central characters articulate what is perhaps as good a summary as any of the rationale for the present volume. His character, Ralph Messenger, summarizes our current dilemma:

> Homo sapiens was the first and only living being in evolutionary history to discovery he was mortal. So how does he respond? He makes up stories to explain how he got into this fix, and how he might get out of it. He invents religion, he develops burial customs, he makes up stories about the afterlife, and immortality of the soul. As time goes on, these stories get more and more elaborate. But in the most recent phase of culture, moments ago in terms of evolutionary history, science suddenly takes off, and starts to tell a different story about how we got here, a much more powerful explanatory story that knocks the religious one for six. Not many intelligent people believe the religious story any more, but they still cling to some of its consoling concepts, like the soul, life after death, and so on. (Lodge 2001, 101)

We saw in chapter 1 that science does indeed tell "a different story about how we got here" with exactly some of the implications that Ralph Messenger suggests. But Messenger seems too casual in his dismissal of the possibility that a way can be found to provide at least some of what people seek when they crave a concept such as soul by using suitably analogous scientific concepts (or, if not analogous concepts, at least concepts that trigger some similar type of emotional resonance). True, after we digest the implications of modern cognitive science and Darwin's universal acid, the term soul can never mean again, to any thinking person, its dictionary definition as "the principle of life, feeling, thought, and action in man, regarded as a distinct entity separate from the body" (*The Random House College Dictionary*, 1975).[1] However, does there remain at least the possibility that we could satisfy these yearnings for uniqueness and transcendence that the concept of soul provides—our yearnings for a certain type of meaning—in a way that might be scientifically respectable? Such a project, even if daunting, does seem worth the effort, given the potential payoff.

There is no question though that the memeplex soul is in for some big changes when, like the scene in The Wizard of Oz, the tricks this memeplex has beguiled us with are unmasked by science. The alternative to embarking on this conceptual restructuring is, if you think about it, rather demeaning. The only alternative to finding meaning through the use of scientific concepts that reflect the way the brain actually works, and how humans actually go about valuing and choosing, is to put our heads in the sand. As Dennett (1995) argues, "the idea that we might preserve meaning by kidding ourselves is a more pessimistic, more nihilistic idea than I for one can stomach. If that were the best that could be done, I would conclude that nothing mattered after all" (22). Dennett is right that it hardly seems an uplifting view of humans to say that we are the most complex creatures on earth, yet we must avert our eyes from the very knowledge our complex brains have revealed—that we must hide from the implications of our discoveries like little children. It is inconceivable that the genies discussed in this book (the replicator/vehicle distinction, different kinds of minds, that TASS lodged in your brain operates without your awareness, the sources of human irrationality, etc.) will be kept in their bottles. The idea that we can choose to ignore them is pure folly.

There is another reason why these genies will not go back in to their bottles. That is because the conceptual insights of evolutionary theory travel on the back of a scientific technology that people do not want to give up. Simply put, science delivers the goods. People want their technological benefits—they want the increased health, they want the cheaper food, they want the increased mobility, they want convenience and comfort. But riding along with the technologies that people want are the conceptual and metaphysical insights that the scientific discoveries trigger: that we are survival machines, that there is no immaterial "mind" where consciousness occurs and where your "self" makes decisions, that the layperson's notion of free will is utterly confused (Dennett 1984; Flanagan 2002). Along with the snowmobiles, the DVDs, microwave ovens, flat screen televisions, cellular telephones, personal data assistants, PET scanners, laproscopic surgery

techniques, the Nintendo games, and the cancer treatments, ride implications that will undermine the most closely and fervently held beliefs that
people have about themselves (as discussed in chapter 1, there is an odd
sense in which fundamentalist theologies recognize this better than liberal
ones).

But short of a totalitarian dictatorship the likes of which we have never
seen, people cannot have one without the other. They can't have the computers and the cancer treatments without undermining their view of just
what humans are in the context of the universe. Science delivers the goods—
but it also destroys many concepts that people find meaningful. Humankind has repeatedly been given the choice between the goods and the meaning in the past—and it has taken the goods every time.

But is this such a bad thing? Perhaps not, if the meaning we are giving
up (perhaps unknowingly) is a chimera, an illusion, a folk tale as the Ralph
Messenger character in Lodge's novel says—concocted to comfort humans
long before they had available the necessary tools (science, rationality) to
explain and control the world. Perhaps now we have the chance to have it
all—the materialist benefits of science plus a rationally reconstructed sense
of the self (in the new sense of an autonomous and unique self—the creator of our own meaning). This will be the positive side of the story to be
told in this chapter—the "upside" so to speak. But there is a dark side as
well—dangers that remain, as we shall see. The modern triumph of materialist science will present continuing stress to even a rationally reconstructed notion of personal meaning. Cognitive science has destroyed the
traditional notion of soul, but a rationally reconstructed sense of meaning
might well, as we shall see, also devour itself. To avoid this outcome, humans will be challenged to develop what might be termed a type of meta-
rationality. But before explaining how this comes about, I will explore
some dead ends in the human search for meaning. Then I will present my
conception of a rationally reconstructed sense of self—the soul in the Age
of Darwin.

Macromolecules and Mystery Juice:
Looking for Meaning in All the Wrong Places

The generation in vitro of self-replicating molecules, the manipulation of DNA to planned socio-genetic purposes—the eradication of
inherited disease, the cloning of armies—are in reach. These developments will necessitate a thorough revision of our conceptual alphabet.
What were, millennially, the building-blocks of all theological and

teleological narratives, the deistic postulate of a universal design by
some supreme architect, the ascription of a personal, singular destiny,
are now being effaced or fundamentally rethought.
                    —George Steiner, *Errata: An Examined Life* (1997, 161)

We found that we could think of ourselves *both* as having emerged,
assisted only by chance, from some primeval, slimy protein soup,
and as close kin to Beethoven and Lincoln. We became able to stop
pretending, like socially insecure upstarts, that our origins are
commensurate with our latterly acquired dignities.
                    —Richard Rorty (1995, 62–63)

People persist in making one particular error in searching for meaning in
human life—they seek it in human origins. The basic thought behind this
particular dead end seems to be that because we value meaning highly and
meaning comes from our origins, then our origins have to come from on
high. Pardon the vernacular but, as novelist Lodge's character Ralph Messenger alluded to previously, when we came up with this we weren't exactly hitting on all cylinders, culturally speaking. It is this type of childlike logic that
plays out in all of the myths of special creation lodged in religions and other
ancient worldviews.

The insights of evolutionary theory have given us a more accurate view
and, with it, a better way to think. As Dennett (1995) explains, one key insight from Darwin's dangerous idea is that meaning "doesn't come from on
high; it percolates up from below, from the initially mindless and pointless algorithmic processes that gradually acquire meaning and intelligence as they develop" (205). The string of self-replicating macromolecules
("macros" as Dennett calls them) that began the evolutionary road that
leads to us are, as Dennett, notes, best viewed as mindless automata—robots. Dennett uses this fact to emphasize our plebeian origins. He asks if you
would want your daughter to marry a robot. His question is a setup for his
humorous observation that "Well, if Darwin is right, your great-great-great
. . . grandmother was a robot! A macro, in fact" (206). So if we are looking
for an uplifting story in our origins to bolster us, we have been looking in
the wrong place—it is macromolecules all the way down.[2]

The other place people look for meaning is inside themselves—in their
own conscious introspections. What people want to find when they contemplate the nature of their own minds is what I called in chapter 2 the Promethean Controller: "We would like to think of ourselves as godlike creators of ideas, manipulating and controlling them as the whim dictates, and

judging them from an independent, Olympian standpoint" (Dennett 1995, 346).

The idea of a Promethean Controller is of course a version of what in psychology is called the homunculus ("little man in the brain") problem, a problem I discussed at length in chapter 2. To the extent that an attempt to explain what goes on in the brain ends up positing what is essentially another "little man in the head" who makes the decisions and pulls the levers, so to speak, it has really explained nothing. Instead, it has left the problem of explaining what goes on inside the brain of the little man. Such explanations have merely regressed the problem.

The model of the mind held by most nonspecialists usually ends up positing a Promethean Controller (the seat of the soul) at the heart of the action, despite the tendency now to pay at least some obeisance to the notion that our minds arise from brain activity. This sop to science usually ends up being perfunctory, however, because such homunculus-based models succeed in maintaining the much desired mystery surrounding consciousness, because the inside of the Promethean Controller's mind remains untouched by science. Thus, the Promethean Controller preserves for most people the sense of transcendent importance—the religious need that they are attempting to reconcile with science. The conscious Promethean Controller inside them (for many people, in direct communication with God) is the thing that represents their essence, their soul—a discrete, unchanging, coherent entity.

The trouble, of course, is that as we saw in chapter 2, consciousness is anything but a discrete, unchanging, coherent entity. Modern cognitive science has exploded virtually every assumption behind the layperson's view of a Promethean Controller.[3] The brain is a complex system that schedules output routines, information gathering, and knowledge restructuring. This can all be explained within the modern conceptual language of executive functioning and the idea of coalitions of subsystems successively controlling action and internal processing.[4] There is no little man in the head, there is no one place where it all comes together, there is no one place where the "I" sits, there is no Promethean Controller inside us that represents, in brain terms, the soul.

My last phrasing ("in brain terms, the soul") was deliberately intended to draw attention to the odd mixture of nonmaterialist metaphysics and science that our folk psychologies have become. People want to acknowledge scientific evidence because they are aware that they live in a scientific society and wish to maintain their intellectual integrity. Thus, they know that the source of their sense of self resides in the structure of their brains and that information about the brain must be incorporated into their models.

Yet at the same time they want to inject a squirt of "mystery juice" into their conceptions. They know that they cannot leave out neuroscience entirely from their conceptions, but they are equally determined that it not be allowed to explain *everything*.

Just like the notion of special creation—that the meaning of our life can be found in the special, magical way we came to be human—the idea that the architecture of the mind reveals a Promethean Controller with magical powers in which resides all that is meaningful is an intellectual mistake. If we look to our origins, all we will find is the replicators—self-copying macromolecules. In turning to the architecture of the mind to find meaning, we will find something equally off-putting (autonomous subsystems that operate beyond our awareness, multiple and changing brain systems involved in the supervisory attentional activities to which we give the label executive control). In modern times, the quest for the soul has been translated into the notion of understanding the self. This has turned out to be a dangerous move, because when science started to analyze the self science began to remove the mystery behind it. Under the scrutiny of cognitive science, the self became transfigured, and the "I," the Cartesian Theater, the Promethean Controller were all dissolved.

Ironically though, later in this chapter I am going to argue that human uniqueness and worth—the closest things that we can call, in a scientific age, our sense of soul—do in fact derive from a feature of the human mind. But that feature is not the one favored by the high priests of popular science—consciousness. It is not the Promethean Controller. It is a more subtle feature of human ability that is often confused with both. Importantly, it is a cognitive feature that relates more to rationality than to consciousness.

Even in the face of evolution, we want to separate ourselves from other beasts. I will argue that we can do that, but that consciousness is not the concept that provides the best separation. Consciousness turns out to be an epiphenomenon of the morally transcendent features of human cognition that really do separate humans from other animals. I will start constructing my argument about just what those features are by asking a question recently posed by philosopher John Searle.

Is Human Rationality Just an Extension of Chimpanzee Rationality? Context and Values in Human Judgment

Searle (2001) opens a recent book on rationality by referring to the famous chimpanzees on the island of Tenerife studied by the psychologist Wolfgang Kohler (1927) and the many feats of problem solving that the chimps displayed that have entered the lore of textbooks. In one situation, a chimp was presented with a box, a stick, and a bunch of bananas high out of reach. The chimp figured out that he should position the box under the bananas, climb up on it, and use the stick to bring down the bananas. Searle asks us to appreciate how the chimp's behavior fulfilled all of the criteria of instrumental rationality—the chimp used efficient means to achieve its ends. The primary desire of obtaining the bananas was satisfied by taking the appropriate action.

Searle uses the instrumental rationality of Kohler's chimp to argue that, under what he calls the Classical Model of rationality, human rationality is just an extension of chimpanzee rationality. The Classical Model of rationality that he portrays is a quite restricted and pinched model, and somewhat of a caricature in light of recent work in cognitive science and decision theory, but my purpose is not to argue that here. Instead, I wish to stress the point on which Searle and I agree (but for somewhat different reasons). Human rationality is not simply an extension of chimpanzee rationality.

Consider a principle of rational choice called the independence of irrelevant alternatives that is discussed by Nobel-prize winning economist Amartya Sen (1993). The principle of independence of irrelevant alternatives (more specifically, property alpha) can be illustrated by the following humorous imaginary situation. A diner is told by a waiter that the two dishes of the day are steak and pork chops. The diner chooses steak. Five minutes later, the waiter returns and says "Oh, I forgot, we have lamb as well as steak and pork chops." The diner says "Oh, in that case, I'll have the pork chops." The diner has violated the property of independence of irrelevant alternatives, and we can see why this property is a foundational principle of rational choice by just noting how deeply odd the choices seem. Formally, the diner has chosen x when presented with x and y, but prefers y when presented with x, y, and z.

Yet consider a guest at a party faced with a bowl with one apple in it. The individual leaves the apple—thus choosing nothing (x) over an apple (y). A few minutes later, the host puts a pear (z) in the bowl. Shortly thereafter, the guest takes the apple. Hasn't the guest just done what the diner did in the previous example? The guest has chosen x when presented with x and y, but has chosen y when presented with x, y, and z. Has the independence of irrelevant alternatives been violated? Most would answer no. Choice y in the second situation is not the same as choice y in the first—so the equivalency required for a violation of the principle seems not to hold. While choice y in the second situation is simply "taking an apple," choice y in the first is contextualized and probably construed as "taking the last apple in the bowl when I am in public" with all of its associated negative utility inspired by considerations of politeness.[5]

What Sen's (1993) example illustrates is that there is more to choice alternatives than just their objective utility, independent of the context in which they are presented. It is true that sometimes it does not make sense to contextualize a situation beyond the consumption utilities involved. In the first example, it did not make sense to code the second offer of y to be "pork chops when lamb is on the menu" and the first to be "pork chops without lamb on the menu." A choice comparison between steak and pork should not depend on what else is on the menu. Sometimes though, as in the second example, the context of the situation *is* appropriately integrated with the consumption utility of the object on offer. It makes social sense, when evaluating the utilities involved in the situation, to consider the utility of the first y to be: the positive utility of consuming the apple plus the negative utility of the embarrassment of taking the last fruit in the bowl.

This example illustrates one way in which human rationality is not like animal rationality—humans code into the decision options much more contextual information. Animals are more likely to respond on the basis of objective consumption utility without coding the nuanced social and psychological contexts that guide human behavior (this of course is not to deny that many animals can respond to contingencies in their social environments, which is obviously true).[6]

It is not difficult to demonstrate that humans incorporate a host of psychological, social, and emotional features into the options when in a choice situation. For example, if I were to offer you $5, no obligations, no strings attached, most people would take it. Why not? If you are bothered by taking money for doing nothing, then give it to a charity that is close to your heart. It's likely though that you would refuse my $5 if I offered it to you in the context of the Ultimatum Game.

The Ultimatum Game has been studied by economists (Davis and Holt 1993; Thaler 1992) and it works as follows. There are two players, one termed the Allocator and the other termed the Recipient. The two players do not communicate with each other. The Allocator is given a stake of $100. The Allocator can divide the stake however he wishes, offering the Recipient part of it. Call the part the Recipient is offered x. The Recipient can decide whether to accept the offer. If the Recipient accepts, then he receives x and the Allocator receives $100 minus x. If the Recipient rejects the offer, then neither gets anything. Pretend you are a Recipient in an Ultimatum game and have been offered $5 by the Allocator. If you are like most subjects who have participated in Ultimatum Game experiments, you would turn it down, thus denying yourself a $5 gift.

Like most people, you are insulted by the greed of the Allocator, and it is clearly worth $5 to you to punish that greed. Thaler (1992) notes that, as

human as this response seems, in a technical sense, it has violated the strictures of economic theory. Thaler jokes that if we take the rationality assumption of the most Panglossian of economic theories to its extreme, the
optimal solution to this problem is that the Allocator should offer one
penny and the Recipient should accept! After all, both are income maximizing responses.

You will not be surprised to learn that virtually no one follows this
purely economic stricture. In one experiment described by Thaler (1992),
the Allocators offered an average of 37 percent of the stake, and the modal
offer was 50 percent of the stake. Offers of anything less than 30 percent of
the stake turn out to be in serious danger of being refused. The subjects in
these experiments demonstrate, in their violation of the maximizing response predicted by the strictly economic theory, the truism that there's
more to life than money. Their responses were determined by more than
just the objectively presented values. Probably in this case we have another
demonstration of how human rationality is not just an extension of animal
rationality. One assumes that most hungry animals would take the single
food pellet offered under similar circumstances.[7] Humans put a utility on
other values that they bring to the situation, such as fairness. Fairness considerations mean incorporating many contextual details about how the offer is made rather than simply the monetary value of the offer. Such considerations represent another case, like that of the fruit in the bowl, where
context matters because of values and social considerations that humans explicitly want to be integrated with the utility of the options.

That virtually everyone in the Ultimatum Game experiments rejects the
strictly maximizing response illustrates what Sen, in his famous paper titled
"Rational Fools," refers to when he says that "the *purely* economic man is indeed close to being a social moron" (1977, 336). The purely economic man
who would accept one penny as the Recipient in the Ultimatum Game
would indeed seem to be a rational fool. Such behavior would suggest that
this person recognized no social considerations, had no values, and coded
only monetary value into his options. That we perceive something inhuman
in such a person's psychological constitution is, I submit, an accurate perception. Rational choices not backed by values, by reflective evaluation in
light of larger life goals, would be like animal rationality. Like Searle (2001),
I intend to argue that human rationality is (and should be) more than that.
However, consistent with my Meliorist tendencies, and with my empirical
research on individual differences in rationality (Stanovich and West 1998c,
1999, 2000), I will argue some additional things as well. In particular I am
going to argue that we all achieve more than chimp rationality—but only

more or less. We could do better. Most of us have a rationality more enriched than that of the chimp but not as rich as it might be.

There's More to Life than Money—
But There's More than Happiness Too: The Experience Machine

In order to demonstrate the truism in the title of this section, I am going to use a thought experiment devised by philosopher Robert Nozick (1974, 42–45) and called the experience machine. I am going to take some liberties with his creation, but the idea is entirely his.

Imagine it is hundreds of years in the future and your son brings home the latest device concocted by the neurophysiological games industry. He plugs a computer into the wall and then puts over his head a helmet (somewhat like those worn in present-day virtual reality demonstrations) connected to the computer and then proceeds to lie on the couch for two hours a day. And then for four hours a day. And then for six hours a day. And then for eight.

It seems that the machine is on its "hedonic setting." It is delivering pleasurable sensations to your son—true happiness—as long as his head is in the helmet. Would you want him to live most of his life inside it?

I detect some hesitation, so I'll ask you: Why not? You want your son to be happy, don't you? You are an advocate of instrumental rationality—maximizing personal utility[8]—aren't you? That's what the machine is doing. Your son couldn't be happier. He reports that being inside the helmet "just feels great." What more could you want for him than that? Nozick asks you to ponder the question "What else can matter to us, other than how our lives feel from the inside?" (43).

Your response might be that you want your son not just to have diffuse "happiness experiences" but to *experience life* and be happy. "You should have said so" says your son, "there is a life experience setting on the machine. See right here. I'll change modes"—at which point your son goes inside the helmet for another sixteen hours in a row.

It seems that the life experience setting simulates in the brain of the user the experience of *actually doing* all of the things the person would like to do in his/her life—making friends, bungee jumping, playing baseball, flying an airplane, skiing, painting a picture, etc., etc.—along with the concomitant happiness induced by those fulfilling activities. Before putting the helmet on, an inexhaustible list of activities is presented and as many activities as the user desires for as much time as the user wants can be chosen. There is nothing that your son could think of to do that wouldn't be presented as an option. Your son, once inside the helmet, is now not missing any *experiences.* He is having the experience of actually doing the things both you and he wanted him to do—along with the happiness you wanted for him as well. There he is, right on the couch, getting everything you could ever want for him. Now there really is no reason for him to ever take the helmet off, right?

Nonetheless, you resist. Somehow, you agree with Nozick (1974, 43) that plugging into the machine would be a kind of suicide. We are concerned with more than just how our time is filled; we are also concerned with *who we are.* In turn, you want your son to be a certain type of person, not just to have pleasant experiences. The experience machine thought experiment demonstrates that you are willing to trade the latter for the former. You have values that can only be fulfilled for your son if he is in actual causal contact with the world—and that efficacious causal contact with the world, hopefully in meaningful activities, is a value that you hold over and above first-order preferences for pleasurable experiences.

The experience machine thought experiment is aimed straight at the advocates of instrumental rationality (particularly those with more simplified hedonistic conceptions of utility) in the thin theory sense that does not critique the content and structure of desires. It is designed to show us that almost no one wants to stop at a thin theory. Only one of Amartya Sen's (1977) "rational fools" would plug him/herself in to the experience machine—only a rational fool would think that there was nothing to life other than hedonic experience.

We all have values—we all want to be a certain type of person. In order to become that person, we need (contra the thin theory) to include an evaluation of the content of our desires (Dworkin 1988; Frankfurt 1971; Taylor 1992). In a broad theory of rationality that does so, the mechanism we would use to critique our desires consists of the higher-order values that we hold.[9] Being a certain type of person is not a singular event that can be discretely experienced like a consumption good. Being a certain type of person, a person congruent with one's deepest values, has symbolic utility. Symbolic utility is utility that we experience by carrying out a symbolic action that stands for the utility of something else. That behavioral scientists should explicitly recognize the concept of symbolic utility has been argued eloquently by Robert Nozick (1993).

Nozick on Symbolic Utility

Nozick (1993) defines a situation involving symbolic utility as one in which an action (or one of its outcomes) "symbolizes a certain situation, and the utility of this symbolized situation is imputed back, through the symbolic connection, to the action itself" (27).[10] Nozick notes that we are apt to view a concern for symbolic utility as irrational. This is likely to occur in two situations. The first is where the lack of a causal link between the symbolic action and the actual outcome has become manifestly obvious yet the symbolic action continues to be performed. Nozick mentions various antidrug measures as possibly falling in this category. In some cases, evidence has accumulated to indicate that an antidrug program does not have the causal effect of reducing actual drug use, but the program is continued because it has become the *symbol* of our concern for stopping drug use. In other cases, the symbolic acts will look odd or irrational if one is outside the networks of symbolic connections that give them meaning and expressiveness. Nozick mentions concerns for "proving manhood" or losing face as being in this category.

Although it would be easy to classify many instances of acts carried out because of symbolic utility as irrational because of a lack of causal connection to the outcome actually bearing the utility, or because the network of social meanings that support the symbolic connection are historically contingent, Nozick warns that we need to be cautious and selective in removing symbolic actions from our lives. Most reactions to the idea of the experience machine indicate that no one wants to live a life without any symbolic meaning. The concern with "being a certain type of person" is a concern for living a life that embodies values that do not directly deliver utilities but are indicative of things that do.

Nevertheless, it is easy to see how symbolic utility can run amok in the escalating and reverberating circuits of "meaning making" in the modern world. As Nozick notes, "conflicts may quickly come to involve symbolic meanings that, by escalating the importance of the issues, induce violence. The dangers to be specially avoided concern situations where the causal consequences of an action are extremely negative yet the positive symbolic meaning is so great that the action is done nevertheless" (1993, 31). In the wake of the destruction of the World Trade Center, Nozick's warning is chilling, especially when combined with our knowledge of the logic of faith-based memeplexes that resist critical evaluation (see chapter 7).

Yet despite these warnings, performing actions for reasons of symbolic utility is an essential component of human conduct and provides critical feedback in our quest to become a certain type of person. Nozick (1993, 49) provides an especially good discussion of how symbolic actions that help maintain a valued concept of personhood are not irrational despite their lack of a causal connection to experienced utility. People may be fully aware that performing a particular act is characteristic of a certain type of person but does not contribute causally to their becoming that type of person. But in symbolizing the model of such a person, performing the act might enable the individual to maintain an *image* of himself. A reinforced image of himself as "that kind of person" might make it easier for the person to perform the acts that *are* actually causally efficacious in making him that type of person. Thus, the pursuit of symbolic utility that maintains the self-image *does* eventually get directly cashed out in terms of the results of actions that are directly causally efficacious in bringing about what the individual wants— to be that sort of person.

For many of us, the act of voting serves just this symbolic function. Many of us are aware that the direct utility we derive from the influence of our vote on the political system (a weight of one one millionth or one one hundred thousandth depending on the election) is less than the effort that it takes to vote (Baron 1998; Quattrone and Tversky 1984), yet all the same we would never miss an election! Voting has symbolic utility for us. It represents who we are. We are "the type of person" who takes voting seriously. Not only do we gain symbolic utility from voting, but it maintains a self-image that might actually help to support related actions that are more efficacious than a single vote in a national election. The self-image that is reinforced by the instrumentally futile voting behavior might, at some later time, support my sending a sizable check to Oxfam, or getting involved in a local political issue, or pledging to buy from my local producers and to avoid the chain stores.

Book buying has symbolic value for many intellectually inclined people, and often the symbolic utility has become totally disconnected from consumption utility or use value. I am like many people in that I buy many books that I will never read (I am fully aware that my "I'll have time for that in retirement" is a pipe dream). Even though my reading of fiction has slipped in recent years, I'll always pick up the latest Booker Prize winner. I'll do this even when the description of the book in reviews makes it seem very unlikely that I'll ever read the book. Despite the fact that my actual consumption of the utility of reading fiction has waned, I'm still "the type of person" who reads quality fiction—or at least I signal so to myself when I buy the latest Booker-winner (even when it is Margaret Atwood's *Blind Assassin*—which I'm *sure* I'll never read!).

Why were all those copies of Stephen Hawking's *A Brief History of Time* on all those coffee tables some years ago, virtually all but unread? Ignoring the utility derived from intellectual status display, many of us were deriving symbolic utility from buying it despite the fact that it would never produce utility as a consumption good. Either because of the magnificence of the topic, respect for the life of the author, or a unique combination of both, this candidate for the biggest unread book of all time was an impressive purveyor of symbolic utility.

"It's a Meaning Issue, Not a Money Issue": Expressive Rationality, Ethical Preferences, and Commitment

We have high expectations for meaning and worth, and we understand challenging ourselves to excel. Maybe we are—some of us at any rate—meaning-in-life overachievers.

—Owen Flanagan, *Self Expressions: Mind, Morals, and the Meaning of Life* (1996, 201)

Human rationality is not an extension of chimpanzee rationality: symbolic utility plays a critical role in human rational judgment, but there is no counterpart in chimpanzee rationality. The representational abilities of humans make possible a level of symbolic life unavailable to any other animal. There is a large literature on how these representational abilities bootstrap knowledge acquisition and behavioral regulation.[11] My focus here is on the related (and I believe profound) question of the implications that these representational abilities have for models of human rationality. In making room for symbolic utility, the evolution of cognition opened new expressive possibilities for humans, new ways of valuing—ways that are not simply an extension of chimpanzee rationality.

The recognition that the expressive function of action often dissociates responses from being understood in means-ends terms is beginning to reverberate in theoretical developments within economics and psychology. For example, Hargreaves Heap (1992) argues for distinguishing what he terms expressive rationality from instrumental rationality. When engaged in expressively rational actions, the agents are attempting to articulate and explore their values rather than trying to fulfill them. They are engaging in expressions of their beliefs in certain values, monitoring their responses to these expressions, and using this recursive process to alter and clarify their desires. Such exploratory actions are inappropriate inputs for a cost-benefit calculus that assumes fully articulated values and a single-minded focus on satisfying first-order preferences (E. Anderson 1993).

Likewise, in his critique of rational choice theory, Abelson (1996) argues the necessity of distinguishing instrumental from expressive motives and defines the latter quite similarly to Nozick's (1993) concept of symbolic utility. Expressive behavior, according to Abelson (1996), is a value-expressive action performed for its own sake. He notes the fervor behind (and reaction to) such symbolic actions as the flying of the Confederate flag in the United States, the protests against fox hunting in England, the demonstrations at abortion clinics, and the protest by the two black sprinters at the 1968 Olympics. A theory of rationality that fails to deal with such actions because they fail to conform with a strictly instrumental calculus is radically incomplete.

Nozick's (1993) concept of symbolic utility also has echoes in the notion of ethical preferences in economics and in Sen's (1977, 1987, 1999) discussions of the need for incorporating a concept of commitment in economic theory. As a counterpoint to the emphasis in economics on the isolated market participant concerned only with his/her own "consumption bundle," Sen (1977) discusses the concepts of sympathy and commitment which he defines as follows. Imagine that the knowledge that individuals are being tortured makes you sick. You are thus motivated to work to stop it and you do so. You have produced an instance of sympathy according to Sen's (1977) distinction. The treatment of other individuals directly affects your own experienced utility, and in aiding the others, you affect not only them but your own directly experienced utility. Sen (1977) contrasts this with another individual whose values also commit her to stopping torture but whose constitution is different from yours—the thought of torture does not *literally* make her sick. Knowledge of torture affronts her values, but does not affect her ongoing visceral feelings. A successful intervention to stop torture is consistent with her values, but she derives no direct experiential utility gain from its stopping. This is an instance of commitment according to Sen (1977).

There is a sense in which commitment is less egoistic than sympathy. A person acting on a commitment is expressing a value but is not seeking an instrumental utility increase. People acting on sympathy are themselves reinforced by an increase in another's hedonic state. Sen (1977) tells a humorous story of two small boys to illustrate the difference between sympathy and commitment. Two boys are presented with a large apple and a small apple from which they are to choose. Boy A says "you choose" and boy B immediately takes the larger apple. Boy A is incensed at this behavior. Boy B is puzzled and asks boy A which one he would have picked if he had gone first. Boy A says "Well the small one of course"—to which boy B responds "Well what's the matter? You got exactly the one you wanted!"

Sen (1977) points out that that would literally be true if boy A's offer of "you choose" had been based on sympathy. That boy A was indignant suggests that his offer for the other one to choose was based on commitment to the principle of politeness and not because he sympathetically identified with boy B's hedonic consequences. Had this been the case, boy A would have gained in utility when B took the larger apple and would have been pleased rather than indignant.

Sen (1977) points out that commitment more than sympathy is a threat to the traditional assumptions of economics because commitment, unlike sympathy, drives a wedge between choice and personal welfare. Choices based on commitment may actually lower the personal welfare (as economists measure it) of the individual, as when we vote for a political candidate who will act against our material interests but who will express other societal values that we treasure. The concept of ethical preferences that has received discussion in the renegade economics literature (Anderson 1993; Hirschman 1986; Hollis 1992) has the same function of severing the link between observed choice (revealed preference in the discipline's parlance) and the assumption of instrumental maximization in the economic literature.

The boycott of nonunion grapes in the 1970s, the boycott of South African products in the 1980s, and the interest in fair-trade products that emerged in the 1990s are examples of ethical preferences affecting people's choices and severing the link between choice and the maximization of personal welfare that is so critical to standard economic analyses. The latter persist in seeing choices as emanating from unanalyzed and unreflective "tastes" (see Hirschman 1986, for a discussion of this assumption in economic theory) that can without complication be collapsed to a unidimensional scale of formalisms that filter the world through a narrow lens of hedonic maximization. Ethical preferences throw a monkey wrench into this worldview. Hollis (1992) notes "the strain put on the notion of utility. An agent who prefers a fifth apple to a fourth pear can be held to have tastes which are given, need no reason, are known introspectively and mean that the fifth apple has greater utility. But what is the common measure when the juiciness of the orange is outweighed by its South African origin?" (309).

Despite the complication entailed for the traditional models, decision theorists have lately more strongly emphasized that symbolic considerations will have to be captured in our models sooner or later—so we might as well start now. In a series of papers on the meaning inherent in a decision, Medin and colleagues (Medin and Bazerman 1999; Medin, Schwartz, Blok, and Birnbaum 1999) have emphasized how decisions do more than convey utility to the agent but also send meaningful signals to other actors and symbolically reinforce the self-concept of the agent. Medin and Bazerman

(1999, 541) point out that the Recipient in the Ultimatum Game who rejects a profitable offer that is considerably under 50 percent of the stake may be signaling that he sees positive value in punishing a greedy Allocator. Additionally, he may be engaging in the symbolic act of signaling (either to himself or to others) that he is not the kind of person who condones greed. Medin and Bazerman discuss a number of experiments in which subjects are shown to be reluctant to trade and/or compare items when protected values are at stake (Anderson 1993; Baron and Leshner 2000; Baron and Spranca 1997). For example, people do not expect to be offered market transactions for their pet dog, for land that has been in the family for decades, or for their wedding rings. Among Medin and Bazerman's subjects, a typical justification for viewing such offers as insults was that: "It's a meaning issue, not a money issue."

Rising Above the Humean Nexus: Evaluating Our Desires

Bringing meaning into the equation severely complicates the instrumental calculus of thin theories of economic rationality. In effect, meaning-based decisions and ethical preferences do more than just complicate first-order preferences (turning "I prefer an orange to a pear" into "I prefer a Florida orange to a pear and a pear to a South African orange"). Instead, the interrelated set of meaning-based concepts discussed in this chapter—symbolic utility, ethical preferences, expressive actions, commitment—represent the mechanisms which can be used to turn traditional thin theories of human rationality (those that take beliefs and desires as given and focus only on the instrumental efficiency of achieving uncritiqued desires) into a broader and more accurate conception of human rationality, one that does not view human rationality as merely an extension of chimpanzee rationality.

These concepts can also be used in an ongoing cultural project of human self-improvement. We can use these higher-order concepts to critique the nature of our own first-order desires and goals. The development of the conceptual tools to aid people in the evaluation of their present wants and desires represents one of the seminal achievements in human cultural history. It is a historically contingent development—an ongoing project in human history.

Recall from chapter 3 the discussion of Hume's famous dictum: "Reason is, and ought only to be the slave of the passions, and can never pretend to any other office than to serve and obey them" ([1740] 1888, bk. 2, part 3, sec. 3). On this narrow instrumental view, the role of reason is only to serve unanalyzed first-order desires. However, as Nozick (1993) says, "if human beings are simply Humean beings, that seems to diminish our stature. Man is the only animal not content to be simply an animal. . . . It is symbolically important to us that not all of our activities are aimed at satisfying our given desires" (138). Nozick suggests that having symbolic utilities is the way to rise above what he terms the "Humean nexus" (preexisting desires in correct causal connection with actions in the instrumentally rational agent). He thinks that it is important to us that we view ourselves as above the Humean nexus. I agree with the thrust of this, but why limit the mechanism of escape to just symbolizing per se?

As was discussed above, much symbolizing activity is rightly viewed as irrational—having come unstuck completely from the actual utility-producing act with which it was once in causal connection. Recalling the discussion in the last chapter, it is certainly the case that much symbolizing represents only the churning and recombining of bad or useless memes. Symbolizing of this type begins to look like little more than the by-product activity of a large-brained mammal with time on its hands. We can do better than that.

Notions of rationality that are caught in the Humean nexus allow for no program of cognitive reform. Humean instrumental rationality takes a person's presently existing desires as given. Why not break the Humean nexus by bringing our analytic processing capabilities for symbolizing to aid in a project of cognitive reform of the Neurathian type discussed in the last chapter? Why not use our symbolizing abilities to leverage ourselves into a reflective evaluation of our presently existing desires?

Specifically, we can enjoy the symbolic meanings experienced in situations, but we can use them in other ways too. As discussed above, expressive actions can be used to aid in the maintenance of self-image that in turn is a mechanism to be used in future goal regulation. Much in the way that cognitive scientists have speculated about auditory self-stimulation leading, in our evolutionary history, to the development of cognitive pathways between brain modules (see Dennett 1991), the feedback from expressive actions can also help to shape the structure of goals and desires. However, in the next section, I will explore a more explicit and reflective way of using our symbolic abilities to rise above the Humean nexus—to engage in a process of evaluating our first-order desires in order to examine whether they really reflect the kind of person we want to be.

Second-Order Desires and Preferences

The Humean thesis fares poorly as a claim about meaning. People think some things worth wanting whatever in fact one wants, and other

things not worth wanting even if in fact one does want them. We seem
to understand these claims, and it would be good to have an analysis
that counts them intelligible.

—Allan Gibbard, *Wise Choices, Apt Feelings* (1990, 12)

Philosophers have long stressed the importance of self-evaluation in the
classically modernist quest to find one's truest personal identity. Charles
Taylor (1989), for example, stresses the importance of what he terms strong
evaluation, which involves "discriminations of right or wrong, better or
worse, higher or lower, which are not rendered valid by our own desires, in-
clinations, or choices, but rather stand independent of these and offer standards by which they can be judged. So while it may not be judged a moral
lapse that I am living a life that is not really worthwhile or fulfilling, to describe me in these terms is nevertheless to condemn me in the name of a
standard, independent of my own tastes and desires" (4; see Flanagan 1996,
for an excellent discussion of Taylor's concept of strong evaluation).

The meaning-based concepts discussed earlier in this chapter (symbolic
utility, expressive rationality, commitment, ethical preferences, etc.) represent some of the mechanisms of strong evaluation. Most people are accustomed to conflicts between their first-order desires (if I buy that jacket I
want, I won't be able to buy that CD that I also desire). However, a person
who forms ethical preferences creates new possibilities for conflict. So, for
example, I watch a television documentary on the small Pakistani children
who are unschooled because they work sewing soccer balls, and vow that
someone should do something about this. I find myself at the sporting
goods store two weeks later instinctively avoiding the more expensive
union-made ball. A new conflict has been created for me. I can either attempt the difficult task of restructuring my first-order desires (e.g., learn to
not automatically prefer the cheaper product) or I must ignore a newly
formed ethical preference. Actions out of kilter with a political, moral, or
social commitment (Sen 1977, 1999) likewise create inconsistency. Note
how the commitment causes "backdraft" on to first-order desires and actions. Values and commitments create new attention-drawing inconsistencies that are not there when one is only aware of the necessity of scheduling
action to efficiently fulfill first-order desires.

In short, our values are the main mechanism we use to initiate an evaluation of desires. Action/value inconsistency can signal the need to initiate
normative criticism and evaluation of both the first-order desires and the
values themselves. Values thus provide an impetus for the possible restructuring of the architecture of desires. They are what allow human rationality
to be a broad rationality—one where the content of desires makes a differ-

ence—in contrast to the thin instrumental rationality characteristic of chimpanzees and other animals.

What I have been calling a critique of one's own desire structure (and Taylor's notion of strong evaluation) can be a bit more formally explicated in terms of what philosopher Harry Frankfurt (1971), in a much-cited article, terms second-order desires—desires to have a certain desire.[12] In the language more commonly used by economists and decision theorists (see Jeffrey 1974), this higher-level state would be called a second-order preference: a preference for a particular set of first-order preferences. Frankfurt speculates that only humans have second-order desires, and he evocatively terms creatures without second-order desires (other animals, human babies) "wantons." To say that a wanton does not form second-order desires does not mean that they are heedless or careless about their first-order desires. Wantons can be rational in the thin, purely instrumental, sense. Wantons may well act in their environments to fulfill their goals with optimal efficiency. A wanton simply doesn't reflect upon his/her goals. Wantons want but they don't care what they want.

To illustrate his concept, Frankfurt (1971) uses as an example three kinds of addict. The wanton addict simply wants to get his drug. That is the end of the story. The rest of the cognitive apparatus of the wanton is simply given over to finding the best way to satisfy the desire (i.e., the wanton addict could well be termed instrumentally rational). The wanton addict doesn't reflect on his desire—doesn't consider one way or another whether it is a good thing. The desire just is. The unwilling addict, in contrast, has the same first-order desire as the wanton addict, but has the second-order desire not to have it. The unwilling addict wants to want not to take the drug. But the desire to desire not to take it is not as strong as the desire to take it. So the unwilling addict ends up taking the drug just like the wanton. But the relation of the unwilling addict to his behavior is different than that of the wanton. The unwilling addict is alienated from his act of taking the drug in a way that the wanton is not. The unwilling addict may even feel a violation of self-concept when he takes the drug. Such a feeling would never occur in the wanton's act of drug taking.

Finally, there is the interesting case of the willing addict (a possibility for humans). The willing addict has thought about his desire for drug taking and has decided that it is a good thing. He actually wants to want to take the drug. Frankfurt (1971) helps us understand this type by noting that the willing addict would, if the craving for the drug began to wane, try to take measures to reinstate the addiction. The willing addict has *reflected* on the addiction, just as the unwilling addict, but in this case has decided to *endorse* the first-order desire.

All three of Frankfurt's addicts are exhibiting the same behavior, but the cognitive structure of their desire hierarchies is quite different. This difference, although not manifest in current behavior, could well have implications for the likelihood of the addiction continuing. The unwilling addict is (statistically speaking of course) the best candidate for behavioral change. He is the only one of the three to be characterized by an internal cognitive struggle. One would posit that this struggle at least has the possibility of destabilizing the first-order desire or perhaps weakening it. The wanton is characterized by no such internal struggle and thus has less of a chance of the first-order desire being disrupted. However, note that the wanton is actually more likely to lose the addiction than the willing addict. The latter has an internal governor keeping the addiction in place, namely the second-order desire to have it. The willing addict would take steps to stop the natural waning of the addiction. A natural waning of the addiction would be unimpeded in the wanton addict, who would simply take up some other activity that is higher in his first-order goal hierarchy. The wanton would not be sad to be rid of the addiction. He would not be happy either—for the logic of his condition is that he does not reflect on the coming and going of his desires.

Achieving Rational Integration of Desires: Forming and Reflecting on Higher-Order Preferences

We are animals. But we are self-creating in certain respects. How self-creation, self-control, and authorship are possible for animals like *Homo sapiens* can be explained. It is a nonmystery that "mysterians" like to spread magic dust over.

—Owen Flanagan, *Self Expressions: Mind, Morals, and the Meaning of Life* (1996, viii)

Of course, as humans, we sometimes act wantonly in Frankfurt's sense and sometimes not. Should we consider it problematic that we sometimes act as wantons? Employing the concepts from the earlier chapters (summarized in figures 7.1 and 7.2 in the last chapter), the situations that are problematic and nonproblematic can be demarcated. It is largely the case that the first-order desires of the TASS subsystems serve the vehicle's overall goals in useful ways. But discussions in previous chapters have illustrated the danger of wanton behavior. Some gene-installed goals that are remnants of the EEA can, in certain situations, thwart the vehicle's interests in the modern environment. Additionally, some desires are memes that have been unreflec-

tively acquired and are actually parasites—serving their own replicative interests but doing little to help their hosts. These two classes of goals motivate the Meliorist admonition that an ongoing reflective critique of first-
order desires is necessary in order for humans to achieve broad rationality.

The logic of this project of cognitive reform based on the concept of
higher-order desires can be made clear using the notation from an article on
second-order evaluation by Jeffrey (1974). He makes use of the preference
relationship used in decision theory that is the basis for the formal axioma-
tization of utility theory (Edwards 1954; Jeffrey 1983; Luce and Raiffa 1957;
Savage 1954; von Neumann and Morgenstern 1944). I will make use of the
preference relation informally and nontechnically here,[13] in order to ease
the verbal convolutions that result when talking about higher-order preferences.

If it is said of a person that he prefers A to B, then the preference relation
expresses this as: A pref B (simply taken to mean that A is preferred to B). As
was mentioned in chapter 3, from this simple relationship and a few basic
axioms, the entire structure of expected utility theory (the theory of instrumentally rational choice) can be built up. But that is not my concern here,
because I simply wish to make use of the notation.

Imagine John, who has the choice between smoking (S) and not smoking (~S). John chooses to smoke. So we have:

John prefers to smoke
S pref ~S

Imagine also that John is a strong evaluator (in Taylor's 1989 sense). He
evaluates and reflects on his first-order desires. Furthermore, John is knowl-
edgeable. He knows that smoking represents a devastating health risk.
He therefore wishes he preferred not to smoke. Like the unwilling addict
discussed above, John wants to want not to smoke. Therefore, John has a
second-order preference:

John prefers to prefer not to smoke
(~S pref S) pref (S pref ~S)

In this scenario, John has a second-order preference that conflicts with his
first-order preference. At the level of second-order preferences, John prefers
to prefer to not smoke; nevertheless, as a first-order preference, he prefers
to smoke. The resulting conflict signals that John lacks what Nozick (1993)
terms rational integration in his preference structure. It is a feature of Noz-

ick's broad theory of rationality that people should want to achieve rational integration. Like the unwilling addict, John is rightly uncomfortable with the mismatch in his desires at different levels of analysis. Nozick's principle (actually principle IV in his strictures for a broad theory of rationality) says only that John should make efforts to resolve this inconsistency. However, it does not state *how* John should resolve the inconsistency.

There was a strong tendency in early philosophical writings on the topic of higher-order preferences (e.g., Taylor 1989) to assume that higher-level desires should always trump lower-level desires. In the previous chapter we have seen why this is a dangerous assumption. Yes, it may well be the case that a TASS-determined preference that is suboptimal for the vehicle should be overridden by a reflectively acquired second-order preference in line with the person's values. But many examples discussed in chapter 7 suggest another possibility. Perhaps a TASS-determined preference that is actually *good* for the vehicle is coming into conflict with an unreflectively acquired meme that has high replicating power and has become part of a person's value structure but does not serve the person well.[14] In this case, it is the second-order preference that is not efficacious for the individual. A bias toward strong evaluation like that emphasized by Taylor (1989) is not always warranted. Evaluation at higher levels should not necessarily have priority over lower-order desires. Instead, the higher-order desires are merely part of the overall structure that a person should try to bring into reflective equilibrium so as to achieve rational integration.

Perhaps we should leave the smoking example, which of course is strongly biased, for a more neutral example. Suppose that:

> Bill prefers Y to X

and that

> Bill prefers that he preferred X to Y

Which to change? There is no hard and fast rule. However, the concepts discussed earlier can help. In particular, critical awareness of the potential mismatch between genetic interests and vehicle interests, as well as potential mismatches between meme interests and vehicle interests, should of course make us ask first if either of the levels of analysis appear to be sacrificing us as vehicles to one of the replicators. Adverse consequences for the vehicle's long-term physical well-being has prima facie validity as at least one criterion that can be employed in an attempt to achieve rational integration. If, as in the smoking example, Y looks like a TASS-conditioned response that harms the vehicle and X represents a reflectively considered act that serves the vehicle's long-term health, then a vehicle well-being criterion might dictate honoring the second-order preference and taking behavioral steps to change the first-order preference.

In contrast, the situation might be one in which Y is an act either useful to the vehicle or neutral with regard to vehicle well-being, but X is a costly physical sacrifice dictated by a faith-based meme. In this case, the burden of proof is reversed. The second-order preference has a double burden of proof to overcome. There are two reasons for the reflective person to be suspicious of identifying with it. First, the second-order evaluation is spawned by a memeplex that disables critical evaluation, and, secondly, it dictates a sacrifice of the vehicle's physical health. These two considerations demonstrate that the support for the second-order preference weakens under reflective scrutiny.

The criterion of vehicle well-being is not always definitive, however. It could, and should, be overridden at times, especially when it is arrayed against a converging set of other criteria and decision rules. For example, moral or ethical considerations could collectively trump vehicle well-being at times, particularly when the sacrifice to the latter is minimal.[15]

In summary, it is rational to want to achieve rational integration of preferences, but no theory of rationality provides rules for how this should be brought about. It is not always wise to bring first-order preferences into conformity with second-order preferences (nor the converse). Instead, a process that is more Neurathian (see chapter 7) must take place—coherence must be achieved by making certain assumptions in order to bring coherence. But subsequently, due to the examination of other, interlocking preferences and goals, those assumptions themselves may be deemed insecure and have to be altered at a later date. What is certain is that every person should be motivated to achieve rational integration, and one way to start to achieve such integration when there is a first order/second-order mismatch is to construct another level.

There of course is no limit to the hierarchy of higher-order desires that might be constructed.[16] But the representational abilities of humans may set some limits. Dworkin (1988) notes that "as a matter of contingent fact, human beings either do not, or perhaps cannot, carry on such iteration at great length" (19). I will deal here with no more than three levels, which seems a realistic limit for most people in the nonsocial domain.

Nozick (1993) argues that it is rational for people to want to achieve rational integration. This means that when a first-order preference is not endorsed by a second-order strong evaluation, then the individual should take steps to reconcile the conflicting first- and second-order preferences. One cognitive tool that we can use is to inquire as to the state of our third-order preferences. One type of third-order preference is structured to evaluate whether a person prefers his or her second-order evaluation to the first-order preference or vice versa. So John, the smoker discussed previously, might realize when he probes his feelings that:

> He prefers his preference to prefer not to smoke over his preference for
> smoking:
> [(~S pref S) pref (S pref ~S)] pref [S pref ~S]

We might in this case say that John's third-order judgment has ratified his second-order strong evaluation.[17] Presumably this ratification of his second-order judgment adds to the cognitive pressure to change the first-order preference by taking behavioral measures that will make it more likely (entering a smoking secession program, consulting his physician, asking the help of friends and relatives, staying out of smoky bars, etc.). On the other hand, a third-order judgment might undermine the second-order preference by failing to ratify it:

> John might prefer to smoke more than he prefers his preference to prefer
> not to smoke
> [S pref ~S] pref [(~S pref S) pref (S pref ~S)]

In this case, although John wishes he didn't want to smoke, this preference is not as strong as his preference for smoking itself. We might suspect that this third-order judgment might not only prevent John from taking strong behavioral steps to rid himself of his addiction, but that over time it might erode his conviction in his second-order preference itself, thus bringing rational integration to all three levels.

The idea of third-order preferences is sometimes difficult to hold in mind, so here are some alternative ways of thinking about it. First, in general, the higher-level judgment might be thought of as asking "am I right to prefer?" the preference judgments at lower levels. Or, another way of conceiving of the higher-order judgment is as asking "do I want to be a strong evaluator of my lower-order judgment?" It is strong evaluation of the first-order preference that first causes a lack of rational integration. A wanton has no problem of rational integration, because he/she does no strong evaluation. Once a strong evaluation has been made, however, one way to conceive of the stance at the third-order preference level is that it is asking your-self the question: Do I want to be a strong evaluator about this? Do I want to ratify the strong evaluation over the first-order preference? Or, do I wish I hadn't made the strong evaluation in the first place, because I'd rather re-tain the first-order preference?

I will now discuss some examples, going up to third-order preferences, that are less biased against the first-order preference than the smoking ex-ample, in order to illustrate how the process of achieving rational integra-tion is an ongoing process of mutual adjustment and constraint satisfaction driven by reflection and evaluation across many different levels of analysis.

Tessa always loved Christmas as a child. She loved decorating her tree and helping her parents string lights on the outside of their house and dec-orating the inside as well. Tessa loved wrapping her presents to others and unwrapping her own. As a thirty-five-year-old adult, Tessa still loves all the accouterments of Christmas: the tree (which she decorates just as assidu-ously as ever), the caroling, performances of *The Christmas Carol* put on by theater groups, all of the Christmas movies and musical programs on tele-vision. So there is no doubt that:

Tessa prefers celebrating Christmas to not celebrating it
XMAS pref NOXMAS

The only problem is that as an adolescent Tessa lost her faith in God. Throughout young adulthood the belief waned even further and now, in her thirties, Tessa has been an atheist for some years. She does wonder whether there isn't something a little strange about an atheist celebrating Christmas. Sometimes she feels there probably is, and thus creeps in the strong evalua-tion, the second-order judgment that:

Tessa prefers to prefer not to celebrate Christmas
(NOXMAS pref XMAS) pref (XMAS pref NOXMAS)

That is, it probably would be better if she really didn't prefer to celebrate Christmas, she thinks sometimes (thus creating a lack of rational integra-tion—a mismatch between her first-order and second-order preferences).

In her thinking about this, Tessa sometimes seeks to resolve the lack of rational integration by making a third-order judgment and ponders whether it is right for her to have this second-order preference. She wonders whether she should be a strong evaluator about this domain of her life. When she thinks about it, she finds reason to question whether she should be worrying herself with strong evaluations about this particular preference of hers. No one is hurt by her behavior (her other family members are either agnostics who also like Christmas or practicing Christians). The joy she gets out of it seems to vastly outweigh the satisfaction of making a point by not celebrating—the point being only a mildly effective public display of her atheism (only mildly effective because most people who know her already know she is atheist). Besides, she thinks, doesn't the much repeated lament that "Christmas is so material" mean that it has really become the perfect holiday for a scientific materialist? And shouldn't the fact that the religion column in her local paper frequently complains that "Christ has gone out of Christmas" mean that, conveniently for her, the holiday has become, in North America at least, a materialist orgy of buying, decorations, song, and celebrations—the perfect holiday for a scientific materialist who has joie de vivre! In light of all these considerations, Tessa decides that her strong evaluation is not ratified by a higher level analysis. Instead,

> Tessa prefers to celebrate Christmas more than she prefers her preference
>   to prefer not to celebrate it
> [XMAS pref NOXMAS] pref [(NOXMAS pref XMAS) pref (XMAS pref
>   NOXMAS)]

In light of this third-order analysis—in light of the failure of her second-order preference to be ratified—it is unlikely that Tessa will change her first-order preference to celebrate Christmas. Instead, it will be more likely that her degree of rational integration will increase when her second-order preference, undermined by the third-order judgment, begins to wane.

There are, however, many cases in which rational integration of conflicting lower-order desires is precluded because there is uncertainty about what the nature of the third-order judgment should be, specifically whether second-order evaluations should be ratified. In fact, many political and moral debates in our society are about the appropriateness of second-order judgments—for example, debates about whether certain ethical preferences are justified. Jim loves to shop at Costco, Wal-Mart, and other discount stores. He loves the bargains and loves beating the price for an object that he had seen at a higher price in another store in town. Jim likes those cheap soccer balls from Pakistan discussed above. In short, Jim much prefers getting cheap stuff to not getting cheap stuff (to missing out on a bargain). For Jim, it's definitely the case that:

> Jim prefers cheap stuff to more expensive stuff
> CS pref ~CS

But since he developed these habits, Jim has become aware of some of the darker sides of global trade. He's seen, for example, a *60 Minutes* special report on the small Pakistani children working brutal hours at low wages so that the ball that Jim buys his son can be purchased cheaply. Over the course of the last couple of years Jim has become very queasy about this whole scene. He's read reports of slave labor in China, child labor throughout the Indian subcontinent, companies trying to reduce labor benefits by threatening to move to the Third World, and the massive pollution and environmental degradation in the maquiladoras (the free-trade districts of Mexico)—and he knows that all these things are linked to making those products in Wal-Mart so cheap. He's now heard about fair-trade products and he's now realized that he has friends who actually buy the more expensive union-made product on purpose. Jim has now started to believe that's it's probably not a good thing that he likes cheap stuff so much. He's begun to develop an ethical preference. In fact, he now prefers that he didn't prefer cheap stuff so much:

> Jim prefers that he did not prefer cheap stuff so much
> ($\sim$CS pref CS) pref (CS pref $\sim$CS)

Jim has mismatched first-order and second-order preferences, and to achieve rational integration he must reverse his first-order preference or withdraw his strong evaluation of the first-order preference. But he finds himself deadlocked at the third-order level. Jim admits that he felt more comfortable when he didn't have a second-order preference. He now wrestles with conflicting desires when he didn't before. He liked it better when he could shop like a wanton—just fulfilling his first-order desires and thinking about nothing else.

Jim's friends tell him that his reflective attitude toward his shopping makes him a better person and that, before, he was not much more than a shopping robot—vacuuming up the last bargain in the bin before it was greedily taken by another shopping automaton. He tells them this may be true, but he was a lot happier as a shopping robot. His friends remind him of the Pakistani children and the inexorable logic of the global economy and the race to the bottom in order to make things cheaper and cheaper by externalizing costs. They tell him to imagine two detergent companies (A and B) who make their detergents for the same cost and of the same quality. They point out that if company A finds a way to deprive its workers of health benefits, then detergent A will show up at Wal-Mart looking cheaper and people like the old Jim will be gobbling it up, forcing company B to try to do the same thing. Just as Jim is about to succumb to these arguments, just as he thinks that it's right for him to be a strong evaluator, just as he's about to believe that

> He prefers his preference not to prefer cheap stuff over his preference for
> cheap stuff
> [(~CS pref CS) pref (CS pref ~CS)] pref [(CS pref ~CS)]

He reads other arguments in *The Economist* and the *Wall Street Journal* that it is wrong to have any second-order preferences at all about his purchases; that not only will he be happier shopping without worry about how the things in his shopping basket are made, but everyone in the world would be better off if he did so. Jim had heard of Dr. Pangloss before, but this seemed to take things to a new level. But these people were serious. They literally believed these things. The editorials of these respected publications told Jim that the Pakistani children deforming their fingers sewing the soccer balls wouldn't have gone to school anyway—their country was too poor. The only way that future children would go to school was if the country's economic productivity was lifted by the sewing children now creating more wealth so that future children could go to school. And as for the workers losing their health benefits and perhaps their jobs when company A cracked down on its workers, well all was for the best because the company would make large profits that would then be fed back through it and the economy and the rising productivity that that entailed would create even better jobs somewhere for the workers that had been fired (well maybe not for *those* workers exactly, but for someone, sometime—in the future).

Amazingly, and this would have taken even Dr. Pangloss aback, all the information in the world Jim could possibly need was summarized in the price of the product—so the only thing he had to look at in order to make the world a better place was the only thing the old Jim looked at anyway: the price. What an astounding argument. It was almost enough to get Jim to think that he really thought that, at a third-order level of judgment, he didn't think that he should be a strong evaluator of his preferences in this domain. But on the verge of this, Jim had realizations that gave him pause— what was right about this new (to him) view in the *Wall Street Journal* seemed to also be what was wrong with it. Only his purchases seem to matter in this Panglossian world. We are what we shop—and the world is what we shop. But this was horrific too, Jim thought. If I don't respond to a price nothing good can happen, these people say. If I stay home and read to my children rather than buy two videos and a new shirt then has the net utility in the world really gone *down*? It was in essence what these people were saying. Jim found this hard to believe. And what about the negative effects of price? Jim had not wanted the downtown stores of his picturesque small town to be gutted by the Wal-Mart, but that is what happened. The constitution of his neighborhood and local downtown was an indirect effect of the prices of the Wal-Mart products, but those products didn't advertise the gutting of his local shops or other of their effects on his environment (increased traffic) as part of their cost (Jim learned that there were even economists who studied such so-called externalities, but they tended not to be quoted in the *Wall Street Journal*).

Jim found that he could not ratify his strong evaluation, but he could not be convinced to overturn it either. He was left like many of us in the modern world—wrestling to bring his higher-order preferences into coherence with his lower. But in fact it is not the achieving of rational integration that is the mark of personal autonomy and identity, as was emphasized in early treatments of the second-order desire concept in the literature; it is merely *engaging* in the second-order evaluative project that is sufficient (see Dworkin 1988; Lehrer 1997). It is the distinctive mark of the human condition to wrestle with such higher-order judgments, not necessarily to reconcile all inconsistencies.

Many political debates wherein the disputants hope to change or reinforce a particular human behavior or choice have the character of debates about whether a certain strong evaluation that people are making is justified. Earlier in history, the cultural trends that led to the overturning of so-called Victorian morality were essentially third-order arguments that certain second-order judgments on our behavior (Victorian admonitions in favor of sexual restraint—preferring not to prefer certain types of sexual practices) were less to be preferred than the first-order preferences themselves (the first-order preference for sex, for instance).

The outcome of any third-order struggle will likely have a large effect on how rational integration will eventually be achieved (if it is). It will bias the analysis in favor of the second-order judgment if the latter is ratified and in favor of the first-order judgment if it is not. But it should not be assumed that the third-order process is determinative—that rational integration is a matter of simply building more levels and mechanically counting up the results. Reflection must be deeper than this, because of course, via the arguments of the last chapter, merely by being higher-order judgments it is possible that both second- and third-order judgments could be derived from the same parasitic memeplex. Nothing in Frankfurt's (1971) notion of higher-order desires guarantees against higher-order judgments being infected by memes that are not efficacious for the vehicle's well-being.

That a Neurathian project of meme evaluation must accompany any cognitive project of rational integration is revealed by comparing two examples. Imagine that John, our smoker above, ratified his second-order desire to prefer not to smoke with a third-order judgment. Imagine also that the two higher-order judgments had enough cognitive weight to lead him to take behavioral steps (therapy, etc.) to overturn his first-order desire and he did so. It is, however, unnerving to realize that John's desire structure and resolution has exactly the same structure as that of the terrorist hijackers who destroyed thousands of lives in the World Trade Center attack.

An (obviously oversimplified) model might be that, like most people, even the hijackers—at least at one time—had a wanton desire for life over any religious martyrdom that they imagined:

> They preferred life to martyrdom
> LIFE pref MARTYR

But at some point in their lives, a faith-based memeplex found them to be good hosts and became the basis of a second-order judgment on their first-order desires. At some point in their lives, although they preferred life to martyrdom, they began to wish that they did not. They began to appreciate people who were terrorist martyrs even though they weren't prepared to be one. They began to wish that they could be like those people:

> They preferred to prefer martyrdom to life
> (MARTYR pref LIFE) pref (LIFE pref MARTYR)

Perhaps the mismatched preference structure created the same discomfort that Jim, who continued to like cheap stuff even after he had decided that he wished he didn't like cheap stuff, had felt. The discomfort of the mismatched preference structure creates motivation to achieve rational integration. This might have spawned a third-order evaluation of the second-order preference—the person might have asked himself: am I right to have this second-order preference to prefer to prefer martyrdom. But since the individual is immersed in the same conceptual community that caused the original second-order judgment to be made—the same community that was the environment for the memeplex that began his stance of strong evaluation—then it is likely that, as in the smoking example, the third-order judgment will ratify the second-order preference:

> They prefer their preference to prefer martyrdom to life over their preference
>   for life
> [(MARTYR pref LIFE) pref (LIFE pref MARTYR)] pref [LIFE pref MARTYR]

But if that, as in the smoking example, begins the cognitive/behavioral cascade that leads to the flipping of the first-order preference, rational integration is achieved by murdering thousands of innocent people.

This unnerving example illustrates that there is no substitute for the Neurathian project of examining—in turn, and recursively—the planks represented by each level of judgment. We achieve coherence, float in our boat for a while, bring a different level of judgment into question, perhaps flip a preference relation causing incoherence that must be brought into reflective equilibrium again, this time perhaps by giving a different level of judgment priority.

My point is that rational integration is not achieved by simply flipping preferences that are in the minority of the count across the levels of analysis; nor can it be achieved by the simple rule of giving priority to the highest level. Part of rational integration is the evaluation of the memes that form the values that are the basis of the higher-order evaluations. Note also that it is not just reflective higher-order preferences that are potentially meme-based. As discussed in the last chapter (see figure 7.2) certain higher-level evaluations can become so practiced that they become part of the reflexive, first-order goals of TASS (Ainslie 1984). This of course is the purpose of the repetition of things like Victorian moral codes. Their promoters are hoping to make these judgments into reflexive responses rather than objects of analytic reflection (where processes of criticism might reject them).

Philosophers have fretted about the potential regress in constructing higher and higher order judgments, and they have tended to bias their analyses toward the highest level desire that is constructed—giving it a unique privilege in defining a person's so-called will.[18] Modern cognitive science, in the form of all of the concepts discussed in this volume—TASS, the representational abilities of the analytic system, memes, rationality both broad and thin—would suggest instead a Neurathian project in which no level of analysis is uniquely privileged. Philosophers have thought that unless we had a level of cognitive analysis (preferably the highest one) that was foundational, something that we value about ourselves (various candidates in philosophical literature have been personhood, autonomy, identity, and free will) would be put in jeopardy. But, as philosopher Susan Hurley (1989) has eloquently argued, we can reconstitute ourselves as valuing human beings without the existence of a foundational form of human cognition.

Hurley (1989) has pointed out that philosophers have been focused on finding the highest point in the regress of higher-order evaluations, using that as the foundation, and defining it as the true self—or doing the same thing by trying to find a viewpoint outside the hierarchy of desires. But Hurley (1989) endorses a Neurathian view in which there does not exist either a "highest platform" or a so-called true-self, outside of the interlocking nexus of desires. She argues that "the exercise of autonomy involves depending on certain of our values as a basis for criticizing and revising others, but not detachment from all of them, and that autonomy does not depend on a regress into higher and higher order attitudes, but on the first step" (364). In short, the uniquely human project of self-definition begins at the first step, when an individual begins to climb the ladder of hierarchical values—when a person has, for the first time, a problem of rational integration.

In Hurley's (1989) description of the Neurathian process, the self is not some elusive vanishing point at the endpoint of higher and higher valuations. Instead, "before we get very far into the regress, the self is already with us. . . . Personhood depends on the capacity for reflection on and evaluation of one's own attitudes, for self-interpretation" (322). Using the Neurathian rebuilding the boat analogy that I introduced in the last chapter, Hurley notes that we can use desires at certain levels to criticize others "depending on certain of them as a basis for criticism and revision of others; but we must always occupy ourselves in the process. Self-determination does not depend on detaching ourselves from the whole of what we *are*" (322).

How do we know that a person is engaging deeply in such a process of self-definition and rational integration? Interestingly, perhaps the best indicator is when we detect a mismatch between a person's first-order and second-order desires with which the person is struggling: The person avows a certain set of values that imply that he should prefer to do something other than he does. For example, the person's shopping *behavior* is consistent with the first-order preference: they prefer cheap stuff; but you know the person's values imply that they would prefer not to prefer cheap stuff—and furthermore this is just what the person avows to you. The struggle between the person's first-order and second-order preferences is apparent to us. However, when we do not detect such a struggle, the state of a person's preference hierarchy is ambiguous. The person might well have engaged in strong evaluation and have a second-order preference that endorses his first-order preference, and thus no internal struggle will be apparent in behavior because there is none. Or, the person could simply be a wanton—a person who experiences no struggle up and down a vertical set of higher-order preferences (the wanton of course might experience a horizontal, first-order struggle among conflicting basic desires). With humans, of course, we can ask. We could, for example, take the word of Frankfurt's (1971) willing addict that he is indeed a willing addict—that he has thought carefully about his addiction and has very reflectively decided that he prefers to prefer the drug.

However, it is crucial to note that we would become suspicious of someone in whom we never ever detected a first-order/second-order struggle—even if this person claimed that she was examining her first-order preferences by forming second-order ones and just finding, by golly, that the two always seemed to coincide. Modern life is just too full of contradictions for someone to claim that his/her behavior was *always* in line with his/her higher-level values. In fact, life is so complex, and full of moral and personal choices that are potentially conflicting, that instead of admiring people who claim perfect congruence between behavior and second-order judgments, we would suspect that such people are more likely to be behaving wantonly. They are simply doing what they want, but in modern society it is socially desirable to be a strong evaluator, so they avow that they have second-order preferences. We are right to be suspicious of this.[19] How we know people are displaying the embarrassing mismatches between their first-order and second-order preferences is when they reveal to us their internal struggle to achieve a rational integration.

The phenomenon of the impostor—someone merely posing as a strong evaluator—reveals the logic behind a perplexing feature of modern moral life. The charge of hypocrisy is a cutting modern moral stricture—a wounding one to a real strong evaluator. The logic of the charge creates some curious consequences. First, only people who make many second-order judgments tend to get accused of hypocrisy. Wantons and impostors escape the charge entirely. Consider Ruth. Ruth is a vegetarian who lives in co-op housing and works as a social worker and has the other interests that might be thought to be associated with this stereotype. She is a strong environmentalist and she has participated in antiglobalization demonstrations. Ruth's uncle Ralph continually needles her and calls her a hypocrite. He delights in pulling out the tags of her clothes to reveal Made in Malaysia ("no union rep there, heh, heh") or Made in China ("probably slave labor, heh, heh"); and he relishes pointing out the convenience foods in her refrigerator that are environmentally unsound and that violate her vegetarianism.

Ruth finds this all very upsetting (it always *has* bothered her when she bought those convenience foods); and it is even more vexing because nothing Ralph seems to do will allow her to return his charge of hypocrisy. He hates paying taxes and, consistently, voted for Ronald Reagan in the 1980s and both Bush the elder and Bush the younger after that. He drives a huge SUV and will drive to the store for a pint of milk, but then again he thinks that environmentalists who worry about fuel use and global warming are "wackos," so there is no inconsistency in his behavior. Ruth finds all this immensely frustrating because she can't find a way to get back at him in the way that he needles her.

What Ruth needs to do is to realize that there are worse things than being a hypocrite. Because she is a strong evaluator trying to achieve rational integration, she is focused on the issue of hypocrisy. She should realize that at least a hypocrite is attempting self-definition via strong evaluation, albeit perhaps failing to achieve rational integration (and perhaps showing moral weakness if few first-order preferences ever reverse). Uncle Ralph is something worse. He is a wanton. Or, to be more accurate, Uncle Ralph is a wanton posing as a strong evaluator, because he claims to make second-order evaluations. In fact, he probably does not have seriously examined second-order preferences at all, because examination of his life from an external, third-person perspective would reveal plenty of strong evaluations that Ralph could make (but doesn't) that would lead to inconsistencies with his first-order preferences.

Consistent with this conjecture is that Ralph does not seem to notice rank inconsistencies among the values he purports to hold, even at the same level of analysis. Uncle Ralph says he's for traditional values, stable families, cohesive communities, and free-market capitalism. But the last of these does not cohere with the previous three. There is no more continuously disruptive force in the world than unrestricted capitalism. The business pages of our newspapers and our corporate-controlled electronic media constantly tell us that capitalism's "creative destruction" (leaving behind old ways of production in order to promote the more efficient and new) is what produces our impressive material wealth, and this may well be true. But among the most notable things that are "creatively destroyed" by capitalism are traditional values, stable social structures, families, and cohesive communities. The temporary employment agency, Manpower, and Wal-Mart, known for its destruction of the commercial life in small downtowns, are two of the largest employers in the United States. Thus, the first-order preferences of Uncle Ralph that are consistent with his support of global capitalism will not cohere with second-order preferences derived from his belief in stable communities and families—or, they would not cohere if he were not blind to this lack of rational coherence.

In being a wanton (or an *unconscious* hypocrite), Ralph is at a level of self-definition *below* that of Ruth. Hypocrisy charges only come into play when people are *attempting* self-definition by making second-order judgments. An additional level of personhood is achieved when, like Ruth, one notices and is bothered by lack of rational integration.

Why Rats, Pigeons, and Chimps Are More Rational than Humans

The many mismatches between Ruth's second-order evaluations and her first-order preferences might well make her choices in the actual world more inconsistent than choices resulting from Uncle Ralph's wanton desires, and this has some startling implications for conceptions of rationality. Because certain types of consistency and coherence are the defining features of the axiomatic approach to instrumental rationality (see chapter 3 and Luce and Raiffa 1957; Savage 1954; von Neumann and Morgenstern 1944), it may be the case that Ruth's choices will violate these more than Ralph's. In terms of the standards of instrumental rationality, Ralph is more rational than Ruth. Can this conclusion be right?

Yes it can. Not only is Ralph probably more rational than Ruth, but rats and pigeons and chimps are probably more rational than Ruth as well. That is, it is possible that animals and other wantons are more rational than most humans. How can this be?

First, it has been established that the behavior of many nonhuman animals does in fact follow pretty closely the axioms of rational choice (Kagel 1987; Real 1991); many animals appear to have at least a fair degree of instrumental rationality. As Satz and Ferejohn (1994) have noted: "pigeons do reasonably well in conforming to the axioms of rational-choice theory" (77). Some investigators have seen a paradox here. They point to the evidence reviewed in chapter 4 indicating that humans often violate the axioms of rational choice. Some (often Panglossian critics of experiments showing human error) have thought that there was something puzzling or even incorrect about these experiments indicating failures of instrumental rationality in humans in light of the findings of high levels of instrumental rationality in lower animals such as bees and pigeons. For example Gigerenzer (1994) draws attention to the fact that "bumblebees, rats, and ants all seem to be good intuitive statisticians, highly sensitive to changes in frequency distributions in their environments. . . . One wonders, reading that literature, why birds and bees seem to do so much better than humans" (142).

However, it is wrong to adopt the assumption that rationality should increase with the complexity of the organism. To the contrary, there is nothing paradoxical at all about lower animals demonstrating more instrumental rationality than humans, because the principles of rational choice are actually *easier* to follow when the cognitive architecture of the organism is simpler.

One reason there is no paradox here—that it is unsurprising that bees show more instrumental rationality than humans—is that the axioms of rational choice all end up saying, in one way or another, that choices should not be affected by irrelevant context. As described in chapter 4, the axioms of choice operationalize the idea that an individual has preexisting preferences for all potential options that are complete, well-ordered, and stable. When presented with options, the individual simply consults the stable preference ordering and picks the one with highest personal utility. Because the strength of each preference—the utility of that option—exists in the brain before the option is even presented, nothing about the context of the presentation should affect the preference, unless the individual judges the context to be important (to change the option in some critical way)— and there is the rub. Humans, being the most complex organisms on the planet, are vastly more sensitive to contextual features. As such, they are more likely to be in situations that render the axioms of choice ambiguous by making their application unclear because it is debatable whether a contextual feature should be coded into the option or not. Part of that debate might be internal (as in Ruth's struggle with her first-order and second-order desires), thus leading to inconsistency across choice situations that violate the axioms.

The previous discussion of the principle of the independence of irrelevant alternatives (Sen 1993) illustrated how the choice axioms preclude context effects. Recall that that principle states that if x is chosen from the choice set x and y, then y cannot be chosen when the choice set is widened to x, y, and z. In the example, it was shown how human social context led to violations of this principle if the choice options were narrowly construed— or were adhered to only if the complexities of human social context were assumed to be rightly coded into the options ("taking the last apple in the bowl when I am in public"). The point is that the principle is one that prohibits the irrelevant contextualization of choice. When x and y are truly the same across situations, one should not switch preferences from x to y when alternative z is added.

Likewise, all of the other principles of rational choice have as implications, in one way or another, that irrelevant context should not affect judgment. Take transitivity again (if you prefer A to B and B to C, then you should prefer A to C). The principle contains as an implicit assumption that you should not contextualize the choices such that you call the "A" in the first comparison "A in a comparison involving B" and the "A" in the third comparison "A in a comparison involving C." Otherwise the rule would put no constraints on your behavior at all, and you could not be conceptualized as maximizing utility. The rule says you should not change the value of A depending upon what it is being compared with.

Other axioms of rational choice have the same implication—that choices should be appropriately decontextualized. Consider another axiom from the theory of utility maximization under conditions of risk, the so-called independence axiom (a different axiom from the independence of irrelevant alternatives, and sometimes termed substitutability, see Baron 1993a; Broome 1991; Luce and Raiffa 1957; Savage 1954; Shafer 1988; Slovic and Tversky 1974). The axiom states that if the outcome in some state of the world is the same across options, then that state of the world should be ignored. Again, the axiom dictates a particular way in which context should be ignored. And just like the independence of irrelevant alternatives example, humans sometimes violate it because their psychological states are affected by just the contextual feature that the axiom says should not be coded into their evaluation of the options. The famous Allais (1953) paradox provides one such example. Allais proposed the following two choice problems:

Problem 1. Choose between:
  A: One million dollars for sure
  B: .89 probability of one million dollars
      .10 probability of five million dollars
      .01 probability of nothing
Problem 2. Choose between:
  C: .11 probability of one million dollars
      .89 probability of nothing
  D: .10 probability of five million dollars
      .90 probability of nothing

Many people find option A in Problem 1 and option D in Problem 2 to be the most attractive, but these choices violate the independence axiom. To see this, we need to understand that .89 of the probability is the same in both sets of choices (Savage 1954). In both Problem 1 and Problem 2, in purely numerical terms, the subject is essentially faced with a choice between .11 probability of $1,000,000 versus .10 probability of $5,000,000 and .01 probability of nothing. If you chose A in Problem 1 the independence axiom dictates that you should choose C in Problem 2 (or you should choose B and D).

Many theorists have analyzed why individuals finding D attractive might nonetheless be drawn to option A in the first problem (Bell 1982; Loomes and Sugden 1982; Maher 1993; Schick 1987; Slovic and Tversky 1974), and many of their explanations involve the assumption that the individual incorporates psychological factors such as regret into their construal of the options. But the psychological state of regret derives from the part of the option that is constant and thus, according to the axiom, shouldn't be part of the context taken into account. For example, the zero-money outcome of option B might well be coded as something like "getting nothing when you passed up a sure chance of a million dollars!" The equivalent .01 slice of probability in option D is folded into the .90 and is not psychologically coded in the same way. Whether this contextualization based on regret is a justified contextualization has been the subject of intense debate (Broome 1991; Maher 1993; Schick 1984, 1987; Slovic and Tversky 1974; Tversky 1975). Unlike the case of "taking the last apple in the bowl when I am in public," in the Allais paradox it is less clear that the .01 segment of probability in the B option should be contextualized with the negative utility of an anticipated psychological state that derives from the consequences in an outcome that did not obtain.

My point here is not to settle the debate about the Allais paradox, which has remained deadlocked for decades. Instead the point is to highlight how humans recognize subtle contextual factors in decision problems that complicate their choices and perhaps contribute to their instability. Since decision theorists debate the rationality of reacting to these contextual factors, it is not difficult to imagine such a debate going on (either explicitly or without awareness) in the mind of the decision maker. Regardless of the outcome of the internal debate, it is nearly certain that such an internal struggle would introduce instability in responses. Such variability would no doubt raise the probability of producing a sequence of choices that violated one of the coherence constraints that define utility maximization under the axiomatic approach of Savage (1954) and von Neumann and Morgenstern (1944).

Note that an agent with a less subtle psychology might be less prone to be drawn into complex cogitation about conflicting psychological states. An agent impervious to regret might be more likely to treat the A vs. B and C vs. D choices in the Allais problem as structurally analogous. Such a psychologically impoverished agent would be more likely to adhere to the independence axiom and thus be judged as instrumentally rational.

My point in considering the principles of independence of irrelevant alternatives, transitivity, and independence is to highlight one common feature of these axioms: they all require the decision maker to abstract away aspects in the contextual environment of the options, as do the other rules of rational choice (see the discussion of descriptive invariance and procedural invariance discussed in chapter 4—as well as others not discussed here such as reduction of compound lotteries). It is this fact, combined with one other assumption, that explains why the finding that lower animals can fulfill the strictures of instrumental rationality better than humans is not only not paradoxical at all but is actually to be expected. Humans are the great social contextualizers (one of the fundamental computational biases discussed in chapter 4). We respond to subtle environmental cues in the contextual surround and are sensitive to social flux and nuance. All of this means that the contextual features humans code into options may lack stability both for good reasons (the social world is not stable) and bad reasons (the cues are too many and varying to be coded consistently each time).

In having more capacity for differential coding of contextual cues from occasion to occasion, humans create more opportunities for violation of any number of choice axioms, all of which require a consistent coding of the options from choice to choice. The more such contextual cues are coded, the more difficult it will be to be consistent from decision to decision. The very complexity of the information that humans seek to bring to bear on a decision is precisely the thing that renders difficult an adherence to the consistency requirements of the choice axioms. The situation with respect to contextual cues is analogous to the struggle of Ruth, described previously. The complexity of her hierarchical goal structure renders her first-order preferences more unstable than they would be if she did not reflect upon them at all. Or to put it another way, because she was concerned that she be broadly rational (not simply accepting her desires as given) she engaged in a critique of her first-order desires. Engaging in this second- (or higher) order critique might actually mean temporarily sacrificing some degree of instrumental rationality because of the desire to seek rational integration across all vertical levels of preference. Any instrumental loss caused by instability in the first-order preferences thus induced is the result of her attempt to engage in the broader cognitive program of critiquing lower-level desires by forming higher-level preferences. Uncle Ralph's instrumental rationality is not threatened by any such destabilizing program.

Escaping the Rationality of Constraint

In this and the previous sections I have identified three reasons why human choices might display less of the coherence and stability that define instru-

mental rationality as it is operationalized in axiomatic utility theory. I will
call them contextual complexity, the strong evaluator struggle, and symbolic
complexity, respectively. The first, contextual complexity, was the subject of
the last section. Due to coding more contextual features into their options,
humans risk exhibiting more trial-to-trial inconsistency (of the type that
leads to violations of rationality axioms) than less cognitively complex animals whose TASS systems respond more rigidly to their stimulus triggers.

The strong evaluator struggle is illustrated by Ruth's tendency to form
many second-order preferences because she holds many values (overcon-
sumption is bad; all of our actions impact the environment; in an intercon-
nected world all choices in the First World affect the destitute in the Third
World, etc.) that dictate a critical stance toward unreflective human responses. These many second-order preferences, on a purely statistical basis,
create many opportunities for first-order/second-order conflicts. These conflicts cause instability in the first-order preferences that lead to choice consistency violations.

Finally, symbolic complexity leads to problems in maintaining instrumental rationality in the same manner that contextual complexity does. To
the extent that options are evaluated partially in terms of symbolic utility,
then social context will importantly affect responses (Nozick 1993, for example, emphasizes the socially created nature of symbolic utility). Assuming that the social cues that determine symbolic utility will be complex and
variable, as in the manner of the contextual complexity hypothesis (indeed,
these two might be considered part of the same category), such variability
could well create inconsistency that disrupts the coherence relationships
that define instrumental rationality.

All three of these mechanisms—contextual complexity, the strong evaluator struggle, and symbolic complexity—lead human behavior to deviate
from instrumental rationality. Many authors have commented on how the
behavior of entities in very constrained situations (firms in competitive
markets, people living in subsistence-agriculture situations, animals in
predator-filled environments) are the entities whose behaviors fit the rational choice model the best (e.g., Clark 1997, 180–84; Denzau and North
1994; Satz and Ferejohn 1994). The harshly simplified environments of
these entities allow only for the valuing of instrumental rationality in the
most narrowly defined, self-interested, and basic-needs satisfaction way (or
else the entities perish). Additionally, these environments are all subject to
evolutionary or quasi-evolutionary (e.g., markets) selection processes. Only
entities who fit the narrow criteria of instrumental rationality are around for
us to study. Tiny charitable contributions aside, corporations are not no-

table for valuing symbolically (any corporation valuing symbolic acts more than the bottom line would be rapidly eliminated in the market); nor do they engage in a struggle between their desire for profits and some higher-order preference. Corporations, like animals in harsh environments, achieve what we might call the instrumental rationality of constraint. The Panglossian economists who tend to extol the rationality of humans tend to analyze situations where there is no choice; or, more specifically, they analyze situations set up to ruthlessly exploit those not making the instrumentally optimal choice.

Most humans do not now operate in such harsh selective environments of constraint (outside of many work environments that deliberately create constraints, such as markets). They use that freedom to pursue symbolic utility, thereby creating complex, context-dependent preferences that are more likely to violate the strictures of coherence that define instrumental rationality. But those violations do not make them inferior to the instrumentally rational pigeon. Degrees of rationality among entities pursuing goals of differing complexity are not comparable. One simply cannot count up the number of violations and declare the entity with fewer violations the more rational. The degree of instrumental rationality achieved must be contextualized according to the complexity of the goals pursued.

In addition to pursuing symbolic rationality, humans engage in the risky project of evaluating their desires by forming higher-order preferences and examining whether they rationally cohere with their first-order preferences. I say risky project because the potential lack of rational integration (conflicts between first- and higher-order preferences) thereby entailed puts instrumental rationality in jeopardy. Optimal levels of first-order desire satisfaction will be disrupted as long as this program of cognitive criticism and rational reintegration continues. This cost in instrumental rationality is the price humans pay for being a species, the only species, that cares about what it cares about (Frankfurt 1982). We are the only species that disrupts the coherence of its desires by destabilizing them through internal cognition directed at self-improvement and self-determination.

Two-Tiered Rationality Evaluation:
A Legacy of Human Cognitive Architecture

People aspire to rationality broadly conceived, not just instrumental rationality. People want their desires satisfied, but they are concerned about having the *right* desires. If those desires are TASS subsystems that serve ancient genetic goals better than current life goals, they need to be overridden by the analytic system pursuing considered, environmentally appropriate long-term goals. If those desires are TASS responses acquired in childhood or overpracticed rules not appropriate to the current situation,[20] then they again need to be overridden by analytic processes in the service of a meme-plex that has been thoughtfully examined. The full robot's rebellion is achieved by pursuing instrumental rationality in the context of a continuing critique of those desires being pursued.

Because humans aspire to rationality broadly rather than narrowly defined, a two-tiered evaluation of their rationality is necessary. As described in the last section, the instrumental rationality we achieve must be evaluated by taking into account the complexity of the goals being pursued and by analyzing the dynamics of the cognitive critique. Or, to put it another way, both thin and broad rationality need evaluation. The rules for examining instrumental rationality are well articulated. The criteria that should be applied when evaluating broad rationality are much more complex and contentious (see Nozick's 1993, discussion of twenty-three criteria for the evaluation of preferences) but would certainly include: the degree of strong evaluation undertaken; the degree to which a person finds lack of rational integration aversive (Nozick's 1993, principle IV) and is willing to take steps to rectify it; whether the individual can state a reason for all second-order desires (Nozick's 1993, principle VII); whether it is the case that a person's desires are not such that acting on them leads to irrational beliefs (Nozick's 1993, principle XIV); whether a person avoids forming desires that are impossible to fulfill (Nozick's 1993, principle X), and others (see Nozick 1993).[21]

Your life plans are the goal structures instantiated within your brain, which include TASS subsystems with genetically short-leashed goals, reflectively acquired memes that determine goal structure, and unreflectively acquired memes that do the same thing. Thus, in order to achieve two-tiered rationality, I have stressed: (1) the importance of selective TASS override by the analytic system; (2) the importance of reflectively acquired beliefs; and (3) the importance of reflectively acquired desires. We can thank a feature of our cognitive architecture for making the last two possible. In holding, contemplating, and evaluating second-order desires that conflict with first-order desires, we are cognitively dealing with a hypothesized mental state—one that is actually not true of us. We are able to represent a state of affairs that does not map into an actual, causally active, mental state of our own. We are able to mark a mental state as not factual. Many cognitive theorists (see note 11) have emphasized the critical importance (and specialness to human mentality) of being able to separate a belief or desire from its coupling to the world (to mark it as a hypothetical state). This is what the representational abilities of the analytic system (vastly augmented by the powerful tool of language) can accomplish. These representational abilities allow you so say to yourself "if I had a different set of desires, it would be preferable to the ones I have now," and they appear to be uniquely human.

These metarepresentational abilities make possible the higher-order evaluations that determine whether we are pursuing the right aims. They make it possible to add symbolic utility to our lives and actions. They provide the distancing from belief that is necessary for meme evaluation. The critiques of the memes we host and of our first-order desires made possible by these representational abilities make it possible to evaluate whether either genes or memes are sacrificing us as human vehicles.

The Spookiness of Subpersonal Entities

What constitutes evolutionary biology's difficulty as a subject . . . is something quite different from what makes difficulty in, say, a physics topic. The difficulty over these social and evolutionary questions is not the rigour of chains of logic or maths, nor complexities of geometry. . . . The problem is rather that of thinking the socially unthinkable.

—W. D. Hamilton, *Narrow Roads of Gene Land* (1996, 14)

The thought of these two entities (genes and memes) optimizing for themselves rather than their hosts (us) is something people find profoundly demoralizing and distressing. It is one reason that the so-called gene's-eye view of life (Barash 2001; Dawkins 1976; Dennett 1995), although implicit in the neo-Darwinian synthesis of the twentieth century for some time, has only been made explicit in the last two decades and has only entered into social discourse in the last decade.

But these distressing facts aren't ghosts that will go away if we don't look at them—they are concomitants of our representational abilities and the cultural achievement of science that have enabled humans to self-examine with a clarity unavailable to more simply constructed creatures. The clarity of human self-examination, for all its benefits, brings with it new and scary notions for humans to face—the spooky notions of the selfish gene and equally selfish meme. It is creepy to think about the notion that these subpersonal entities both construct and constitute our bodies and minds and that they are not necessarily only in it for us (or, to put it another way, are not optimizing *for* people but are optimizing *across* people). Here's how things get creepy:

Creepy Fact #1: There is no "I" in the brain who is aware of everything going on and who controls everything.

Discussions of the nonconscious mind well predate Freud. What modern cognitive science has done is to flesh out the details of how the brain processes of which we are unaware do their work. What this research has uncovered is that there is no single place in the brain that can be identified as the "I"—the seat of the soul. What we experience as the "I" is simply what it feels like to be the inside of a supervisory attentional system that is actually distributed throughout the brain and is attempting to optimally use and schedule the outputs of TASS systems whose operations yield no conscious experience (e.g., Norman and Shallice 1986; Shallice 1988). This interacts with:

Creepy Fact #2: Our brains were built by entities not *exclusively* concerned with instantiating goals that were good for us.

What is particularly disturbing is contemplating how these two Creepy Facts interact. The reflex-like operation of TASS means that there is an autonomous part of the brain that may be more tightly keyed to the ancient goals of the replicators than to the ongoing goals we have as people living in a complex modern world. Because of Creepy Fact #1, it could be said that TASS system outputs come "bubbling up" from below to offer the analytic system scheduling options. But the spooky part is that what comes bubbling up actually comes from subsystems designed not with the vehicle in mind but largely in the interests of replicator success.[22]

In chapter 1 I discussed the disturbing logic dictating how the only replicators that are around and currently building bodies are those that, throughout evolutionary history, sacrificed the vehicle's interests to the replicator's interests when the two came in conflict. Any nonselfish replicator who chose the vehicle over replication when the two were conflict is no longer around to tell the tale. Short-leashed TASS goals always contain this danger—the danger of a replicator/vehicle conflict in goals. Why this is scary is the mindless way that TASS goes about its business.

I discussed in chapter 2 how mindless creatures doing complex things (recall the Sphex from that chapter) are creepy because they tempt us to attribute all kinds of valued properties to them (intelligence, awareness, thought, etc.), but when we open them up we find nothing but the blind mechanical logic of an automaton—true robots doing what they do with no consciousness of what they are doing. Likewise, the mindless complexity of

TASS is unnerving, particularly since it resides within our own brains! (Recall Clark's [1997] use of the phrase "The Martian in John's brain" discussed in chapter 2.) It is spooky to think that there are things inside you—not just the temporary viruses that give you a cold, but deep in your brain—that are controlling your body and that: (a) you are not aware of them; (b) they might not be acting specifically in your interests; (c) the "I" that you think represents you isn't your whole brain nor does it have control of your whole brain; and that (d) this is because the purpose of your brain is not to serve the "I" (your brain was built instead to serve the subpersonal replicators).

As we have seen, the only escape hatch from some of these unnerving conclusions—the only way for a vehicle to rebel against the genes' using it as a survival vehicle—is for the analytic system to instantiate goals that serve the long-term interests of the vehicle and then use those goals to monitor the automatic outputs of the TASS system and override and redirect them accordingly by exercising supervisory attentional control. More simply put, the analytic system must exercise its capabilities to override TASS. But that leads us to:

Creepy Fact #3: There is another subpersonal replicator that constitutes the software that the analytic system must use to monitor TASS, and this subpersonal replicator, like the gene, may sometimes have interests that conflict with vehicle well-being.

Earlier in this chapter and in the previous one I discussed the Neurathian program of cognitive evaluation that must be undertaken in order to prevent adverse consequences arising from Creepy Fact #3. In this section, however, I want to draw attention to one common aspect of the creepy facts that cause us distress: We shrink from anything that does not have human consciousness at the center of the action.[23] We find it difficult to acknowledge the fundamental importance of anything that does not put humans (or more specifically human consciousness) center stage (probably because our folk psychologies have not yet accommodated a view of mind without a homunculus).

TASS firing in the absence of conscious control; genes working their will; memes caring only about replication whether or not they help us—all these disturb our sense of the rightness of a "humans first" view of the universe. And we are right to be concerned. If human well-being is to be thought of as one of the supreme values, then we are right to be concerned about subpersonal entities that are not optimizing at the single human level.

Modern culture can be viewed as an attempt to thwart the untoward effects of subpersonal entities when they are discovered to be antithetical to human interests. Genetic engineering was characterized as the ultimate robot's rebellion in chapter 1 because it represents—for the first time in the history of the planet—a vehicle using the replicator for the vehicle's purpose rather than reverse. Likewise, cultural institutions sometimes evolve to guard against selfish memes that might hurt their hosts. Some governments make pyramid sales schemes (a really bad parasite meme) illegal to protect their citizens from catching a meme virus that will damage their personal well-being through financial loss. This book represents a cultural product aimed at raising awareness of how subpersonal entities can undermine personal well-being.

Desires Connected to Dollars:
Another Case of Spooky Subpersonal Optimization

However, cultural evolution throws up new challenges all the time, and we may be approaching one now. We may be reaching a stage of social, economic, and technological development where a kind of meta-rationality is needed if our hard-won ability to value the vehicle over the ends of subpersonal entities is to be preserved.

Decision theorists and cognitive psychologists have studied extensively so-called Prisoner's Dilemmas and commons-dilemma situations (Colman 1995; Hargreaves Heap and Varoufakis 1995; Komorita and Parks 1994). Without going into detail, the essential features of these multiple-agent interaction situations are these. First, for each single agent, trying to maximize his own utility, there is one response that according to a principle of decision theory (usually the dominance principle) is the rational response (call it NR for narrowly rational). However, the game is interactive—payoffs depend on the responses of other individuals—and if every player makes the NR response, the payoff for all is low. The alternative response (C, for cooperative) is dominated in the technical sense (whatever the other agents do, you are worse off for having chosen C), but if everyone responds C everyone is much better off than had they chosen NR.

Littering has this logic. I gain a lot by driving through a far-away city and throwing an inconveniently full paper cup of beverage out the window of my car. Since I will never see the cup again, it will have no negative utility for me because it will never litter my landscape. In a very narrow sense it is rational for me to throw it out—it is an NR response. The problem is that it is narrowly rational for each driver to reason the same way. However, the result of everyone making the NR response is a totally trashed landscape that we all hate to view. Had we all sacrificed the small convenience of not throwing our cups (made the C response), we would all enjoy the immense benefit of an unlittered landscape. The C response is better in the collective sense, but notice the pernicious dominance of the NR response. If you all cooperated and didn't throw your cups, then if I throw mine, I get the benefit of an unlittered landscape but I also get the convenience of throwing my cup (I do better than had I responded C). If the rest of you all threw your cups, then I am better off throwing (NR) because had I not thrown, the landscape would have still been littered but I would have forgone the convenience of throwing my cup. The problem is that everyone sees the same dominance logic and hence everyone throws their cups and we are all less happy than we would have been if everyone had responded C.[24]

What Prisoner's Dilemmas and commons dilemmas show is that rationality must police itself. We must constantly ask ourselves the question of whether it is rational to be (narrowly) rational. It is possible that world history has entered a stage where rationality itself may be changing the environment in such a way as to present a series of very special types of Prisoner's Dilemmas that will require almost what might be termed meta-rationality—using rational judgment to control rationality itself. If we do not rise to this level, we might risk being demoralized by other types of sub-personal optimization that are in one sense a product of rationality.

Numerous social commentators have described the paradoxical malaise that has descended upon the affluent, successful majority in Western societies.[25] We seem to have a surfeit of goods which we certainly show no signs of wanting to give up, but we detect that the other aspects of our environment are deteriorating and we do not know what to do about it. Commuting times have doubled in many North American municipalities in just the last ten years; childhood has been transformed into one long consumption binge; small communities die out as young adults see that in order to have a good job one must live in a large urban conurbation; music and movies become coarser and coarser, and children are exposed to this at earlier ages; food and water poisoning incidents increase in frequency and scope; we wait in lines of automobiles as we attempt to visit sites of natural beauty; hard-working youth are marginalized in the economy because they lack the requisite educational qualifications; asthma and autoimmune diseases increase in frequency; obesity among young children is at an all-time high; our rural areas and small towns are defaced by proliferating "big box" stores and their ugly architecture; libraries that were open full-time thirty years ago, when we were less rich, now cut back their hours; smog alerts during the summer increase yearly in most North American cities—all while we enjoy a cornucopia of goods and services that is unprecedented in scope.

Several authors have written eloquently about the Prisoner's Dilemma logic that causes many of these problems and is the cause of the seeming paradox (e.g., Frank 1999; Frank and Cook 1995). Others have discussed

how the effects of concentrations of wealth on the democratic process distort social priorities and contribute to some of these problems (e.g., Greider
1992; Lasn 1999; Lindblom 2001). These types of analyses (not mutually
exclusive) are no doubt both partially correct, but it is not my purpose to recapitulate them here. Instead I wish to draw an analogy and to reveal the demoralizing logic of a process that—analogous to evolution—optimizes not
at the level of the individual person but across individuals. Surprisingly,
market societies have this logic at the level of personal desires—a logic that
may be contributing to our present malaise.

In this book, phrases like maximizing utility and optimality have been
used to discuss certain aspects of rationality, and similar phrases are used to
characterize the workings of economic markets when operating properly.
Markets are said to be efficient, to optimize, to satisfy wants, to maximize
the satisfaction of preferences, or finally and more colloquially, to "give
people what they want."[26] For example, the MIT Dictionary of Modern Economics, in defining the term optimum tells us that "much of economics is
concerned with analyzing how groups or individuals may achieve optimal
arrangements. . . . Generally, it is assumed that the satisfaction of individual
desires is the objective of the economic system" (Pearce 1992, 315–16). This
all seems pretty innocuous until we read further down in the definition and
find the following qualification: "In attempting to attain an optimum we are
usually constrained by the fundamental scarcity of goods and resources—
individuals are constrained by their income" (316).

What could this possibly mean that "optimal arrangements" in terms of
satisfaction of people's desires is "constrained by an individual's income"?
Yale University economist Charles Lindblom (2001) unpacks this a little for
us by admitting that the claim that the market system is efficient because it
responds to popular preferences is, to use his words, overstated. Instead, he
points out that "at best it responds only to such preferences as can be expressed by voluntary offers of skills and assets. Change the allocation of
skills and assets, and the preferences to which the market responds will
change correspondingly" (168). This hints at what the MIT Dictionary
means when it speaks of the constraint of income. The optimal allocation
that markets achieve is the satisfaction of desires that are backed by assets.

It takes a philosopher, in this case Elizabeth Anderson of the University
of Michigan, to say baldly what all the economic lingo obscures: "The market is a want-regarding institution. It responds to 'effective demand'—desires backed by the ability to pay for things" (1993, 146). This is what the
MIT Dictionary's genteel phrase "constrained by income" and Lindblom's
phrase "expressed by voluntary offers of assets" means. I suggest we say it

even more baldly. With apologies to my non-American readers, it makes the logic of the situation much clearer if we say that markets optimize the satisfaction of *desires connected to dollars*. This way of thinking about markets makes closer contact with the belief/desire analyses of cognition in philosophy and psychology and makes it easier to evaluate, from the standpoint of these disciplines, what is going on here and why it might be related to the malaise of modernity referred to above.

An individual desire is a subpersonal entity, and it is this that the market optimizes across people. The optimization that the *MIT Dictionary* talks about is not an optimization of *people's* satisfied desires. Thinking so represents a misunderstanding actively fostered by market advocates. The layperson would naturally take the repeated admonition in the popular press that "markets give people what they want" to mean that markets produce the maximum number of satisfied *people.* But we have just seen that this is what economic efficiency is *not* about. The phrase "people's desires" in the language of free markets means desires *aggregated across people.* And since the only desires that get satisfied in markets are those connected to dollars, we see that what markets actually optimize is a subpersonal quantity—desires connected to dollars aggregated across people (who of course differ vastly in the dollars they have available to connect with their desires).

The perverse consequences of subpersonal optimization are often not recognized because we are so used to thinking, when we make attempts to optimize allocations, in terms of allocations that satisfy people. Fairness considerations, for example, dictate a focus on the individual person as the unit of analysis. When the Halloweeners come to my door, I tell each to choose two candy bars. When coaching a children's soccer team, the coach tries to give all the players roughly equal individual instruction time. Both the coach and I feel that things have worked out pretty optimally if all the players received roughly equal instruction time and if all the Halloweeners received two candy bars. Just as in markets, there were constraints in both cases—the number of candy bars and the amount of time. Optimal allocation to the layperson in these situations is each person getting an equal chance to satisfy a desire (for candy and for coaching).

The natural inclination of most people when thinking about allocation issues is to think in terms of a whole person as the relevant unit in determining the efficacy of the outcome. The notion of market efficiency overturns this default assumption in a very radical way.[27] Utilitarian moral theory encounters well-known difficulties because of its analogous prescription of evaluating outcomes in terms of an aggregation across people.[28] The doctrine of the greatest good for the greatest number has trouble dealing with cases, both actual and imaginary, in which the separateness of persons seems to matter. Take Nozick's (1974, 41) description of a thought experiment involving a so-called "utility monster." The utility monster has the property of gaining enormous utility benefits from watching the suffering of others. When five people suffer and lose utility, the utility monster gains so much that his gain in utility is greater than the loss of the five. A utilitarian calculus which merely summed up the utility across people would dictate that the five be sacrificed to the utility monster because that would result in greater overall utility.

It would seem bizarre to feed the utility monster, because it is bizarre to aggregate human satisfaction across humans without recognizing the separateness of people. But this is just what market equilibriums accomplish. They optimize advantageous trades among desires connected to dollars without recognizing any differences between two desires within the same person and two desires each within a separate person. In a market economy some people end up with a status somewhat like the utility monster in Nozick's example—they have monstrous amounts of money connected to their desires and hence in a market system can make the world do their will.

Author Wallace Shawn, in his play *The Fever,* a continuous first-person narrative, captured this logic eloquently. He begins by describing how, on any given day in history, there is a fixed capacity in the world to do things— a fixed amount of desire satisfaction that will take place:

> At each particular moment I can see that the world has a certain very particular ability to produce the things that people need: there's a certain quantity of land that's ready to be farmed, a certain particular number of workers, a certain stock of machinery. . . . And each day's capacity seems somehow so small. It's fixed, determinate. Every part of it is fixed. And I can see all the days that have happened already, and on each one of them, a determinate number of people worked, and a determinate portion of all the earth's resources was drawn up and used, and a determinate little pile of goods was produced. So small: across the grid of infinite possibility, this finite capacity, distributed each day. (Shawn 1991, 63–65)

Of course, the fixed capacity of the Earth implies a fixed capacity to satisfy people's desires by doing different things every day. Shawn asks, in his monologue, how this is determined ("And of all the things that might have been done, which were the ones that actually happened?") given that there is an infinity of different combinations of things that could be done and hence an infinite number of sets of desires that could be satisfied on any given day. The answer is:

The holders of money determine what's done—they bid their money for the things they want, each one according to the amount they hold—and each bit of money determines some fraction of the day's activities, so those who have a little determine a little, and those who have a lot determine a lot, and those who have nothing determine nothing. And then the world obeys the instructions of the money to the extent of its capacity, and then it stops. It's done what it can. The day is over. Certain things happened. If money was bid for jewelry, there was silver that was bent onto the shape of a ring. If it was bid for opera, there were costumes that were sewn and chandeliers that were hung on invisible threads.

   And there's an amazing moment: each day, before the day starts, before the market opens, before the bidding begins, there's a moment of confusion. The money is silent, it hasn't yet spoken. Its decisions are withheld, poised, perched, ready. Everyone knows that the world will not do everything today: if food is produced for the hungry children, then certain operas will not be performed; if certain performances are in fact given, then the food won't be produced, and the children will die. (65)

With the telling image of the money speaking or the money remaining silent, Shawn has accomplished the critical reversal of figure and ground for us—the critical backgrounding of our tendency to see the individual human actor as center stage. But this tendency must be suppressed if we are to understand how the proportionality in markets actually works. What is proportionally determining what gets done in the world is not people but money. Because the world responds only to desires connected to dollars, it is wrong to say that in markets "*people* get what they want." It is instead, to use the anthropomorphic language that I used earlier in the context of speaking of genes and memes, the desires connected to dollars that get what they want. Putting it this way highlights the analogy with genetic optimization which, as was discussed in chapter 1, occurs across people—with no concern for equality of effects from person to person. Likewise, markets do not care whether the desires connected to dollars that it satisfies are within one person or spread out among many. How much each *person* gets desires satisfied is irrelevant to market optimization.

   People sense in an inchoate way, but do not know how to articulate (other than in hackneyed phrases that market advocates have succeeded in discrediting and marginalizing), their concerns that there is something about the wealth of our now numerous multi-billionaires (*Forbes* magazine lists 497 as of March 18, 2002; Kroll and Goldman 2002) that is distasteful. I suggest that what people are sensing but unable to articulate is that a multi-billionaire is analogous to a utility monster. As in Shawn's example, compared to the average person in the world (even compared to the average person in the United States!) his desires affect the world (because they are connected to dollars) in such godlike disproportion that it makes a mockery of our belief in equal human worth.

Whatever a multi-billionaire wants done in the world gets done, daily— today, tomorrow, throughout his natural life—to the tune of whatever multi-billion dollar amount the net worth is. And the insight Shawn's piece has triggered—better than any economic treatise—is that this is true whether the billionaire actually buys anything (in the sense that the person in the street views buying and selling) or not. Suppose, for example, that throughout tomorrow a billionaire owns $10 billion worth of United States Treasury Bills. The desire that the billionaire would be fulfilling there would be: I want to keep my money (and get a fair return on it—although we would presume that the return, at his level, would be less important than the keeping). The billionaire could affect the world greatly and immediately by, tomorrow, shifting his $10 billion to the government bonds of Bangladesh. By doing so he would actually even earn a higher return. However, he would incur somewhat greater risk to his money by doing so. From the fact that he does not do so, we would infer that he wants to keep his money more than he wants an extra return. The desire, which thus actually seems to be "I *really* want to keep my money," is better fulfilled by the U.S. Treasuries than the government bonds of Bangladesh.

Note that buying the Bangladesh notes would—just as does a utility monster's actions—create ripple effects across the world and affect many people. The excess demand for $10 billion of Bangladesh notes would inevitably lower the interest rate Bangladesh would need to offer on world markets, lower its debt load, and benefit a country of millions. But the bonds weren't bought, there was no excess demand, and nothing in Bangladesh changed. Thus does a monster in a market affect what gets done every day all over the world in the way of Shawn's scenario. The alienation that even affluent people feel in market economies is thus in part caused by another type of subpersonal optimization (like that of the gene and meme) that seems distasteful to us because it ignores the boundary between humans that is so important to most people. Situations like the Bangladesh example reveal the logic of extremely skewed distributions of desires connected to dollars, and this logic follows through regardless of the morals and personal propensities of the individuals holding the dollars, who may or may not be taking steps to ameliorate these effects. The inherent logic of the situation, not personal morality, is my focus here.

Optimizing goal fulfillment based on desires connected to dollars actually does two demoralizing things. I have been focusing on the first: it opti-

mizes for an entity other than the individual person. But the second demoralizing feature is equally important and interacts with the first. It is that
markets recognize no values that are not expressed in actual market choice
behavior. Hausman and McPherson (1994) discuss how most economic
theory specifically eschews interest in the program of cognitive reform that
I have outlined earlier in this chapter—the attempt to achieve rational integration by seeking consistency among first-order and higher-order preferences. Economic theory treats preferences as givens that are not subject to
theoretical inquiry (Hirschman 1986). How these preferences arose and
whatever values are or are not behind them are irrelevant to markets and
economic theory. As Hausman and McPherson (1994) note, "markets are
not a political forum in which one's *reasons* matter" (264).

People find this a profoundly demoralizing aspect of the markets that
now dominate modern life. As discussed above, people attach utility to the
symbolic, they have ethical second-order preferences, they take action for
expressive reasons, and they make meaning-based judgments. Yet none of
these mean a thing to a market unless they are cashed out in terms of an actual consumer choice. But we value things that we will never purchase—
public goods like clean air and parks for example. Additionally and often
forgotten are the many nonpublic goods that we value yet will never purchase and hence have no effect over. I value the idea that low-cost housing
should be provided for those with modest incomes. But, being affluent, I
will never purchase low-income housing myself and thus will never prod
the market by helping to create a demand for it. The market provides no
mechanism for me to express this value. The value has no impact because it
is not connected to the dollars that act to stimulate market changes.

Of course this is what governments and democracy are for. There, on a
one-person one-vote basis, we express our values by participating in the political process. (For reasons of brevity, I will very charitably bypass the obvious five-hundred-pound gorilla that market power acts to distort the per-
son-based political process as well; see Greider 1992; Johnson and Broder
1996; Lindblom 2001.) Democracies are one way that people express their
values in a manner that more explicitly recognizes the importance of people
by at least attempting to more equally weight their views. But nearly every
commentator on the social and economic trends of the last thirty years is in
agreement that the impact of governments has waned and the impact of
markets has strengthened.

Thus, our ability to express values not directly connected to our consumer preferences has waned precisely as we have progressed materially,
and this in part accounts for the malaise and alienation from life felt in the
early twenty-first century. Even people who lack any economic vocabulary

to express the changes that they feel sense, in the churning so-called "turbo-capitalism" (Luttwak 1999) of our time, that as the pace of life and quantity of acquisition ratchets up, something is being optimized, but whatever it is, it appears not to be the conditions of individual people and what they value. Something out there *does* seem to be working with ever greater efficiency— it just doesn't seem to be optimizing the right things.

As mentioned above, economics vehemently resists the notion that first-order desires are subject to critique. Since, as I have argued in this chapter, engaging in such a critique is one way that people find meaning in their lives and utilize their unique cognitive powers, it is no wonder that people find a world dominated by markets alienating and demoralizing. Economic analyses (the formal and highly quantitative analyses of optimality for example) are predicated on the assumption that humans are wantons. Such analyses are totally undermined by the notion that humans are strong evaluators, that they can prefer to have preferences other than those they have, that they value things they do not consume, or that some desires are, when fully considered, irrational.

Hirschman (1986, 145–47) discusses how economists resist the notion of second-order preferences and find especially distasteful (because it upsets their elegant models) the notion that first-order preferences might change due to a systematic attempt to achieve rational integration by reconciling first-order and higher-order preferences. In fact, Hirschman (1986) argues, economists actually prefer the assumption that first-order preference change is "inscrutable, often capricious" (146). This is because if preference change is as wanton as the pursuit of first-order desire satisfaction, then all is well with the economic assumptions of self-interest and maximization and the optimization analyses they support (a logic analogous to saying that if we assume Johnny had gotten an A instead of a C and a B instead of a D then, gosh isn't it great to think that Johnny would have been on the honor roll).

People feel differently than economists, however. It is a recognition that markets will not respond to values that are not translated into desires connected to dollars that led to the movement for ethical consumerism that gained strength in the 1990s. Products such as ethical mutual funds and fair-trade coffee gained popularity, and the concept of ethical production allowed people to have their values reflected in the marketplace. The ethical consumer movement represents an explicit attempt to insert into the market process the values whose absence from the process has led to the paradoxical feeling engendered by materialism that things are getting worse as, just as certainly, they are getting better. The paradox is resolved if we recognize that the markets that now dominate our lives are indeed powerful mechanisms for satisfying first-order desires connected to dollars but that they totally ignore a host of other desires and values—among them, values not connected to marketplace behavior and first-order desires that are not connected to dollars because the individual lacks financial assets.

One aspect of markets that interacts strongly with the previous discussion of strong evaluation is how markets bias the process of rational integration. It was emphasized in that discussion that in a Neurathian view of the process of rational integration, all levels of representation are equally open for critique. No priority should be given to higher-level representations of preferences, contrary to some earlier views in the philosophical literature. However, in direct contrast to the Neurathian stricture, markets radically privilege first-order desires. Second-order desires are not expressed in market behavior and are thus invisible to markets. As mentioned previously, economists view desires as tastes—unanalyzeable givens. They prefer to view changes in desires as capricious and unpredictable changes in tastes rather than the outcome of a process of rational deliberation and integration of higher-order judgments. Viewing humans in markets as pure wantons aids the formal quantitative modeling of human behavior, and thus economists are prone to concern themselves only with first-order desires.

It is of course highly debatable how widely applicable is the assumption of capricious preference change. However, there is a more important point to appreciate: markets act so as to make the assumption true. That is, markets act so as to encourage wanton behavior. In her analysis of value expression and economics, Anderson (1993) points out that economics "assumes that people adequately express their valuations of goods only through satisfying their unexamined preferences. . . . Markets are responses only to given wants, without evaluating the reasons people have for wanting the goods in question" (194).

There are a number of ways to illustrate how markets recognize only the wanton side of human behavior. One way is to recall Jim, the unwilling cheap-stuff addict in the example earlier in the chapter. Jim had a first-order preference: he preferred cheap stuff. But Jim was an unwilling addict because he also was characterized by a second-order preference: Jim prefers that he did not prefer cheap stuff so much. However, imagine that Don is another individual who also is characterized by preferring cheap stuff. But unlike Jim, Don is a wanton. He has never given a single thought to his preference. Both act, for now, on the first-order preference. Jim is a candidate for change in the future because he lacks rational integration of his desires and could well overturn his first-order preference at a later point in time. As a wanton, Don is much less likely to change his behavior. The psychological structures of Jim and Don are quite different.

The important point to realize here is that the market does not distinguish between Jim and Don. As long as Jim does not succeed in overturning his first-order preference to achieve rational integration, the market is indifferent to his second-order preference and to his psychological struggle to achieve rational integration. It is in fact even worse than that. Because Jim's first-order preference coincides with that of the majority of other consumers, the market adjusts daily (by economies of scale, by advertising, by using popular preferences as "loss leaders," and many other economic and sales mechanisms) to make it easier and easier for Jim's first-order preference to be satisfied, thus instantiating it as an even stronger habit and making it harder to overturn by higher-order ethical judgments.[29]

Another way that markets recognize only the wanton side of human behavior is by conspiring (tacitly of course) with another subpersonal entity. Many short-leashed genetic goals are instantiated in TASS. These goals are nearly universal and thus they create many desires connected to dollars. Not surprisingly, the market has reacted to these genetically created interests. Economies of scale make fulfilling these desires cheap and easy. Indeed, their near universal presence makes priming them (in order to jump in and fulfill them at a profit) irresistible for a market. University of Virginia economist Steven Rhoads (1985) points out that it is inevitable that corporations will advertise barbecue grills and *Dallas* (he was writing in the 1980s) more than history books and Shakespeare, and this is "not because they are inherently biased against high culture, but because more persuasion, and thus more money, is needed to induce people to buy the latter" (158). Short-leashed genetic goals create many easily primed desires connected to dollars, and the inexorable logic of markets will make the fulfilling of these goals cheap. The cheapness of fulfilling these goals then becomes part of a positive feedback loop—other things being equal, people prefer more cheaply fulfilled desires because they leave more money for fulfilling other desires.

Everyone has short-leashed genetic goals, whereas only a few people are strong evaluators who have second-order desires. Of course, people can express more considered, higher-order preferences through markets too (e.g., free range eggs, fair-trade coffee) but they are statistically much smaller, harder to trigger via advertising, and lack economies of scale. The positive feedback loop surrounding unconsidered, TASS-based desires can even affect people's second-order judgments—"Well if everyone is doing it, it must not be so bad after all." Many symbolic and ethical choices must be developed in opposition to short-leashed TASS goals that markets are adapted to fulfilling efficiently. Such symbolic and ethical preferences are much less universal and are not even continuously active in those that have them (much of the time, even strong evaluators are wantons—unless second-order judgments are practiced enough to become part of TASS). They are not as widespread, so the market does not make them cheap desires to have.

The positive feedback loop that makes TASS-based desires cheap is related to the discussion of so-called adaptive preferences (we might alternatively call them convenient preferences) by Elster (1983). Adaptive preferences are those that are easy to fulfill in the particular environment in which the agent lives. The market automatically turns widespread, easily satisfied desires into adaptive preferences. If you like fast-food, television sitcoms, video games, recreating in automobiles, violent movies, and alcohol, the market makes it quite easy to get the things you want at a very reasonable cost because these are convenient preferences to have. If you like looking at original paintings, theater, walking in a pristine wood, French films, and fat-free food, you can certainly satisfy these preferences if you are sufficiently affluent, but it will be vastly more difficult and costly than in the previous case. So preferences differ in adaptiveness, or convenience, and markets accentuate the convenience of satisfying uncritiqued first-order preferences. Short-leashed TASS preferences that are not reflected upon will invariably be convenient because they are universal, and since only a minority will overturn them via a second-order evaluation they will be widely held and hence cheap to satisfy. Thus does the market make difficult the critique of first-order desires that, it was argued earlier in this chapter, is part of the essence of our personhood. Neither your genes nor the market care about your symbolic utility or your second-order desires.[30]

The Need for Meta-Rationality

> The role of values is extensive in human behavior, and to deny
> this would amount not only to a departure from the tradition of
> democratic thought, but also to the limiting of rationality. It is the
> power of reason that allows us to consider our obligations and ideals as
> well as our interests and advantages. To deny this freedom of thought
> would amount to a severe constraint on the reach of our rationality.
> —Amartya Sen, *Development as Freedom* (1999, 272)

From the preceding discussion, it seems as if the dominance of markets in our lives may have already begun to become a threat to our broad rationality. There is a deep irony in this, because Western market-based societies are the embodiment of the instrumental rationality that is a defining characteristic of modernism. Earlier in this chapter, I discussed the rationality of constraint—how entities whose behaviors fit the rational choice model the best often exist in highly constrained environments with strong selection pressures. Firms show more instrumental rationality than individuals in part because in the former case there is an awareness of possibilities of goal conflicts with subentities (workers), whereas in the human case the idea of goal conflicts with subentities (TASS) has only recently been the subject of attention.

The mismatch in the degree of rationality at the level of individual and corporation can of course be exploited by the latter. Such exploitation is well known and is responsible for much of the effectiveness of advertising, for example. Markets move quickly to exploit human irrationality for profit (e.g., the interest rates on credit card debt). Of course, as market exploitation of the irrationality of individuals has become more a part of the general environment, it becomes a behavior-shaping force.

Nozick (1993) discusses how such institutional rationality differentially rewards behavioral traits in ways that permanently change society. Discussing Max Weber's (1968) work on the social shift from gemeinschaft (community and personal relationships) to gesellschaft (formal regulations and institutions), Nozick stresses how institutionally embodied principles of instrumental rationality have allowed the memeplex of rationality to extend its domain. In the form of institutions that have as explicit goals the extension of their scope (e.g., corporations), a narrow view of instrumental rationality

> is proceeding now to remake the world to suit itself, altering not only its own environment but also that in which all other traits find themselves, extending the environment in which only it can fully flourish. In that environment, the marginal product of rationality increases, that of other traits diminishes; traits that once were of coordinate importance are placed in an inferior position. This presents a challenge to rationality's compassion and to its imagination and ingenuity: can it devise a system in which those with other traits can live comfortably and flourish. (Nozick 1993, 180)

Interestingly and ironically, among those other traits are those that reflect concerns about broad rationality—those that critique first-order desires by invoking values, meaning, and ethical preferences. The power of instrumentally oriented institutions to shape the world (including individuals) in their own image actually poses a threat to our ability to attain an even higher form of rationality—one that makes full use of the representational powers of humans to reshape their own values and goals.

I do not wish to be misunderstood here. A major theme of this book is that rationality (and its embodiment in institutions) provides a means of creating conditions that optimize at the level of people rather than the genes—the beginning of the robot's rebellion. However, another set of self-insights that humans have even more recently achieved (the implications of the second replicator—the meme) immensely complicates the picture. The goals to be identified with the person—those that serve to define the success of a vehicle-optimizing process of instrumental rationality—should not just be taken as given, or else we again simply give ourselves up to replicator interests. They need to be critiqued by a Neurathian process.

It is the latter program of cognitive reform that is threatened by the current dominance of institutions that seemingly force us into a concern for instrumental rationality in the narrowest sense. As I have alluded in the title of this section, I feel that we will need a form of meta-rationality to avoid this outcome. Rationality will have to critique itself and, importantly, attempt to figure out when the powerful program of optimizing instrumental rationality should be allowed to work its will and, conversely, when it threatens to confine us by short-circuiting our attempts at goal evaluation and should be tempered accordingly.

The necessity of this kind of recursive examination of rationality is hinted at in some of the phrasing in the quotation from Nozick (1993) given above. Note the following wordings: "is proceeding now to remake the world to suit itself," "altering not only its own environment," "extending the environment in which only it can fully flourish," "this presents a challenge to rationality's compassion," "can it devise a system." Without using the term, Nozick (1993) is treating instrumental rationality as a memeplex—a matrix of interlocking ideas with a life of its own. As this memeplex has co-evolved with the social and institutional memeplex "market capitalism," it cannot itself be expected to engage in a self-critical project. Instead, what it can be expected to do is to replicate. It has been selected for its replicative success in the memosphere, not for reining itself in when it encounters other memeplexes that the host values. Instead, it is the *host* that must engage in a Neurathian project that uses parts of rationality to critique rationality itself.

Not only is it the host who must engage in the process of cognitive reform (not the memeplex), but the memeplex, if it needs to be restrained (that is, if instrumental concerns are to be trimmed back temporarily so that rational integration of desires can take place), cannot be expected to co-operate. We use this memeplex for our ends, yes; but it is not *there* for our ends. As illustrated in the last chapter, this co-evolved memeplex can be expected to protect itself—indeed to inoculate itself against reflective examination once it has been installed in a host brain. For example, Anderson (1993) has argued that people do not find a view of themselves as instrumental maximizers in a market society reflectively endorsable, yet they cannot seem to escape defaulting to just this image when thinking and behaving nonreflectively. Anderson speculates that "this is due to the special salience markets have in our lives, as well as to the ideologies that promote their nearly unlimited expansion. . . . Markets keep a crucial evaluative question outside their decision frame: whether it makes sense to govern our conduct with respect to a particular good by market norms at all" (219).

Of course, as discussed in chapter 7, many successful memeplexes include strategies for precluding their own evaluation. That this strategy is operating to protect the co-evolved memeplexes of narrow instrumental rationality pursued through markets is perhaps indicated by the hostility this memeplex displays toward any program of broad rational integration (critique of first-order desires) that humans pursue. For example, the movements toward ethical consumerism, fair-trade products, and ecologically sound production are regularly excoriated by the *Wall Street Journal*, business advocates, trade representatives, and market boosters. On the surface this seems puzzling. After all, ecologically sound production is still production, fair trade is still trade, and ethical consumerism is still consumerism. One suspects that the memeplex guards against a slippery slope that might erode certain assumptions built into it that make it easier for the memeplex to expand its domain. For example, these reforms complicate markets by undermining the assumption that all relevant information is summarized in a commodity's price. They also challenge the assumption that the only way to act in markets is by seeking in the most efficient way to fulfill one's unexamined first-order desires—that desires are simply tastes to be treated as givens. All of these reforms would change the memeplex by making it much more complex, less adaptable to all contexts, and thus perhaps would hurt its powers of replication. Perhaps this is the explanation for the resistance which, on the surface, seems more vehement than these market-preserving reforms should warrant.

The Formula for Personal Autonomy
in the Face of Many Subpersonal Threats

> The ethical progress of society depends, not on imitating the cosmic
> process, still less in running away from it, but in combating it.
> —T. H. Huxley, *Evolution and Ethics* ([1894] 1989, 141)

The full program of Neurathian self-examination that is necessary for personal autonomy, as I view it, now lays revealed. Instrumental rationality is a necessary but not sufficient requirement for full personal autonomy. Yes, we want to be instrumentally rational, but as we have seen, even bees can achieve that. Broader programs of cognitive reform are necessary if humans are to achieve an autonomous rationality superior to that which is achievable by nonhuman animals. For example, in addition to pursuing instrumental rationality, it is essential that humans also pursue a program of improving epistemic rationality. If we rationally pursue our ends from the wrong premises about the way the world is, we will fall short of satisfying our desires. The memes that are our beliefs are themselves replicators with their own interests and thus must be evaluated.

Likewise with our desires. First, we must be careful not to identify our interests with the first-order desires lodged in TASS. These may be weighted in favor of ancient genetic interests and may not be serving our present long-term goals. But the long-term plans utilized by our analytic systems likewise must be evaluated because, again, many of these are memes.

Despite this danger, it is essential to engage in a program of strong evaluation—a program of evaluating first-order desires. When mismatches between our first-order preferences and second-order evaluations occur, we must engage in a Neurathian program of rational integration. Like all Neurathian programs, it is risky—we may start out standing on a rotten plank.

All of this is quite far from the rationality of the bee, or that of the chimpanzee for that matter. It is rightly called a program of meta-rationality, where rationality is used to evaluate itself and to evaluate institutions in which the cultural products of rational evolution are also embedded. The creativity and openendedness involved in achieving full human rational integration—in escaping the Humean nexus—has perhaps been underestimated. The task seems daunting. Are we up to it?

Are We Up to the Task?
Finding What to Value in Our Mental Lives

Based on the arguments of the previous section, it may be that in terms of the cultural evolution of human rationality, the narrow instrumental rationality of market societies is perhaps a local maxima in the evolution of ideas. Is there reason to believe that humans are up to the task of pursuing a broader concept of rationality? I would argue that we are in fact up to the task. The reason for my optimism about our broader evaluative abilities stems from the intensity of our drive to deploy our representational abilities.

The human desire to use our representational powers to engage in second-order evaluation is strong. It is apparent in a telling and poignant vignette that author Jonathan Franzen (2001) relates about his father. His family took his father, who had suffered from Alzheimer's disease for some time, from his nursing home to their house for Thanksgiving dinner. By that time in the progression of his father's dementia, Franzen tells us that "a change in venue no more impressed my father than it does a one-year-old" (89). He describes the sadness of the pictures taken of his father slumped in his chair and his father's silences and shrugs in response to his wife. Franzen's father did manage to thank his other sons for their telephone calls, but that was about it. But after dinner, and in front of the nursing home as Franzen brought his father back to it, a remarkable thing happened. Franzen's wife ran into the nursing home to get a wheelchair, and Franzen and his father sat looking at the entrance to the home, when his father suddenly said: "Better not to leave than to have to come back."

Franzen was stunned at this indication that his father, at least in that moment, indicated an awareness of his situation and his evaluation of it. Franzen says "I'm struck, above all, by the apparent persistence of his *will*. I'm powerless not to believe that he was exerting some bodily remnant of his self-discipline, some reserve of strength in the sinews beneath both consciousness and memory, when he pulled himself together for the statement he made to me outside the nursing home. I'm powerless as well not to believe that his crash on the following morning, like his crash in his first night alone in a hospital, amounted to a relinquishment of that will" (89). I would argue that Franzen is exactly right here; that there is no discontinuity at all between his view—emotionally immersed in this poignant moment—and the most detached and cold-hearted view from the standpoint of cognitive science. What does the latter say about the incident? First of all, as a victim of Alzheimer's, Franzen's father is suffering the indignity caused by the mindless and uncaring logic of the replicators called genes—they care not a whit for the survival machine once its reproductive prospects are over. They are happy for the survival machine to fall apart in whatever way it will, once the object of their concern, replication, has long since been determined. Our creators value not at all the thing we value most—our sense of self. As Franzen and other commentators (e.g., Shenk 2001) have argued, the particular horror of Alzheimer's is that the self dies long before the physical body.

Brain deteriorization as Alzheimer's progresses is the reverse of a child's development—the last cognitive abilities acquired are the first to be lost. In advanced stages, the Alzheimer's patient becomes a wanton—like a baby, re-

sponding only to immediate desires. Franzen's shock at this incident occurred because he had become used to his father's status as a wanton—
someone who did not make a second-order evaluation of his situation.
However, here is where Franzen's poignant description and analysis of this
incident gets it just right. His father *was* exercising his so-called will, if we
view that old fashioned word as indicating the cognitive effort involved in
running the serial virtual machine on our brain's parallel hardware. And
what his father did immediately upon getting his analytic processor running—the last time he ever did so (further severe dementia and death followed this incident)—was to use it to make a strong evaluation.

Admittedly taking some license with this example, and exercising some
charity in describing it, we might say that although he had enjoyed his
Thanksgiving dinner, and no doubt wanted to attend the festivities, Franzen's father wished that he preferred not to leave the nursing home. With
the last bit of cognitive power at his disposal—the cognitive power necessary to sustain the representational abilities that make second-order
judgment possible—Franzen's father fought to be something other than a
wanton. TASS runs on in the Alzheimer's patient but, contrary to what
champions of the "gut instincts" think, it does not sustain the representational abilities that are most central to the self and that make possible the
rebellion against the replicators' interests that is human culture.

With his last bit of cognitive will, Franzen's father rebelled against his
uncaring replicator creators. With this last cognitive exertion he asserted the
human agent's right to judge the world from his own perspective. Franzen's
father carried this rebellion to the last, failing, limits of his cognitive power.
Replicators do not judge the world—they just are. Only we do that—and
the anecdote about Franzen's father demonstrates how tenacious is our
drive to do so.

I think there might be implications in this anecdote for the ongoing cultural conversations about where to look for human significance in a world
without God and in which we are beginning to face the stark implications
of the Darwinian insight. When seeking the source of human uniqueness,
the popular science writer's favorite candidate is consciousness. However,
the status of the concept of consciousness in cognitive science is a mess, to
say the least. There are influential schools of thought in cognitive science
which show that, basically, we simply do not know what we are talking
about when we talk of consciousness in our natural language (Churchland
and Churchland 1998; Dennett 1991, 2001; Wilkes 1988). Much specula-
tion in neuroscience about consciousness still hides background assumptions of a dualistic nature that would be embarrassing if brought out into

the open. Homunculi that are way too smart lurk in the wings of many theories of consciousness. We have yet to find a coherent way of talking about the concept of consciousness (if there is one) in the context of current theories of distributed processing, modularity, and cognitive control.

Of course there are many aspects of consciousness discussed in the literature.[31] On some, there has been, by at least modest scientific agreement, some progress in understanding. Components of the multifarious concept of consciousness that are interpreted as aspects of selective attention, of awareness, of executive control, and of consciousness of memory access have yielded to varying degrees of scientific understanding. But oddly, these are not the aspects of consciousness that commentators focus on as the features of cognition that define human uniqueness. Puzzlingly, commentators tend to focus on the feature that philosophers call qualia—the so-called "raw feels" or interior experience of consciousness, what it "feels like from the inside," so to speak (see Dennett 1988, for an extremely skeptical stance toward qualia). I say puzzling because this is precisely the aspect of consciousness on which there is no scientific or philosophical agreement at all. Some scholars feel that they can "prove" the inherent unexplainability of qualia—others that the former group has played a semantic trick by setting up the problem in such a way that it cannot be solved (for representations of both positions, see Chalmers 1996; Dennett 1988, 1991, 2001; McGinn 1999; Shear 1998).

Flanagan (1992) has coined the term mysterians for those who feel that this aspect of consciousness (qualia) will resist scientific explanation. Puzzlingly to myself and others (e.g., Churchland 1995; Dennett 1995), the mysterians do not—as you perhaps would expect scientists and philosophers to do—sadly lament something that will elude explanation by human reason. Instead, they appear to gleefully accept the conclusion, suggesting to some a hidden agenda in their position (see Dennett 1995, on cranes and skyhooks). Additionally, the very mysterians who wish to place qualia outside the scope of science tend to extol it as an important source of human uniqueness and significance (they feel that qualia are defining in some way, yet that science can tell us nothing about them). They represent a tendency, common throughout history, for people to seek meaning in mystery (see Dennett 1995; Raymo 1999).

None of this controversy and confusion would be so disturbing if it were not the case that we have defaulted so often to calling consciousness the defining feature of humanness. What various authors are in effect saying in their paeans to consciousness is that the thing that makes humans unique is something which we are unsure even that we can talk about correctly.

My intention is not to resolve these disputes about qualia here, but simply to redirect our attention to other important aspects of human cognition. Whether or not these "what it feels like to be" properties of cognition submit to scientific explanation, we might take the lesson of Nozick's (1974) experience machine and consider that there may be more to life than interior experience.

I want to suggest here that the broad conceptions of rationality that I have explored in the final chapters of this book provide useful benchmarks of a meaningful cognitive life. I am not suggesting that broad rationality is the only marker of meaning or significance—only that consciousness is not the only indicator of human uniqueness; and, in fact, it may not be the best. In any case, the situation is not a zero-sum game—there may be many markers of human uniqueness, and evidence for one does not negate the importance of the others. Exploring other concepts as markers of meaning in life does not detract from the importance of ongoing studies of consciousness.

Human values often play out in the form of critiques of our first-order preferences. Thus, the struggle to achieve consistency between our first-order preferences and higher-order preferences is a unique feature of human cognition. It separates us from other animals much more discretely than any other feature of mentality, including phenomenal consciousness which is much more likely to be distributed in continuous gradation among brains of various complexities across the animal kingdom.

Broad rationality is something that we at least can talk coherently about. I am not minimizing the complexities surrounding models of both narrow and broad rationality, but I am suggesting that it might be better to focus on a defining feature of humans that at least seems to be amenable to scientific investigation. The cognitive mechanisms that support the pursuit of narrow instrumental rationality are the subject of many progressive research programs.[32] Philosophically coherent discussions of broad rationality hold at least the hope of progress toward a notion of what makes a valued life for a human being.[33] The cognitive architecture that makes strong evaluation—self-evaluation of mental states—possible is the subject of progressive debate and experimentation.[34] In short, cognitive science has begun to uncover the internal logic of an agent that values its own evaluative autonomy. The uniqueness of the mechanisms that make this possible might be a useful focus when searching for what might be a scientifically plausible concept of self in the age of Darwin and neuroscience. This seems like a more self-respecting direction to look than the slightly insulting tendency to hide behind the so-called mystery of consciousness (qualia). Jonathan Franzen's father might have had better conscious experiences as a wanton, but clearly those around him would have preferred that he maintain the ability to sustain second-order evaluations. Again, in a version of Nozick's experience machine thought experiment, would any of us prefer to turn ourselves into a wanton (give up the ability to form second-order evaluations) if we were guaranteed pleasurable conscious experiences as a wanton?

It is at least moderately uncontentious that our internal experiences result from cognitive activities inside the massively complex causal nexus of the brain. Our experiences result from a massive brain system monitoring itself, responding to sensory activity, exercising supervisory attention, representing at a high level of abstraction, and carrying out other important cognitive activities. Maybe we need to refocus on the activities themselves rather than the experience of these activities. Perhaps the importance that we desire to assign to human mental life should be assigned to those activities themselves rather than to the internal experiences that go with them. Another way to phrase it is that maybe there are things that are important to do other than to experience the activities. Perhaps as well it is important to *do* them (again, recall your response to the experience machine).

Perhaps we have overvalued consciousness in the same way educators have misperceived the importance of self-esteem in the learning process. An extremely popular hypothesis in the 1990s was that school achievement problems were the result of low self-esteem. It turns out that the relationship between self-esteem and school achievement is more likely to be in the opposite direction from that assumed by school personnel. It is superior accomplishment in school (and in other aspects of life) that leads to high self-esteem and not the reverse.[35] Self-esteem simply rides along with the productive activities in which the individual is engaged. At least in part, consciousness rides along in an analogous way. Important and valued cognitive activities—both perceptual and representational—are likely to cause conscious states. Of course, to attribute all of the causal power and value in the activities to the experiential aspect of consciousness is just the homunculus error again in another guise.

The value we attach to cognitive activities should not be solely attributed to experiential correlates. Part of the value rightfully belongs to the activities themselves, especially activities such as strong evaluation of first-order desires and the self-critical evaluation of memes. Perhaps we might benefit from reorienting our search for meaning away from consciousness and internal feels (qualia) and toward the evaluative activities that make us autonomous and unique agents in the world. We create meaning when we work to make second-order evaluations; work to achieve rational integration in our preference hierarchies; attempt to achieve consistency among our first-order preferences; are alert to symbolic meaning in our lives; value ourselves as a vehicle and do not let genetic proclivities in TASS sacrifice our interests in a changing technological environment. All of these activities define what is really singular about humans: that they gain control of their lives in a way unique among lifeforms on Earth—by rational self-determination.

Chapter One

1. To be precise, what Dawkins (1976) meant by the use of this term was "the potential near-immortality of genes in the form of copies" (34).

2. Dawkins (1976) of course did not invent the concepts discussed in *The Selfish Gene*, but instead synthesized the work of Hamilton (1964), Maynard Smith (1974, 1976), Trivers (1971, 1974), Williams (1966), and many others—work which is now part of mainstream evolutionary science. Dawkins's synthesis emphasized the relatively recent insight sometimes termed the "gene's-eye view"—that to fully understand evolution's effects we must focus not only on the individual organism or the species but additionally on the most fundamental unit, the gene, and on the mechanisms by which it can increase its reproductive success relative to its competing alleles. Hamilton (1964), Williams (1966), and Dawkins (1976) have taught us that those mechanisms are essentially threefold—as a gene you might directly make copies of yourself; you might help to build an organism (the so-called vehicle, see next section) and then help it to survive and reproduce the genes that helped build it; or you might cause the organism you build to help other organisms that contain copies of yourself (see Badcock 2000; Barrett, Dunbar, and Lycett 2002; and Buss 1999, for introductory discussions). For readable discussions that include the gene's-eye view, but emphasize multilevel views of natural selection, see Sober and Wilson 1998; and Sterelny and Griffiths 1999.

3. Radcliffe Richards (2000) notes that "the most important current debates are not between pro- and anti-Darwinians, but among people who have crossed the Darwinian threshold but disagree about how far beyond it Darwinian explanations can take them" (26).

4. At the extremely gross level at which I am using the concepts, nothing in my argument depends on the outcome of any specific controversy in biological science. As Dennett (1995) notes, "there are vigorous controversies swirling around in evolutionary theory, but those who feel threatened by Darwinism should not take heart from this fact. Most—if not quite all—of the controversies concern issues that are 'just science'; no matter which side wins, the outcome will not undo the basic Darwinian idea. That idea, which is about as secure as any in science, really does have far-reaching implications for our vision of what the meaning of life is or could be" (19). Likewise, none of the primary conclusions discussed in the present volume are at all influenced by the outcome of the many micro-controversies in evolutionary biology. For other readable

summaries, see Cairns-Smith 1996; Eigen 1992; Maynard Smith and Szathmáry 1999; Ridley 2000; and Szathmáry 1999. For discussions of the issues that are less "gene-centric" and that stress multilevel selection theory, see Gould 2002; Sober and Wilson 1998; and Sterelny and Griffiths 1999.

5. It should be understood that anthropomorphic descriptions of replicator activity are merely a shorthand that is commonly used in biological writings. So for example the statement here referring to "replicators developed protective coatings of protein to ward off attacks" could be more awkwardly stated as "replicators that built vehicles with coatings became more frequent in the population." I will continue the practice here of using the metaphorical language about replicators and genes having "goals" or "interests" in confidence that the reader understands that this is a shorthand only. As Blackmore (1999) notes, "the shorthand 'genes want X' can always be spelled out as 'genes that do X are more likely to be passed on'" (5) but that, in making complicated arguments, the latter language becomes cumbersome. Thus, I will follow Dawkins (1976) in "allowing ourselves the licence of talking about genes as if they had conscious aims, always reassuring ourselves that we could translate our sloppy language back into respectable terms if we wanted to" (88). Dawkins points out that this is "harmless unless it happens to fall into the hands of those ill-equiped to understand it" (278) and then proceeds to quote a philosopher smugly and pedantically admonishing biologists that genes can't be selfish any more than atoms can be jealous. I trust, Dawkins's philosopher to the contrary, that no reader needs this pointed out.

6. In subsequent publications, Dawkins (see especially 1982, 15–19) has clarified the misunderstandings of his writings engendered by the use of the term robot. He of course uses the term in its technical incarnation from cybernetics—not the folk usage common among the public. It is now more well-known than it was in the year *The Selfish Gene* was written (1976) that "robotic" in cybernetics does not connote inflexibility. To the contrary, much high-level computer programming is precisely about how to equip robots and artificially intelligent entities with the ability to respond flexibly to changing environmental input. Furthermore, it is now more commonly recognized that a modern robot is not controlled by its programmer on a moment-by-moment basis as it traverses its environment, but that the programming must be completed *before* the robot begins to perform in its environmental setting. The analogy here is that the genes too must code for the production of a vehicle without knowing exactly what it will encounter in its environment. We will see in this section that this property (lack of foresight in the genes themselves) can prompt the creation of vehicles that, in effect, slip the genetic leash.

7. See Hull (1982, 1988, 2000, 2001) and others on this same theme, e.g., Plotkin 1994; Sterelny 2001b; Sterelny and Griffiths 1999; Williams 1985.

8. Dawkins (1976), Dennett (1984), and Plotkin (1988) all discuss the Mars explorer logic.

9. Dennett, in his short but provocative book *Kinds of Minds* (1996; see also, Dennett 1975, 1991, 1995), describes the overlapping short-leashed and long-leashed strategies embodied in our brains by labeling them as different "minds"—all lodged within the same brain in the case of humans, and all simultaneously operating to solve problems. The minds reflect increasingly powerful mechanisms for predicting the future world. The minds also reflect decreasing degrees of direct genetic control. The Darwinian mind uses prewired reflexes and thus produces hardwired phenotypic behavioral patterns (the genes, when they built such a mind, have "said" metaphorically "do *this* when x happens because it is best"). The Skinnerian mind uses operant conditioning to shape itself to an unpredictable environment (the genes have "said" metaphorically

"learn what is best as you go along"). The Popperian mind (after the philosopher Karl Popper) can represent possibilities and test them internally before responding (the genes have "said" metaphorically "think about what is best before you do it"). The Gregorian mind (after the psychologist Richard Gregory) exploits the mental tools (see Clark 1997) discovered by others to aid in the pretesting of responses (the genes have "said" metaphorically "imitate and use the mental tools used by others to solve problems"). Godfrey-Smith (1996) and Dennett (1975) discuss the circumstances under which it benefits the genes to construct a vehicle with a flexible intelligence and, alternatively, when a more rigid response pre-programming is efficacious (see also, Papineau 2001; Pollock 1995; Sterelny 2001b).

10. The term Darwinian mind is from Dennett's (1991, 1995, 1996) Tower of Intellect model.

11. Although it might seem strange, for a creature as simple as a bee, to talk of "interests" separate from strictly genetic goals, Sen (2002) has argued that we do, in fact, make quality of life judgments for nonhuman animals and that a certain degree of anthropocentrism is inevitable in such judgments. The mildest form of such anthropocentrism is represented in the judgment that an animal is better off alive than dead.

12. I will risk the possibly irritating reiteration of the point that this language is of course metaphorical. The language of intentionality is used to describe gene effects because it is a convenient shorthand for convoluted phrases such as "a gene increasing the frequency of behavior x will increase in frequency over its allelic rival that does not act to increase the frequency of behavior x." No biologist who uses the language of intentionality means to imply that genes consciously calculate. Instead, they are simply saying that the gene pool becomes filled with genes that influence bodies in such a way that they behave as if they had made such calculations. Again, no conscious calculation is implied here. The notion of genes having "interests" is no more strange than speaking of the interests of Ford Motor Company or the interests of the city of Minneapolis. Genes, like these examples, are distributed groupings of entities and the collectivity is acting not consciously but "as if" it were trying to maximize some quantity. Thus, "the biologist's convention of personifying the unconscious actions of natural selection is taken for granted here" (Dawkins 1993, 22–23). Writer Robert Wright (1994, 26) argues that such anthropomorphic metaphors will help us come to moral terms with Darwinism's stark implications, and I agree that the metaphors serve this purpose.

13. The literature on kin selection and altruism contains many classic works in evolutionary biology—e.g., Cronin 1991; Hamilton 1964, 1996; Maynard Smith 1975; Williams 1996.

14. See Dawkins 1976, 1982; Sterelny 2001a; Sterelny and Griffiths 1999; and Williams 1992. Sterelny (2001a) points out that segregation distorters are an example of outlaw genes—so-called because they promote their own replication at the expense of the other genes in the organism's genome—and that outlaw genes are rare. However, such outlaw genes are theoretically important way beyond their numbers because they lay bare the subpersonal logic of evolution. The existence of such effects as segregation distortion demonstrate the necessity of taking a gene's-eye view of life. Certain odd or deleterious effects on individual organisms come about because the organisms are being constructed for the benefit of the genes—not for the organism's own benefit.

15. The literature on the problems with group selection is large—e.g., Cronin 1991; Dawkins 1976, 1982; Hamilton 1996; Maynard Smith 1975, 1998; Williams 1966, 1996. For discussions of the circumstances under which group selection effects do occur, see Sober and Wilson 1998, and Sterelny and Griffiths 1999.

16. Consider another example of the problems with the "for the good of the species" or "the good of the group" notion: the phenomenon of the constancy of the sex ratio in sexually producing animals. In many harem-forming animals such as elephant seals, the male can produce enough sperm to service a harem of one hundred females. The vast majority of males never mate. The excess bachelor males use up a panoply of environmental resources that could be made available to the breeding part of the population. If biological conditions were "for the good of the species," then harem-forming animals would not display 50/50 sex ratios. Only a gene's-eye view of biology explains why there is a 50/50 ratio of elephant seals despite the inefficiency of this ratio from the standpoint of the species: in game-theory terms (Maynard Smith 1974, 1976, 1998; Skyrms 1996), a gene's bet on the "sure thing" female has the same expected value as the "risky" bet on the male. (For a discussion of sex ratios that is less gene-centric and instead focuses on multilevel selection theory, see Sober and Wilson 1998.)

17. Strictly speaking, there are two conceptually different subspaces within area B. There are goals that are currently serving genetic fitness that are antithetical to the vehicle's interests, and there are goals within this area that serve neither genetic nor vehicle interests. The reason there are the latter is because genetic goals arose in the ancient environment in which our brains evolved (the environment of evolutionary adaptation, EEA). Environments can change faster than evolutionary adaptations, so that genetic goals may not always be perfectly adapted to the current environment. Whether these goals currently facilitate genetic fitness—or only facilitated reproductive fitness in the past—is irrelevant for the arguments of this book. In either case, goals which diverge from vehicle goals reside in the brain because of the genes. For example, whether the consumption of excess fat serves current reproductive fitness or not, it is a vehicle thwarting tendency (for most of us!), and it is there because it served reproductive fitness at some earlier point in time.

18. On the so-called "genetic fallacy" in early sociobiology, see Symons 1992. Evolutionary psychology's major impact on psychology in the past decade has been documented in a variety of sources—e.g., Barrett, Dunbar, and Lycett 2002; Bjorklund and Pellegrini 2002; Buss 1999, 2000; Cosmides and Tooby 1992, 1994b; Geary and Bjorklund 2000; Geary and Huffman 2002; Over 2003a, 2003b; Pinker 1997, 2002; Tooby and Cosmides 1992.

19. On reasoning errors, see Kahneman and Tversky 1984, 1996, 2000; Gilovich, Griffin, and Kahneman 2002; and chapter 4 of this volume. The literature on evolutionary explanations is now extensive—e.g., Brase, Cosmides, and Tooby 1998; Cosmides and Tooby 1996; Fiddick, Cosmides, and Tooby 2000; Gigerenzer 1996b; Over 2003a; Rode, Cosmides, Hell, and Tooby 1999.

20. To be precise, I am doubting whether there are people who say they value their genome *and have an accurate view of what they are valuing* when they say this. For example, in such a case, the person would have to be absolutely clear that valuing your own genome is not some proxy for valuing your children; be clear that having children does not replicate one's genome; and be clear about the fact that the genome is a *superpersonal* entity.

21. Indeed, genetic engineering and gene therapy for purposes of human health and longevity represent perhaps the ultimate triumph of Dawkins's (1976) so-called survival machines. With the technology of genetic engineering, we, who were built by the replicators to serve as their survival machines, use *them* for our *own* goals—goals that are not the genes' goals (e.g., survival past our reproductive years). Williams (1988) uses such an example to counter Stent's (1978) argument against Dawkins (1976) that

rebelling against one's own genes is a contradiction. Williams (1988) notes that Stent "apparently missed the relevance of major technologies (hair dyeing, tonsillectomy, etc.) based on just such rebellion" (403).

Chapter Two

1. The extent to which this distinction should be taken as strictly categorical or as reflective of a continuum of different processing styles is a matter of dispute. Speaking in terms of clearly differentiated types of cognition will facilitate the explication of issues in this volume, but little except communicative ease would be lost if a more continuous conception were adopted.

2. The differences between the algorithmic level of analysis and the intentional level of analysis (which is understood in terms of sets of goal hierarchies) will be discussed in chapter 6. It should be noted though that the idea that the brain is composed of many different subsystems has recurred in conceptualizations in many different disciplines—from the society of minds view in artificial intelligence (Minsky 1985); to Freudian analogies (Ainslie 1982; Elster 1989; Epstein 1994); to discussions of the concept of multiple selves in philosophy, economics, and decision science (Ainslie 1992, 2001; Bazerman, Tenbrunsel, and Wade-Benzoni 1998; Dennett 1991; Elster 1985; Hogarth 2001; Loewenstein 1996; Medin and Bazerman 1999; Parfit 1984; Schelling 1984).

3. The notion of many different systems in the brain is, by no means new. Plato (1945) argued that "we may call that part of the soul whereby it reflects, rational; and the other, with which it feels hunger and thirst and is distracted by sexual passion and all the other desires, we will call irrational appetite, associated with pleasure in the replenishment of certain wants" (137). What is new, however, is that cognitive scientists are beginning to understand the biology and cognitive structure of these systems (Goel and Dolan 2003; Harnish 2002; Kahneman and Frederick 2002; Lieberman 2000; Metcalfe and Mischel 1999; Pinker 1997; Sloman 1999; Sloman and Rips 1998; Slovic et al. 2002; Smith, Patalino, and Jonides 1998; Sternberg 1999; Willingham 1998, 1999; T. Wilson 2002; Wilson and Keil 1999) and are beginning to posit some testable speculations about their evolutionary and experiential origins (Barkow, Cosmides, and Tooby 1992; Carruthers 2002; Carruthers and Chamberlain 2000; Evans and Over 1996; Mithen 1996; Pinker 1997; Reber 1992a, 1992b, 1993; Shiffrin and Schneider 1977; Stone et al. 2002).

4. On compositionality, see Fodor and Pylyshyn 1988; Pinker 1997; and Sloman 1996. On the link between analytic processing and intelligence, see Engle, Tuholski, Laughlin, and Conway 1999; Kane and Engle 2002; and Stanovich and West 2000. There is a large literature linking analytic processing to inhibitory control–e.g., Barkley 1998; Case 1992; Dempster and Corkill 1999; Dienes and Perner 1999; Harnishfeger and Bjorklund 1994; Kane and Engle 2002; Norman and Shallice 1986; Zelazo and Frye 1998.

5. On autonomous systems, see Anderson 1998; Bargh and Chartrand 1999; Baron-Cohen 1995, 1998; Carr 1992; Coltheart 1999; Cosmides and Tooby 1994b; Fodor 1983; Hirschfeld and Gelman 1994; Lieberman 2000; Logan 1985; Navon 1989; Pinker 1997; Reber 1993; Rozin 1976; Samuels 1998; Shiffrin and Schneider 1977; Sperber 1994; Stanovich 1990; and Uleman and Bargh 1989.

6. On modularity, see Anderson 1992; Baron-Cohen 1998; Coltheart 1999; Fodor 1983; Gardner 1983; Hirschfeld and Gelman 1994; Samuels 1998; Karmiloff-Smith 1992; Sperber 1994; Thomas and Karmiloff-Smith 1998.

7. The unintelligent nature of modular processes (e.g., Dennett 1991; Fodor 1983, 1985; Karmiloff-Smith 1992) and the evolutionary importance of rapidly executing processes (e.g., Buss 1999; Cosmides and Tooby 1992, 1994b; Pinker 1997) have been much discussed.

8. My concept of TASS is defined more loosely than are Fodorian (1983) modules. TASS is composed of more than just innate mechanisms. My concept also loosens criteria 4 and 5 considerably to the much weaker property that most TASS subsystems are only *relatively more* encapsulated and impenetrable than is analytic processing. The latter stratagem removes the TASS concept from the controversies over whether particular modules meet or do not meet a discrete criterion of complete encapsulation. Indeed, many other theorists (e.g., Karmiloff-Smith 1992; Tsimpli and Smith 1998) have been exploring the consequences of loosening the Fodorian strictures on the definition of a module, particularly that of encapsulation. For example, Tsimpli and Smith (1998) have termed theory-of-mind mechanisms a quasi-module because they appear neither fully impenetrable nor fully encapsulated (see also Carruthers 1998; Baron-Cohen 1995; Sperber 1994, 1996).

9. TASS thus encompasses *both* the Darwinian mind of built-in reflexes based on past evolutionary history discussed in the previous chapter and the reflex-like perceptional and cognitive responses acquired during the individual's own lifetime.

10. For discussions of several of the modules listed in table 2.2, see Atran 1998; Barrett, Dunbar, and Lycett 2002; Baron-Cohen 1995; Carey 1985; Carruthers 2002; Carruthers and Chamberlain 2000; Cosmides and Tooby 1992; Hirshfeld and Gelman 1994; Leslie 1994; Mithen 1996; Pinker 1997; and Tooby and Cosmides 1992.

11. A similar metaphor—that of the adaptive toolbox—is used by Gigerenzer and Todd (1999): "Just as a mechanic will pull out specific wrenches, pliers, and spark-plug gap gauges for each task in maintaining a car's engine rather than merely hitting everything with a large hammer, different domains of thought require specialized tools. This is the basic idea of the adaptive toolbox: the collection of specialized cognitive mechanisms that evolution has built into the human mind for specific domains of inference and reasoning" (30).

12. Some evolutionary psychologists, most notably Cosmides and Tooby (1992 1994b), in their desire to advance the view of mind as an evolutionary adaptation, opt for a totally modular view of mind because they believe that evolution could only have produced modular structures. Other evolutionary theorists disagree, however; see Foley 1996; LaCerra and Bingham 1998; Over 2000, 2002, 2003a, 2003b; Samuels 1998; Smith, Borgerhoff Mulder, and Hill 2001; Sterelny and Griffiths 1999.

13. If we can decompose a smart homunculus into simpler and simpler homunculi that are increasingly nonmysterious in their behavior, then we have made real scientific progress—"intelligence could be broken into bits so tiny and stupid that they didn't count as intelligence at all" (Dennett 1995, 133).

14. For starters, try Baars 1997; Baddeley 1996; Baddeley, Chincotta, and Adlam 2001; Harnish 2002; Kimberg and Farah 1993; Miyake and Shah 1999; Pennington and Ozonoff 1996; Shallice 1988, 1991; and West 1996—all of which indicate that there are scientifically respectable ways to talk about concepts such as a central executive processor, strategic control, attentional processing, and decision-making without succumbing to the homunculus problem.

15. There is a direct analogy here to my caveats in the previous chapter about the use of the language of intentionality when talking about genes. To use the phrase "genes want" is not to imply that inside a gene is a conscious desire. There is no harm done in

the anthropomorphic language of convenience as long as everyone is aware of how it might be cashed out in an actual biological explanation. Likewise, to talk of executive control or analytic system override of TASS output is not to imply the existence of a ghostly homunculus in the brain making explicit decisions. Such phrases can be cashed out in terms of mechanistic models of cognitive control. For some time now, there have existed viable models of how the sequencing of control can be maintained in a cognitive architecture where distributed systems are sharing a common workspace that stores the intermediate products of their computations (Anderson 1983; Baars 1997; Johnson-Laird 1988; Lindsay and Norman 1977; Newell 1990; Selfridge and Neisser 1960). Nothing is wrong with using the language of executive control when talking of organisms that can be modeled by such systems, because all of the higher-level control language can be cashed out in an appropriately scientific fashion without positing a homunculus that is embarrassing in its intelligence.

16. The variations on the prototypical mechanistic model I have in mind can be identified in the sources upon which I draw heavily—Anderson 1983; Baars 1997; Baddeley 1996; Carruthers 2002; Clark 1997; Dennett 1991; Evans and Over 1996; Johnson-Laird 1988; Miyake and Shah 1999; Newell 1990; Norman and Shallice 1986; Perner 1998.

17. The notion that analytic processing acts by somehow integrating information across more domain-specific TASS modules has many incarnations in cognitive science (Clark and Karmiloff-Smith 1993; Karmiloff-Smith 1992; Mithen 1996; Pinker 1997; Rozin 1976). Rozin's (1976) discussion is perhaps the classic version of this idea, setting the stage for its more recent incarnations. His view is that behavioral flexibility, and thus intelligence, increases as the outputs of the peripheral modules of TASS become available (through multiple connections) to more central systems and more output systems. Dennett's conception is different from Rozin's in that, for the latter, the brain becomes rewired (a hardware change), whereas in Dennett's view, the computational power of the massively parallel TASS systems is recruited to run a different software having a different logic (serial, analytic, compositional) than TASS.

18. Individual differences in the ability to sustain serial simulation may be related to the reason why intelligence correlates with reaction time and other speeded tasks. In a monumental review of this literature, Deary (2000) uncovers the theoretical paradox that it turns out to be difficult to explain why elementary information processing tasks correlate with intelligence at all. I have conjectured (Stanovich 2001) that it is not because these tasks measure some inherent "mental speed" at all (Deary reviews evidence indicating that the RT-IQ relationship is virtually unchanged when differences in nerve conduction speed are partialed out). Rather, they all may serve as indirect indicators of the computational power available in the brain's connectionist network—computational power that is available to sustain the simulation of a serial processor. Of course, there are other more direct indicators of the computational power available to sustain serial simulation, such as working memory, and not surprisingly these indicators show larger correlations with intelligence. On such a view, general intelligence could be described as encompassing individual differences in two fundamental properties (that perhaps map into the fluid/crystallized distinction from the Horn/Cattell model, Horn 1982; Horn and Cattell 1967). First, there is the computational power of the parallel network to sustain the serial simulation (this is probably closer to fluid intelligence in the Horn/Cattell model of intelligence). The second major factor is the power of the cultural tools used during serial simulation—the Gregorian mind in Dennett's (1991) Tower of Intellect model (individual differences in this factor might relate to variance in crystallized intelligence in the Horn/Cattell model).

19. Language input can also serve a rapid, so-called context-fixing function (Clark 1996) in a connectionist network (see Rumelhart, Smolensky, McClelland, and Hinton 1986). Where it might take an associationist network dozens of trials and a considerable length of time to abstract a prototype, a linguistic exchange can activate a preexisting prototype in a single discrete communication. Clark (1996) discusses how context fixers are "additional inputs that are given alongside the regular input and that may cause an input that (alone) could not activate an existing prototype to in fact do so" (117). He argues that "linguistic exchanges can be seen as a means of providing fast, highly focused, context-fixing information" (117).

Finally, it should be noted that there is no inconsistency between positing an important role for language in analytic system functioning and the fact that most theorists also posit a module for language in TASS. Carruthers (2002) discusses how there is nothing unusual about a central system coopting a TASS module for nonmodular purposes. It is generally accepted that when we visualize something in our imagination, we coopt the operation of vision modules in TASS (Carruthers cites evidence indicating that some of the same areas of the visual cortex are activated in actual seeing and imaging). Carruthers argues that, likewise, peripheral language modules can be coopted for certain kinds of nonmodular reasoning and problem-solving.

20. The confabulatory tendencies of the analytic system, as well as its tendency toward egocentric attribution, have been much discussed—e.g., Calvin 1990; Dennett 1991, 1996; Evans and Wason 1976; Gazzaniga 1998b; Johnson 1991; McFarland 1989; Moscovitch 1989; Nisbett and Ross 1980; Nisbett and Wilson 1977; Wegner 2002; T. D. Wilson 2002; Wolford, Miller, and Gazzaniga 2000; Zajonc 2001; Zajonc and Markus 1982.

21. Aspects of hypothetical thinking have been discussed by a host of theorists—e.g., Atance and O'Neill 2001; Bickerton 1995; Carruthers 2002; Clark and Karmiloff-Smith 1993; Cosmides and Tooby 2000a; Currie and Ravenscroft 2002; Dennett 1984; Dienes and Perner 1999; Evans and Over 1999; Glenberg 1997; Jackendoff 1996; Lillard 2001; Perner 1991, 1998; Sperber 2000b, 2000c; Sterelny 2001b; Suddendorf and Whiten 2001; M. Wilson 2002. Many theorists have argued that a key early focus of hypothetical thinking might have been the minds (hypothesized intentional states) of other individuals—e.g., Baldwin 2000; Bogdan 2000; Davies and Stone 1995a, 1995b; Dennett 1996; Humphrey 1976; Perner 1991; Tomasello 1998, 1999; Wellman 1990; Whiten and Byrne 1997; Zelazo, Astington, and Olson 1999. There are many subtleties surrounding the concept of metarepresentation–see Dennett 1984; Perner 1991; Sperber 2000b; Suddendorf and Whiten 2001; Whiten 2001.

22. The literature on perception without awareness is large–see Cheesman and Merikle 1984, 1986; Greenwald 1992; Holender 1986; Marcel 1983; Merikle, Smilek, and Eastwood 2001; Purcell, Stewart, and Stanovich 1983; as is the literature on semantic priming without awareness–see Fischler 1977; Masson 1995; Neely and Keefe 1989; Stanovich and West 1983; Stolz and Neely 1995.

23. Gazzaniga's writings are extensive—see Gazzaniga 1989, 1997, 1998a, 1998b; Gazzaniga and LeDoux 1978; Wolford, Miller, and Gazzaniga 2000. It should be noted that Gazzaniga (1998b, 24) posits that interpreter processes can occur both within and outside of conscious awareness. However, in this book I will emphasize interpreter processes carried out explicitly—that is, as analytic system processes.

24. On Capgras syndrome, see Davies and Coltheart 2000; Stone and Young 1997; and Young 2000; and on biased processing in order to maintain a hypothesis, see Mele 1997, 2001.

25. I am not advocating here a simple Platonic view where the lower mind (TASS) should always be ruled by the higher (the analytic system). As we shall see in chapters 7 and 8, sometimes we should decide to identify with TASS in cases of conflict between it and the analytic system. There is a philosophical literature (e.g., Bennett 1974; MacIntyre 1990) on cases like that of the Mark Twain character, Huckleberry Finn, who experienced a conflict between feelings that he should help a slave friend escape and an analytic judgment that helping slaves escape was morally wrong.

26. That many useful information processing operations and adaptive behaviors are carried out by TASS (e.g., Cosmides and Tooby 1994b; Evans and Over 1996; Reber 1993; Gigerenzer 1996b; Gigerenzer and Todd 1999; Pinker 1997) and that much of mental life is infused by TASS outputs (e.g., Cummins 1996; Evans and Over 1996; Hilton 1995; Hogarth 2001; Levinson 1995; Myers 2002; Reber 1993; Wegner 2002; T. D. Wilson 2002) are certainly the default assumptions in cognitive science generally. Also, as will be emphasized in chapters 7 and 8, not all conflicts between TASS and the analytic system should be resolved in favor of the latter.

27. In Pollock's (1991, 1995) model, emotions are conceived as Q&I modules for practical (i.e., means/ends) reasoning, an idea which has been explored by numerous theorists (e.g., Cacioppo and Berntson 1999; Damasio 1994; de Sousa 1987; Frank 1988; Johnson-Laird and Oatley 1992, 2000; Oatley 1992, 1998; Simon 1967). Of course, the idea that conscious rational thought should be used to trump the irrational emotions goes back to Plato (see Nathanson 1994). But this is not the only way that irrationality due to emotional factors can arise. There are actually two different ways in which emotional regulation can go awry. They roughly correspond to the crude distinction of having too little emotional involvement in behavioral regulation and too much. The former view, that Q&I emotion modules may not be active enough—a view amply represented in cognitive science (Damasio 1994; de Sousa 1987; Oatley 1992)—remains under-represented in folk psychology. Several architectures in cognitive science would, in contrast, recognize both possibilities (Damasio 1994; Johnson-Laird and Oatley 1992; Loewenstein, Weber, Hsee, and Welch 2001; Oatley 1992; Pollock 1991, 1995; Stanovich 1999). More importantly, there is empirical evidence for rationality failures of the two different types. Dorsolateral prefrontal damage has been associated with executive functioning difficulties (and/or working memory difficulties) that can be interpreted as the failure to override automatized processes being executed by TASS (Dempster 1992; Dempster and Corkill 1999; Duncan et al. 1996; Harnishfeger and Bjorklund 1994; Kane and Engle 2002; Kimberg, D'Esposito, and Farah 1998; Norman and Shallice 1986; Shallice 1988). In contrast, ventromedial damage to the prefrontal cortex has been associated with problems in behavioral regulation that are accompanied by affective disruption (Bechara et al. 1994, 1997, 2000; Damasio 1994). Difficulties of the former but not the latter kind are associated with lowered intelligence (Damasio 1994; Duncan et al. 1996)—consistent with the idea that psychometric intelligence is more strongly associated with analytic processing than with the operation of TASS (Stanovich 1999; Stanovich and West 2000).

28. Although analytic system overrides of TASS processing will most often be efficacious, this is not necessarily the case. It certainly is possible for metacognitive judgment to go awry and override TASS processing when in fact the latter would have served the person's goals even better than the overriding systems. An example is provided by the work of Wilson and Schooler (1991). They had subjects rate their preferences for different brands of strawberry jam and compared these ratings to those of experts (the ratings in *Consumer Reports*). They found that a group of individuals who were encouraged to be

analytic about their ratings made ratings less in accord with the experts than a group who were not given encouragement to be analytic. Thus, it is possible that cognitive systems can make the error of having too low a threshold for TASS override. Such an overly low threshold for override might be conceived as a Mr. Spock-like hyper-rationality that could actually be deleterious to goal achievement in some situations. Similar arguments hold in the moral domain (recall the Huck Finn example mentioned in note 25).

29. The caveat in note 17 of chapter 1 is relevant here as well. When, in this volume, I label something a genetic goal, it does not necessarily mean that the goal is *currently* serving the interests of reproductive fitness—only that it did so sometime in the past in what is called the environment of evolutionary adaptation, or EEA.

30. The literature on the selection task is vast (for reviews, see Evans, Newstead, and Byrne 1993; Manktelow 1999); and many different response tendencies have been documented (Cummins 1996; Dawson, Gilovich, and Regan 2002; Gebauer and Laming 1997; Hardman 1998; Johnson-Laird 1999, 2001; Liberman and Klar 1996; Manktelow and Evans 1979; Newstead and Evans 1995; Sperber, Cara and Girotto 1995; Stanovich and West 1998a). On the matching bias, see the work of Evans 1984, 1989, 1995, 1998, 2002b; but note that there are several other models of how heuristic, TASS-based processing could lead to the PQ choice (e.g., Oaksford and Chater 1994, 1996; Sperber, Cara, and Girotto 1995).

31. On the current interpretation of representativeness as a form of attribute substitution, see Kahneman and Frederick 2002. Other TASS heuristics have been proposed as the cause of the conjunction error in this problem—many involving conversational assumptions that are automatically applied to the problem. These will be discussed in chapter 4.

32. Note that identifying with one's gut instincts is what creates the area labeled D in figure 2.2—goals in the interests of the genes and not the vehicle that are pursued consciously and reflectively.

Chapter Three

1. One generic reason for the separation of evolutionary adaptation and rationality is that natural selection works on a "better than" principle (see Cosmides and Tooby 1996, 11) and rationality is defined in terms of maximization (Gauthier 1975). Ridley (2000) spins this point another way, calling evolution "short-termish" because it is concerned with immediate advantage rather than long-term strategy. Human rationality, in contrast, must incorporate long-term interests (Ainslie 2001; Haslam and Baron 1994; Loewenstein 1996; Parfit 1984; Sabini and Silver 1998).

2. Philosopher Daniel Dennett (1995,) notes, humorously, that "it isn't clear that the members of any other species *have* an outlook on life" (339). He points out that our outlook, however, can lead to various behaviors that totally subvert genetic goals, such as pledging to celibacy, proscribing the eating of certain nutritious foods, and regulating certain types of sexual behavior.

3. Instrumental rationality concerns the rationality of actions as opposed to the rationality of belief, often termed epistemic rationality (to be discussed below). The terms instrumental rationality, practical rationality, pragmatic rationality, and means/ends rationality have all been used to characterize this aspect of rationality (Audi 1993b, 2001; Evans and Over 1996; Gauthier 1975; Gibbard 1990; Harman 1995; Nathanson 1994; Nozick 1993; Schmidtz 1995; Sloman 1999; Stich 1990), and these terms will be used interchangeably in this book.

4. For the purposes of the discussion here, I am interpreting Hume's "passions" to

correspond to previously existing desires. It should be noted that this quotation is routinely used by decision scientists to characterize instrumental rationality, but no further inferences about Hume's philosophy are meant to be derived from it. Specifically, it is not used to imply that Hume was an exponent of grasping selfishness. To the contrary, social virtues and concern for community were central for Hume.

5. The literature on utility theory in decision science is large–see Baron 1993, 1999; Broome 1991; Dawes 1998; Fishburn 1981, 1999; Gardenfors and Sahlin 1988; Gauthier 1975; Jeffrey 1983; Kahneman 1994; Kahneman and Tversky 2000; Kleindorfer, Kunreuther, and Schoemaker 1993; Luce and Raiffa 1957; McFadden 1999; Pratt, Raiffa, and Schlaifer 1995; Resnik 1987; Savage 1954; Starmer 2000.

6. Another technical reason to reject the thin theory is that, as several theorists have argued, in its failure to address epistemic considerations, it ends up by undercutting itself. These theorists have argued that if we put no constraints on people's representations of problems—if we simply apply the axioms of rational choice to *whatever* are their beliefs about the task situation—we will end up with no principles of instrumental rationality at all (Broome 1990; Hurley 1989; Schick 1987, 1997; Sen 1993; Shweder 1987; Tversky 1975). This issue is discussed in detail in chapter 4 of Stanovich 1999.

7. For examples of models of rationality that deal with complexities beyond our scope here, see Bratman 1987; Bratman et al. 1991; Newell 1982, 1990; Pollock 1991, 1995; Schmidtz 1995; and Stich 1996.

8. A utility function to a decision scientist is a numerical representation of a preference ordering (Luce and Raiffa 1957; Starmer 2000). As Kahneman (1994) has noted, the notion of utility used by economists has "been thoroughly cleansed of any association with hedonistic psychology, and of any reference to subjective states" (20), but modern cognitive science no longer demands this "cleansing" (Diener, Suh, Lucas, and Smith 1999; Kahneman, Diener, and Schwarz 1999; Mellers 2000).

9. Moving to a broad theory of rationality—one that encompasses epistemic evaluation as well as a substantive critique of desires—is not without a cost. It means taking on some of the knottiest problems in philosophy (Earman 1992; Goldman 1986; Harman 1995; Millgram 1997; Nozick 1993; Richardson 1997; Schmidtz 1995, 2002; Schueler 1995). One problem is that concerns about practical rationality can always seem to trump epistemic rationality in a way that would seem to render evaluation of the latter virtually impossible. Foley (1991) has discussed cases where epistemic and practical reasons for belief become separated. So a simultaneous consideration of practical and epistemic rationality will certainly complicate the application of rational standards (Pollock 1995).

10. Instrumental rationality is a necessary but not sufficient condition for the robot's rebellion. In fact, two conditions are needed: the individual's goals must be optimally satisfied, but those goals must also be keyed to the life interests of the vehicle. The second condition is critical because in theory, without it, one could conceive of an organism with only short-leash genetic goals that were maximally achieved. Such an organism would be deemed instrumentally rational, but would have achieved this status only because it has instantiated no goals specifically keyed to the interests of the vehicle. Many Darwinian creatures are no doubt instrumentally rational in this sense—as evolutionary psychologists have emphasized (the paradox of rational lowly creatures will be discussed specifically in chapter 8). With the advent of analytic processing, there is more potential for nonoverlapping replicator/vehicle interests when the long-leash goals of the analytic system are pursued. Thus, the evaluation of those goals (what is meant by a broad theory of rationality) becomes an issue along with the assessment of instrumental rationality. This issue is discussed extensively in chapter 8.

11. For the conceptual arguments that human irrationality is impossible, see Cohen 1981; Henle 1962; and Wetherick 1995, and for the refutation of these arguments, see Stanovich 1999; Stein 1996; and Shafir and LeBoeuf 2002.

Chapter Four

1. Classically, subjective expected utility (SEU) theory defines the expected utility of an action as the sum of the probabilities of each possible outcome multiplied by the utility of each outcome.

2. There is a substantial literature on violations of transitivity—e.g., Dawes 1998; Mellers and Biagini 1994; Slovic 1995; Starmer 2000; Tversky 1969; Tversky and Kahneman 1986.

3. See Kahneman and Tversky (1973, 1979, 1982, 1984, 1996, 2000) and Tversky and Kahneman (1973, 1981, 1982, 1983, 1986) and the many reviews of research in the heuristics and biases tradition; see Baron 2000; Connolly, Arkes, and Hammond 2000; Dawes 1998; Evans 1989; Evans and Over 1996; Fischhoff 1988; Gilovich, Griffin, and Kahneman 2002; Kunda 1999; Manktelow 1999; McFadden 1999; Medin and Bazerman 1999; Shafir and LeBoeuf 2002; Shafir and Tversky 1995; Stanovich 1999.

It is important to emphasize here that the problems in reasoning discussed in this chapter are not merely confined to the laboratory or to story problems of the type I have been presenting. They are not just errors in a parlor game. We will see in other examples throughout this book that the errors crop up in such important domains as financial planning, medical decision making, career decisions, family planning, resource allocation, tax policy, insurance purchases and many other practical domains. The extensive literature on the practical importance of these reasoning errors is discussed in sources such as Baron 1998, 2000; Bazerman 1999, 2001; Bazerman, Baron, and Shonk 2001; Belsky and Gilovich 1999; Dawes 2001; Fridson 1993; Gilovich 1991; Hastie and Dawes 2001; Kahneman and Tversky 2000; Margolis 1996; Myers 2002; Russo and Schoemaker 2002; and Welch 2002.

4. For an introduction to this literature, see Anderson 1990; Evans 1989, 2002a, 2002c; Evans and Over 1996, 1999; Friedrich 1993; Johnson-Laird 1999, 2001; Klayman and Ha 1987; McKenzie 1994; Mynatt, Doherty, and Dragan 1993; Oaksford and Chater 1994; Over and Green 2001; Wasserman, Dorner, and Kao 1990.

5. Recall from chapter 2 that the fact that subjects seem to be thinking about their answers to problems such as this (and the next one we shall discuss) does not contradict the theory that their incorrect response tendencies are due to an automatic TASS response bias. Research has indicated that much of the thinking is really rationalization of the initial responses primed by TASS (Evans 1996; Evans and Wason 1976; Roberts and Newton 2001).

6. There is a large body of work which questions the assumption that people have stable, well-ordered preferences. See Camerer 1995; Dawes 1998; Fischhoff 1991; Gilovich, Griffin, and Kahneman 2002; Hsee, Loewenstein, Blount, and Bazerman 1999; Lichtenstein and Slovic 1971; McFadden 1999; Payne, Bettman, and Schkade 1999; Shafer 1988; Shafir and LaBoeuf 2002; Shafir and Tversky 1995; Slovic 1995; Tversky, Sattath, and Slovic 1988.

7. This literature has been reviewed in several publications; see Kahneman and Tversky 2000; Payne, Bettman, and Johnson 1992; Slovic 1995. There is a large literature on preference reversals and the so-called constructed preference view of its causes; see Camerer 1995; Grether and Plott 1979; Hausman 1991; Kahneman 1991; Lichtenstein and Slovic 1971; Shafir and Tversky 1995; Starmer 2000; Tversky 1996b; Tversky and

Kahneman 1981, 1986; Tversky, Slovic, and Kahneman 1990; Tversky and Thaler 1990. It is not clear whether all of the performance anomalies discussed in this chapter and in chapter 2 are due to TASS tendencies that are evolutionary adaptations. Some may be due to acquired TASS tendencies.

8. Such patterns of choice and pricing imply that there is a sum of money, $M, such that A is preferred to $M, and $M is preferred to B. But, of course, the subject prefers B to A, so an intransitivity results.

9. Cognitive scientists have uncovered some of the reasons why these violations of procedural invariance occur (see Slovic 1995). For example, one reason is due to response compatibility effects. When the response mode is pricing this means that the monetary outcomes in the contracts will be given more weight than they are in the choice situation (so that, in this situation for example, the $2,500 is weighted more highly in a pricing decision than in a choice decision). Also, anchoring and adjustment, discussed in chapter 2, comes into play. If pricing leads people to anchor on the vivid monetary amounts and adjust downward based on the time, this could account for the higher pricing of Contract A, because it is known from anchoring and adjustment studies that people often adjust too little (Mussweiler, Strack, and Pfeiffer 2000).

10. There exist many extensive reviews of the literature on experimental studies of instrumental and epistemic rationality; see Baron 2000; Camerer 1995; Connolly, Arkes, and Hammond 2000; Dawes 1998, 2001; Evans 2002c; Evans and Over 1996; Gilovich, Griffin, and Kahneman 2002; Kahneman and Frederick 2002; Kahneman and Tversky 2000; Kuhberger 2002; Kuhn 1991, 2001; Manktelow 1999; Shafir and LeBoeuf 2002; Stanovich 1999; Stanovich and West 2000.

11. It should be strongly emphasized that the term bias is used here to denote "a preponderating disposition or propensity" (*The Compact Edition of the Oxford Short English Dictionary,* 211) and not a processing *error.* That a processing bias does not necessarily imply a cognitive error is a point repeatedly emphasized by the critics of the heuristics and biases literature (Funder 1987; Gigerenzer 1996a; Hastie and Rasinski 1988; Kruglanski and Ajzen 1983), but in fact was always the position of the original heuristics and biases researchers themselves (Kahneman 2000; Kahneman and Tversky 1973, 1996; Tversky and Kahneman 1974). Thus, the use of the term bias in this volume is meant to connote "default value" rather than "error." Under the assumption that computational biases result from evolutionary adaptations of the brain (Cosmides and Tooby 1994b), it is likely that they are efficacious in many situations.

12. The pervasiveness of TASS is something that evolutionary psychologists and their critics agree on; see Cosmides and Tooby 1992; Evans 2002a, 2002c; Evans and Over 1996; Over 2003a; Stanovich 1999, 2003.

13. There is now a huge literature on aspects of the social intelligence hypothesis— e.g., Baldwin 2000; Barton and Dunbar 1997; Blackmore 1999; Bogdan 2000; Brothers 1990; Byrne and Whiten 1988; Bugental 2000; Caporael 1997; Cosmides 1989; Cosmides and Tooby 1992; Cummins 1996, 2002; Dunbar 1998; Gibbard 1990; Gigerenzer 1996b; Goody 1995; Humphrey 1976 1986; Jolly 1966; Kummer, Daston, Gigerenzer, and Silk 1997; Levinson 1995; Mithen 1996, 2000, 2002; Sterelny 2001b; Tomasello 1998, 1999; Whiten and Byrne 1997.

14. Many of the investigators who have suggested that pragmatic inferences lead to seeming violations of the logic of probability theory in the Linda problem have analyzed the problem in terms of Grice's (1975) norms of rational communication; see Adler 1984, 1991; Dulany and Hilton 1991; Hertwig and Gigerenzer 1999; Hilton 1995; Hilton and Slugoski 2000; Macdonald and Gilhooly 1990; Mellers, Hertwig, and

Kahneman 2001; Politzer and Noveck 1991; Slugoski and Wilson 1998. The key to understanding the so-called Gricean maxims of communication is to realize that to understand a speaker's meaning the listener must comprehend not only the meaning of what is spoken but also what is implicated in a given context assuming that the speaker intends to be cooperative. And Hilton (1995) is at pains to remind us that these are rational aspects of communicative cognition. Clearly, in the view of these theorists, committing the conjunction fallacy in such contexts does not represent a cognitive error. Many theorists have linked their explanation of Linda-problem performance to the automatic socialization of information (fundamental computational bias 2 listed above).

15. In many probability learning situations, both nonhuman animals and humans generally approximate probability matching (Estes 1964, 1976, 1984). There are many different human probabilistic reasoning paradigms; see Fantino and Esfandiari 2002; Gal and Baron 1996; Tversky and Edwards 1966; and West and Stanovich 2003.

16. Although evolutionary psychologists sometimes claim that alternative representations can virtually eliminate cognitive errors (Brase, Cosmides, and Tooby 1998; Cosmides 1989; Cosmides and Tooby 1996; Gigerenzer 1991, 2002; Gigerenzer and Hoffrage 1995), it now appears that more "evolutionarily friendly" representations attenuate such cognitive errors but do not remove them entirely; see Brenner, Koehler, Liberman, and Tversky 1996; Cheng and Holyoak 1989; Evans, Simon, Perham, Over, and Thompson 2000; Girotto and Gonzalez 2001; Griffin and Varey 1996; Harries and Harvey 2000; Lewis and Keren 1999; Macchi and Mosconi 1998; Manktelow and Over 1991, 1995; Mellers and McGraw 1999; Mellers, Hertwig, and Kahneman 2001; Sloman, Over, and Stibel 2003; Tversky and Kahneman 1983.

17. For an updated interpretation of heuristics as involving attribute substitution, see Kahneman and Frederick 2002.

18. Schooling and decontextualization have been linked by numerous theorists— e.g., Anderson, Reder, and Simon 1996; Bereiter 1997, 2002; Cahan and Artman 1997; M. Chapman 1993; Donaldson 1978, 1993; Floden, Buchmann, and Schwille 1987; Hirsch 1996; Luria 1976; Neimark 1987; Paul 1984, 1987; Piaget 1972; Siegel 1988; Sigel 1993.

Chapter Five

1. For statements of the massive modularity thesis, see Tooby and Cosmides 1992 and Cosmides and Tooby 1994b, and for discussions and critiques of the thesis see Over 2003b; Samuels 1998; and Samuels, Stich, and Tremoulet 1999.

2. A variety of sources discuss intelligence as a predictor of social status, income, and employment; see Brody 1997; Gottfredson 1997; Hunt 1995; Lubinski and Humphreys 1997; Schmidt and Hunter 1992, 1998. Other works discuss heritable personality variables as predictors; see Austin and Deary 2002; Buss 1999, 394; Matthews and Deary 1998.

3. Although the outcome of disputes about whether general intelligence is a by-product or adaptation does not alter our argument, it should be noted that many theorists (e.g., Foley 1996; LaCerra and Bingham 1998; Smith, Borgerhoff Mulder, and Hill, 2001) argue that the changing online requirements of the ancestral hominid environment would, contra the massive modularity thesis, have required a flexible general intelligence (see also, Nozick 1993, 120, for a philosophically oriented version of a similar argument).

4. Evans et al. (2000; see also Macchi and Mosconi 1998) have shown that many of the so-called frequentistic versions of base-rate problems are computationally simpler.

This confound means that the facilitative effect of frequentistic representations in some previous research has been overestimated.

5. Likewise, as noted in the discussion above, Gigerenzer's repeated refrain that cognitive illusions (violations of rational reasoning principles) "disappear" with more evolutionarily propitious problem representations implies that these violations are of no concern. Since we know that these axioms of rationality are of concern to the vehicle (a vehicle who does not follow them does not maximize utility), we can only conclude that what these authors want us to infer is that evolutionary rationality is the only rationality that need concern us—precisely Cooper's (1989) point (although the point is much more subtle and somewhat hidden in the writings of the ecological theorists).

6. Several comprehensive summaries of the emergence of evolutionary psychology as a major force in contemporary psychology exist; see Badcock 2000; Barkow, Cosmides, and Tooby 1992; Barrett, Dunbar, and Lycett 2002; Bjorklund and Pellegrini 2000; Buss 1999; Cartwright 2000; Cosmides and Tooby 1994b, 2000b; de Waal 2002; Geary and Bjorklund 2000; Kenrick 2001; Pinker 1997, 2002; Plotkin 1998; Tooby and Cosmides 1992; Wright 1994.

Chapter Six

1. There is now a substantial literature on how the laboratory effects studied by the heuristics and biases researchers occur in real life. The examples in this paragraph are taken from a variety of sources. See Arkes and Ayton 1999; Baron 1998; Bazerman, Baron, and Shonk 2001; Belsky and Gilovich 1999; Dawes 1988, 2001; Fridson 1993; Gilovich 1991; Kahneman 1994; Kahneman and Tversky 2000; Margolis 1996; Piattelli-Palmarini 1994; Plous 1993; Redelmeier, Shafir, and Aujla 2001; Redelmeier and Tversky 1990, 1992; Russo and Schoemaker 2002; Saks and Kidd 1980; Shermer 1997; Sternberg 2002; Sutherland 1992; Thaler 1992; Welch 2002; and Yates 1992.

2. In chapter 3, I drew upon this literature—e.g., Baron 1993; Harman 1995; Jeffrey 1983; Kleindorfer, Kunreuther, and Schoemaker 1993; Nathanson 1994; Nozick 1993.

3. Philosophers use the term intentional somewhat differently than the layperson uses the term, but in this case a nontechnical usage will not lead us too far astray.

4. Indeed, Anderson (1990) uses the term rational level for this level of analysis. In contrast to the terms for the physical/biological level and algorithmic level which are uncontroversial, terms used to describe the third level are variable and controversial. Borrowing from Dennett (1987), I have termed this level of analysis the intentional level. Anderson's (1990) alternative term—rational level—indicates correctly the conceptual focus of this level of analysis. The term is appropriate for Anderson's (1990) project—which is to construct adaptationist models based on an *assumption* of rationality at this level of analysis. But it is inappropriate here because, as outlined in detail in chapter 4, descriptive models of humans' rational level indicate that they are not always rational! Thus, the term intentional level is more appropriate. For general discussions of levels of analysis in cognitive science, see Anderson 1990, 1991; Dennett 1978, 1987; Horgan and Tienson 1993; Levelt 1995; Marr 1982; Newell 1982, 1990; Oaksford and Chater 1995; Pollock 1995; Pylyshyn 1984; and Sterelny 1990.

5. The basic cognitive capacities that underlie intelligence have been extensively investigated. See Ackerman, Kyllonen, and Richards 1999; Carpenter, Just, and Shell 1990; Deary 2000; Engle, Tuholski, Laughlin, and Conway 1999; Fry and Hale 1996; Hunt 1978, 1987; Lohman 2000; Sternberg 1982, 2000; Vernon 1991, 1993.

6. Terminology surrounding the notion of thinking dispositions is remarkably varied in psychology. Some theorists prefer terms such as "intellectual style" (Sternberg

1988, 1989), "cognitive emotions" (Scheffler 1991), "habits of mind" (Keating 1990), "inferential propensities" (Kitcher 1993, 65–72), "epistemic motivations" (Kruglanski 1990), and "constructive metareasoning" (Moshman 1994); see also Ennis 1987; Messick 1984, 1994; and Perkins 1995. I will focus in this volume on "thinking dispositions" and "cognitive styles" that operate to foster rationality. I have previously focused attention on this type of thinking disposition (Sá, West, and Stanovich 1999; Kokis, Macpherson, Toplak, West, and Stanovich 2002; Stanovich 1999; Stanovich and West 1997; Toplak and Stanovich 2002), as have other investigators (Baron 1985a, 2000; Cacioppo, Petty, Feinstein, and Jarvis 1996; Kardash and Scholes 1996; Klaczynski, Gordon, and Fauth 1997; Kruglanski and Webster 1996; Kuhn 1991; Perkins, Jay, and Tishman 1993; Schoenfeld 1983; Schommer 1990, 1994; Sinatra and Pintrich 2003; Sternberg 1997b, 2001).

7. However, it should be noted that the intentional-level goal states of TASS are more likely to be attributed, whereas those of the analytic system are more likely to be actually represented in cognitive architecture (that is, realized as high-level control states in the organism). See McFarland (1989) on goal-seeking versus goal-directed systems.

8. Basic TASS processes such as face recognition, syntactic processing, and frequency detection show little variability (Hasher and Zacks 1979; Pinker 1994, 1997). On the issue of how evolutionary adaptation eats up variability, see Tooby and Cosmides 1990; and Wilson 1994. On why the algorithmic level of the analytic system does show variance, see Over and Evans 2000.

9. Research indicates that in order to predict reasoning performance, the individual's goals and epistemic values (intentional-level constructs) must be assessed as well as their computational capacity; see Kardash and Scholes 1996; Klaczynski, Gordon, and Fauth 1997; Kokis et al. 2002; Sá, West, and Stanovich 1999; Schoenfeld 1983; Sinatra and Pintrich 2003; Stanovich 1999; Stanovich and West 1997, 1998c; Sternberg 1997b; Toplak and Stanovich 2002.

10. That the assessment of rationality must be conditionalized on the cognitive resources of the brain is an argument that has been made by many authors; see Cherniak 1986; Gigerenzer and Goldstein 1996; Goldman 1978; Harman 1995; Oaksford and Chater 1993, 1995, 1998; Osherson 1995; Simon 1956, 1957; Stich 1990.

11. As the Tetlock and Mellers (2002) quote indicates, the debate has been quite contentious; see Cohen 1981; Dawes 1983, 1988; Evans 2002a; Gigerenzer 1996a; Jepson, Krantz, and Nisbett 1983; Kahneman 1981; Kahneman and Tversky 1983, 1996; Krantz 1981; Kuhberger 2002; Samuels, Stich, and Bishop 2002; Samuels, Stich, and Tremoulet 1999; Shafir and LeBoeuf 2002; Stanovich 1999; Stanovich and West 2000; Stein 1996; Vranas 2000.

12. This large literature is reviewed in many useful sources; see Baron 2000; Dawes 1998, 2001; Evans and Over 1996; Gilovich, Griffin, and Kahneman 2002; Johnson-Laird 1999; Kahneman and Tversky 2000; Klaczynski 2001; Oaksford and Chater 2001; Over 2002; Shafir and LeBoeuf 2002; Stanovich 1999; Stanovich and West 2000.

13. Certain types of base-rate problems and the so-called false consensus effect fall into this category; see Birnbaum 1983; Dawes 1989, 1990; Koehler 1996; Krueger and Zeiger 1993; Mueser, Cowan, and Mueser 1999; Stanovich 1999.

14. Numerous studies have documented the imperfect correlation between intelligence and rational thinking dispositions; see the research cited in notes 6 and 9.

15. Although the literature on instruction in rational thinking is scattered, some progress has been made; see Adams 1989; Adams and Feehrer 1991; Baron and Brown

1991; Beyth-Marom, Fischhoff, Quadrel, and Furby 1991; Lipman 1991; Mann, Harmoni, and Power 1991; Nickerson 1988; Nisbett 1993; Paul, Binker, Martin, and Adamson 1989; Perkins and Grotzer 1997; Sternberg 2001; Swartz and Perkins 1989; Tishman, Perkins, and Jay 1995; Williams, Blythe, White, Li, Sternberg, and Gardner 1996.

16. A number of commentators from a variety of disciplines have argued that such a situation might be characteristic of modern life; see Clark 1997, 180–84; Denzau and North 1994; Hargreaves Heap 1992; Nozick 1993, 124–25, 180–81; Satz and Ferejohn 1994; Thaler 1987.

17. In several notorious cases corporations have, in effect, let their chief executive officers determine their own compensation—usually with disastrous results for the corporation and the rest of its employees.

Chapter Seven

1. The complication was that, as discussed in chapter 2, the analytic system, through practice, can instantiate in the algorithms of TASS acquired responses that serve more long-term goals (smile when the boss walks past).

2. Although my definition of the meme follows from Aunger's (2002) discussion, precision of definition is not necessary for my purposes here. The contributors in a volume edited by Aunger (2000a) discuss these and other related definitions (see also Blackmore 1999; Dennett 1991, 1995; Dugatkin 2000; Lynch 1996). The term meme is also sometimes used generically to refer to a so-called memeplex—a set of co-adapted memes that are copied together as a set of interlocking ideas (see Blackmore 1999; Speel 1995). On co-adapted sets of memes see Dennett 1993, 204–5.

Aaron Lynch (personal communication) has suggested that the more neutral term cultural replicator be used instead of the term meme because the latter has been tainted by several popularizations. Also, it should be noted that the term meme, for some scholars, carries with it connotations that are much stronger than my use of the term here. For example, Sperber (2000a) uses the term meme not as a synonym for a cultural replicator in general, but as a cultural replicator "standing to be selected not because they benefit their human carriers, but because they benefit themselves" (163). That is, he reserves the term for category 4 discussed below. In contrast, my use of the term is more generic (as a synonym for cultural replicator) and encompasses all four categories listed below.

3. A meme is both a replicator and an interactor. Here, Hull's (1982) term interactor is more appropriate, because memes do not build vehicles, as do genes. Instead, memes use brains as hosts and as mechanisms for sending signals which start replication reactions in other brains. Templates for future meme signals are stored in books, computers, and other media (see Aunger 2002).

For the arguments of this chapter, it is only necessary to establish that the meme concept fulfills the minimal requirements of a replicator. It is not necessary to establish that every property of a gene finds an analogy in a property of the meme. Blackmore (1999) warns us not to expect a memetic analogy to every genetic concept. All of the major theorists who have commented on the meme concept have drawn attention to meme and gene dissimilarities (Aunger 2000b, 2002). For example, two important differences are, first, that memes have lower copying fidelity than genes, and, second, vehicles *deliberately* create memetic mutations but not genetic mutations (Blackmore 1999; Runciman 1998; Sperber 1985; however, see Hull 2001). The issue of low copying fidelity has received particular attention; see Blackmore 1999; Boyd and Richerson 2000; Dawkins 1982, 1999; Sperber 1985, 1996.

None of these dissimilarities undermine the status of the meme as a true replicator, however. First, Hull (2001) presents some incisive arguments indicating that gene/ meme disanalogies have been exaggerated. Furthermore, Hull (2000) argues that the science of memetics does not rely on a gene analogy at all but instead on the selection principles of universal Darwinism at their most general level: "Memetics does not involve analogical reasoning at all. Instead, a general account of selection is being developed that applies equally to a variety of different sorts of differential replication" (45–46). That we use the gene as a benchmark is only because we are biased toward privileging it because it appeared on the scene first. To be a true replicator, an entity need not recapitulate in detail each of the particular properties of the gene—which simply happened to be the first replicator discovered. As Dennett (1991) notes, "just as genes for animals could not come into existence on this planet until the evolution of plants had paved the way . . . so the evolution of memes could not get started until the evolution of animals had paved the way by creating a species—*Homo sapiens*—with brains that could provide shelter, and habits of communications that could provide transmission media, for memes" (202). To whit, memes may not be the last replicators to be discovered (see Aunger 2002).

    4. See Dennett 1991, 204–5.

    5. Several, for example, are represented in the volume edited by Aunger (2000a).

    6. See Blackmore (1999), Dennett (1991, 1995), and Lynch (1996) for taxonomies of reasons for meme survival. Of course, it is recognized that a meme may reside in more than one category. A meme might spread because it aids the vehicle *and* because it fits genetic predispositions *and* because of its self-perpetuating properties. However, as Dennett (1995) has argued, mismatching replicator/vehicle interests create conditions of particular theoretical interest; hence my emphasis in this chapter. Likewise, Blackmore (2000a) admits that in her 1999 book she chose to emphasize disadvantageous and dangerous memes out of proportion because they highlight the contrast between the perspective of memetic theory and that of sociobiology and the gene-culture coevolution theorists (e.g., Cavalli-Sforza and Feldman 1981; Durham 1991; Lumsden and Wilson 1981). Finally, for a discussion of the evoked culture that is emphasized by evolutionary psychologists, see Sperber 1996; Atran 1998; and Tooby and Cosmides 1992. As I mentioned in note 2, my use of the term meme does not necessitate the denial of evoked culture.

    7. Lynch (1996) and Blackmore (1999) discuss many of these subcategories.

    8. Dawkins (1993) and Blackmore (2000a) have contentiously claimed that most religions are essentially "copy-me" memeplexes, "backed up with threats, promises, and ways of preventing their claims from being tested" (Blackmore 2000a, 35–36).

    9. Below, in the section on the co-adapted meme paradox and in chapter 8, I will discuss the really Alice-in-Wonderland feature here—that those very replicators are *us* (or at least the brain processes that we tend to call "me").

    10. My meme evaluation criterion 1 relates to principle V of Nozick's (1993) taxonomy of principles of rational preference formation. He notes that it is natural that, all other things being equal, people prefer the preconditions needed for forming *any* preferences. These preconditions would be the obvious candidates: life, freedom, nutrients, absence of excruciating pain, etc. Thus, principle V states that "the person prefers that each of the preconditions (means) for her making any preferential choices be satisfied, in the absence of any particular reason for not preferring this" (142). The contradicting case is of course suicide, but the only thing the principle precludes is suicide without a

reason, so it is not really a contradiction. The principle does define as irrational the wish for death without a reason.

11. Although the principles discussed here do emphasize vehicle-centered meme evaluation, this does not mean that it is rational to ignore society. Neither this rule nor those that follow preclude altruistic or social desires as being rational parts of a goal structure. In fact, such goals may actually be rationally prescribed in some circumstances. See Baron 1998; Flanagan 1996; Gauthier 1986, 1990; Nozick 1993; Parfit 1984; Rawls 1971; Scanlon 1998; Schmidtz 1995; Skyrms 1996; Wolfe 1989.

12. TASS modules that are evolutionary adaptations execute based on contingencies in the environment from long ago. They represent rigid statistical guesses based on the logic of signal detection theory (Swets, Dawes, and Monahan 2000). The criterion for a stimulus triggering a TASS module is set, not according to a rule of maximizing truth, but instead according to an aggregate calculation encompassing the relative benefit of a "hit" (responding correctly to the right stimulus) and the relative costs of a false positive (responding to the absence of the correct stimulus) and a false negative (failing to respond to the presence of the correct stimulus). In short, many TASS modules were designed to maximize genetic fitness in the EEA, not to maximize true beliefs as theories of rationality require (Cooper 1989; Skyrms 1996; Stein 1996; Stich 1990).

13. Many examples of such memes are contained in Dawkins 1993; Dennett 1991, 1995; and Lynch 1996.

14. The literature on this meme-induced hysteria is large; see Dawes 2001; Garry, Frame, and Loftus 1999; Loftus 1997; Loftus and Ketcham 1994; Pezdek and Banks 1996; Pezdek and Hodge 1999; Piper 1998; Shermer 1997; Spanos 1996.

15. The term memosphere is Dennett's (1991, 1995). For examples of how the critical thinking literature stresses skills of detachment see Baron 2000; Paul 1984, 1987; Perkins 1995; and Stanovich 1999. Belief bias has been studied in many different ways by psychologists; see Baron 2000; Evans, Barston, and Pollard 1983; Goel and Dolan 2003; Nickerson 1998; Stanovich and West 1997.

16. Memes in a mutually supportive relationship within a memeplex would be likely to form a structure that prevented contradictory memes from gaining brain space for some of the same reasons that genes in the genome are cooperative (Ridley 2000). Vehicles tend to be defective if any new mutant allele is not a cooperator because the other genes in the genome demand cooperation. Resident memes are also selecting for a cooperator—someone advantageous *to them*. This could be the source of the ubiquitous belief bias effect observed in experiments in cognitive psychology (e.g., Evans, Barston, and Pollard 1983; Nickerson 1998)—memes contradicting previously residing memes are not easily assimilated.

An account in terms of memetics might also explain a puzzling individual difference finding—that there seems to be a large amount of domain specificity in belief bias (Sá, West, and Stanovich 1999; Toplak and Stanovich 2003). A person showing high belief bias in one domain is not necessarily likely to show it in another. On the other hand, there are large differences between domains in how much belief bias people display in them (Toplak and Stanovich 2003). Thus, it is not *people* who are characterized by more or less belief bias but *memes* that differ in the degree of belief bias they engender—that differ in how strongly they are structured to repel contradictory memes that are competitors for that "locus" in the memosphere. Of course, one way that memes differ from genes is that memes have no fixed locus for alleles and no necessarily finite number of alleles if there were. But perhaps this is only generally true. Perhaps something more like loci and alleles do exist in the domains of evoked culture which the evolutionary psychologists like to stress, as opposed to so-called epidemiological culture—evolutionarily fixed domains that provide slots that then are filled in by the local environments, things like kinship and perhaps religiosity or belief in supreme beings (master worldviews). Thus, there is no general tendency for a *person* to have high or low belief bias, but certain memeplexes might resist conflicting memes better than others.

17. We might well have an innate mechanism for seeking causal explanations, but mystical beliefs, if picked up from the culture, might fit the free parameters of the evolved mechanisms as well as science (see McCauley 2000, D. S. Wilson 2002, and several of the contributions in Keil and Wilson 2000). However, these two memeplexes may have very different effects on the vehicle.

18. One example of a disarming strategy that indefensible memes use is to hide behind the modern shibboleth "everyone has a right to his/her opinion." If taken literally, this claim is trivial. An opinion is merely an idea—a meme. We are not a totalitarian society. In free societies like ours, it is taken for granted that people are allowed to host whatever memes they want as long as they do not lead the host to harm others, in which case the person will be sanctioned. No one has ever denied this so-called "right to hold an opinion." So why is it that we so often hear people demanding this right, when in fact virtually no participant in discussions in free societies wants to exercise mind control—wants, literally, to deprive someone of his/her right to an opinion? Why people continue to utter the phrase "everyone has a right to his/her opinion" when in fact the statement is utterly trivial raises suspicions that people actually do intend to gain some nontrivial consequences from the utterance (see Ruggiero 2000).

By uttering the "right to my opinion" phrase, the person is additionally signaling that no defense of the idea need be presented, so that it is impervious to critique. Thus, it is important to understand that what many people want is much *more* than simply the right to declare and hold their opinions. What they really want is to suppress critique of their opinions. At the very least they are asserting that they are asking their interlocutor to desist from requesting that the meme be rationally defended. The "right to an opinion" disabling trick has symbiotically attached to our concept of tolerance so that it often works quite effectively in our society. It is thought impolite to persist in the request for opinion justification once the person has held up the "right to an opinion" shield. Thus, a very useful inoculating strategy for a meme is to get itself labeled as an opinion and to brandish the "right to an opinion" trump card when its logic or empirical support is weak.

19. One might ask, however, why *shouldn't* we identify our beliefs with our self? And the answer is that of course to some extent we should. Some beliefs should be considered an essential part of the self—particularly the memeplexes that have undergone selective tests and have thus been acquired reflectively. Blackmore (1999) treats the notion of the self as itself a memeplex, but often adopts a hostile stance toward this so-called memeplex self. I am more positively disposed toward this memeplex and its potential for human benefit (memeplex self might help leverage the stance necessary for critiquing TASS responses that are thwarting vehicle-level goals). But as Blackmore (1999) argues, perhaps some housecleaning is in order. A pruning of some of our unsupported memes might leave room for new, life-enhancing ideas to take up residence in our brains. It seems that we have little reason to identify with nonreflectively acquired memes—those, for instance, that are the accidental characteristics of our birth, our parents, and our early environment over which we had little control. Some of our present TASS responses—responses that trigger rapidly and automatically—are accidental memetic features of our backgrounds that were then practiced enough to be compiled as

automatic TASS micromodules. There seems little reason to let this happen again, in our adult lives, when we can exert reflective control over the memes that enter our brains and are resident long enough to become habitual features of our behavior.

Developing the skill of distancing ourselves from currently resident memes may be difficult, however, because several authors have suggested that we have been genetically selected to be uncritical believers (see Boyer 1994, 2001; Dawkins 1995a; McCauley 2000). Nevertheless, my research group has shown that there are at least modest individual differences in the stance people take toward their resident beliefs and toward unfamiliar memes (Sá, West, and Stanovich 1999).

20. As Churchland (1989, 1995) has emphasized, an advancing cognitive science might well have a profound effect on how people examine their cognitive processes and talk about them. Indeed, past advances in psychological understanding have had this effect. People routinely speak of things like introversion and extroversion, and examine their own cognitive performance using terms like short-term memory—all linguistic tools for self-analysis that were unavailable one hundred years ago. It is important to realize that our folk language of the mental evolves (Churchland 1989; Gopnik 1993; Stich 1983; Wilkes 1988)—in part in response to the diffusion of scientific knowledge. "Introversion," "short-term memory," "Freudian slip," "inferiority complex," and "repression" have been woven into the current folk theory of the mental (D'Andrade 1987), as have notions of information processing from computer science and cognitive psychology (Stanovich 1989; Turkle 1984). Not all of these concepts are perspicuous or useful; but some are—and by using them we become more reflective and self-aware cognitive beings.

21. Of course, I do not mean to imply that the critique in this section applies to all evolutionary psychologists. For example, Pinker (1997) does not endorse the culture-on-a-short-leash view and explicitly recognizes the implications of the differing interests of the replicators and the vehicle: "Genes are not puppetmasters; they acted as the recipes for making the brain and body and then they got out of the way. They live in a parallel universe, scattered among bodies, with their own agendas" (401).

Nevertheless, a particular type of error is fostered by the tendency for evolutionary psychologists to attack what Tooby and Cosmides (1992) call the Standard Social Science Model (SSSM). Tooby and Cosmides feel that the default assumptions of the SSSM of most social scientists are that "the individual is the more or less passive recipient of her culture and is the product of that culture" (1992, 32). Tooby and Cosmides wish to diminish the importance of what they term epidemiological culture and emphasize the importance of so-called evoked culture and its assumption that "our developmental and psychological programs evolved to invite the social and cultural worlds in, but only the parts that tended, on balance, to have adaptively useful effects" (87; see Sperber 1996, for a related discussion of the epidemiology of representations). It is for this reason that evolutionary psychologists are hostile to the concept of the meme.

In contrast, memetic theorists have tended to emphasize the importance of the epidemiological properties of memes, although none of them denies the importance of evoked memes (see Boyd and Richerson 2000, for a discussion). Science writer Robert Wright (1994) paraphrases the Tooby and Cosmides position as essentially saying that ideas must "have a kind of harmony with the brains they settle into" (366). However, unlike Tooby and Cosmides (1992), Wright realizes that there are implications that follow from becoming aware of this fact *in the context of replicator/vehicle distinction*. So after noting that ideas must have a kind of harmony with the brains they settle into, Wright (1994) warns that *"that doesn't mean they're good for those brains in the long run"* (366, italics added). A brain that realizes this startling (and still underappreciated) fact might begin the (admittedly difficult) process of slipping culture off the leash of the genes when the culture is dysfunctional for the person. Evolutionary psychologists seem to have underestimated the power of the memes to break the meme/gene linkage.

22. Those not committed a priori to a relativistic denial of the notion of cultural advance might well argue that the history of civilization reflects just this trend. The emancipation of women and the control of our reproductive lives come immediately to mind.

23. Recall the Huckleberry Finn example from chapter 2. Huck's thought that it was morally wrong for slaves to run away and for whites to help them was a meme—a meme in conflict with TASS feelings of friendship and sympathy. Note how in this case we hope that Huck will not be "captured" by this meme, but will instead follow his TASS-based feelings.

Blackmore (1999, 42–46; 2000b) discusses what should and should not be considered a meme. For example, she points out that subjective experiences, sensory experiences, and innate behaviors are not memes. Although some of the short-leashed genetic goals in TASS are not properly viewed as memes, many of the long-leashed goals of the analytic system would be memes ("education will open up many opportunities in future life," "a marital relationship is to be desired," "it is important to be fashionable in one's dress"). These issues are still highly contentious, however. A chapter by Hull (2000) contains a particularly useful discussion of the conceptual distinctions in the science of memetics.

24. Philosopher Daniel Dennett (1991) is referring to the co-adapted meme paradox when he argues that we should not portray the situation as one of "the memes versus us" because our conscious selves are in large part installed (as so-called virtual machine software) memeplexes. Dennett warns that "the 'independent' mind struggling to protect itself from alien and dangerous memes is a myth" (207). Dennett (1993) uses the point in the last quote to critique Dawkins's (1993) rather more optimistic view that the scientific memeplex can help us evaluate our other memes. I agree with Dennett (1993) that Dawkins (1993) has not demonstrated a foundationalist solution to this problem. But I think too that Dawkins is correct to extol the memeplexes of science and rationality as critical to a Neurathian restructuring of our own goal hierarchies in ways that better serve the host.

Chapter Eight

1. In his fine book, *The Concept of the Soul* (2002), which I became aware of after this book was completed, Owen Flanagan argues that the modern concept of soul encompasses four ideas: a nonphysical mind, prospects of immortality, soul as our personal essence, and the idea of free will. Flanagan explores, with great erudition, many of the themes I have dealt with in this volume.

2. I am referring here to a joke that has circulated widely. A version of it is in Steven Hawking's *A Brief History of Time* (1988). One version has philosopher William James after a lecture on the solar system being approached by a determined elderly lady. "We don't live on a ball rotating around the sun," she says. "We live on a crust of earth on the back of a giant turtle." James decides to be gentle: "If your theory is correct, madam, what does this turtle stand on?" The lady says "The first turtle stands on the back of a second, far larger turtle, of course." "But what does this second turtle stand on?" says James. The old lady crows triumphantly. "It's no use, Mr. James—it's turtles all the way down!"

3. The literature on aspects of modern cognitive science that conflict with the

layperson's introspections about mind and consciousness is quite large—e.g., P. M. Churchland 1989, 1995; P. S. Churchland 1986, 2002; P. M. Churchland and P. S. Churchland 1998; Clark 1997, 2001; Coltheart and Davies 2000; Crick 1994; Dennett 1991, 1998; Gazzaniga 1998a; Harnish 2002; Minsky 1985; Ramachandran and Blakeslee 1998.

4. As I discussed in chapter 2, there are respectable and unrespectable ways to discuss executive control in cognitive science. To the extent that the term implies a homunculus sitting there making decisions, then it is little more than a joke from a scientific standpoint. To the extent that it refers to capacity-demanding scheduling computations that are necessitated when the outputs of more automatic mental systems conflict or are indeterminate, then the positing of such an entity is scientifically legitimate. There are many examples of scientifically sound concepts of executive functioning and supervisory attentional systems (Norman and Shallice 1986).

5. The issue of how the alternatives should be contextualized is a version of the construal problem discussed in chapter 4 of Stanovich (1999), and by Hurley (1989, 55–106) as the problem of the eligibility of interpretations (see also Moldoveanu and Langer 2002).

6. For example, a chimp might well refrain from taking a single apple in the presence of a dominant individual (although some animals might not refrain—many of our pets get bitten by other animals for making just this mistake). The only claim being made here is that humans process more, and more complex, contextual information than other animals.

7. Of course, the joke here is that if animals could understand the experiment, they would be nearly human and probably respond as humans do. There is a sense in which the argument to be presented in this chapter, relying as it does on the representational abilities of the analytic system, is an argument for just that. Additionally, it should be noted that in repeated iterations of such a game, it is an open question whether any higher primate might be able to learn to use punishing refusals to shape the partner's response. I know of no animal work on the Ultimatum Game.

8. Recall from chapter 3 that, to a decision theorist, utility is a numerical representation of a preference ordering, and it is traditional to purge it "of any association with hedonistic psychology" (Kahneman 1994, 20). However, recent cognitive science has been more open to re-importing subjective notions of well-being and hedonic intensity into psychology (Diener, Suh, Lucas, and Smith 1999; Kahneman, Diener, and Schwarz 1999; Mellers 2000).

9. The symbolic value of being a certain type of person has been emphasized in numerous philosophical discussions (Dworkin 1988; Flanagan 1996; Frankfurt 1971; Gewirth 1998; Nozick 1981, 1993; Schick 1984; Taylor 1992; Turner 2001), but is only slowly being recognized in economics (Hirschman 1986; Sen 1977, 1987, 1999) and decision theory (Abelson 1996; Beach 1990; Hargreaves Heap 1992; Medin and Bazerman 1999; Medin, Schwartz, Blok, and Birnbaum 1999).

10. This is by analogy to the way that utility is imputed back to actions in causal expected utility theory. In the classical view, the utility of the consequence is imputed back to the action that leads to the consequence due to the causal connection between the two. For example, I assign utility to the act of turning on the radio when I choose to turn it on, but it is not really the turning on that has the utility, it is the music that comes out. Any expected utility I assign to the act of turning the radio on is imputed back to the act from the consequence of listening to the music itself. Symbolic utility involves a kind of "imputing back" but one relying on a symbolic rather than a causal connection. Nozick

(1993) discusses how there are two possible views of instances of symbolic imputing back—one in which it is literally utility that is imputed back and another, the expressive action view, in which "the symbolic connection of an action to a situation enables the action to be expressive of some attitude, belief, value, emotion, or whatever. Expressiveness, not utility, is what flows back. . . . Expressing this has high utility for the person, and so he performs the symbolic action" (28). For example, Nozick discusses how the instrumentally irrational cooperative response in the one-shot Prisoner's Dilemma game might result from symbolic concerns that one present oneself to oneself as being a cooperative person. The two situations are often hard to distinguish, however, and the distinction is not critical for the present discussion.

11. This literature is summarized in many sources—e.g., Atance and O'Neill 2001; Carruthers and Chamberlain 2000; Clark 2001; Cosmides and Tooby 2000a; Currie and Ravenscroft 2002; Davies and Stone 1995a, 1995b; Deacon 1997; Dennett 1984; Dienes and Perner 1999; Donald 1991, 2001; Glenberg 1997; Karmiloff-Smith 1992; Malle, Moses, and Baldwin 2001; Mithen 1996; Nelson 1996; Perner 1991, 1998; Sperber; 2000b, 2000c; Suddendorf and Whiten 2001; Tomasello 1999; Wellman 1990; Zelazo, Astington, and Olson 1999.

12. Frankfurt (1971) distinguishes second-order desires from second-order volitions. The latter occur when the individual wishes a certain desire to become his will. I will ignore this philosophical distinction as well as several others in the discussion that follows. There are, in fact, many important philosophical debates surrounding Frankfurt's paper that I will not be taking up here because they are tangential to the issues that I wish to raise (see Harman 1993; Scanlon 1998). For example, Frankfurt related the second-order concept to issues of free will that are beyond the scope of this volume (see Harman 1993; Lehrer 1990 1997; Watson 1975). Below, I have followed Jeffrey (1974) in using the preference relationship to illustrate second-order judgments in order to signal my sidestepping of these issues. The reader is directed to Perner (1998) for an excellent discussion of the psychological issues involved in taking a stance toward one's own goals.

One additional caveat that deserves mention, however, is that I do not wish to *identify* values with second-order desires. For the discussion here, values are viewed as one thing that prompts the formation and evaluation of states that can be described as second-order preferences or desires; but higher-order preferences do not necessarily arise from values. Not all second-order judgments need have a moral basis. There are many higher-level evaluative states that represent personal or egocentric concerns rather than moral values. As Frankfurt (1982) notes, to critique our first-order desires we use a multitude of specific ideals such as family tradition, the pursuit of mathematical elegance, and devotion to connoisseurship—many of which are not in the moral domain (and many of which are memeplexes).

13. Hurley (1989, 57) discusses the slightly different notions of the preference relation held in different disciplines—for example that it is variously held as relating objects (in economics), actions (in decision theory), and propositions (in Jeffrey's 1983, system). There is a technical literature in decision theory on the nature of the preference relationship (Broome 1991; Maher 1993). I use the preference terminology here loosely, as a way of talking about the strength of desires. There has been considerable theoretical debate about the idea of second-order preferences (Harman 1993; Lehrer 1997; Lewis 1989; Maher 1993; Sen 1992) and some empirical work (e.g., Sahlin 1981).

14. Note that it is not to be assumed that first-order preferences always derive from TASS, nor that they always are short-leashed genetic goals. Likewise it is not to be assumed that all second-order judgments derive from analytic processing (some second-

order judgments have been practiced and automatized to the extent that they become part of TASS). As already explicated in earlier chapters, there is a much wider potential combinatorial mix of possibilities.

15. Another way that second-order preferences can be used in achieving coherence in a preference network is by helping to resolve conflicting first-order preferences (see Lehrer 1997). Imagine that there are two acts, A and B, and that a person prefers doing act A to not doing it and prefers doing act B to not doing it. The only trouble is that A and B are in conflict. They derive from incompatible desires and they are conflicting acts—it is impossible to do both, and furthermore the strength of the person's first-order preference for each is the same. What should this person do? Perhaps engaging in a strong evaluation of the first-order preferences would help. The person might consult the structure of her second-order preference for guidance. Suppose that the person prefers not to prefer A, but that she actually prefers to prefer B. Here, the second-order preferences aid in resolving the conflict. The greater rational integration with respect to B would argue for resolving the conflict in its favor.

16. The hierarchy of higher-order desires (Dworkin 1988; Frankfurt 1971; Hurley 1989; Lehrer 1997; Zimmerman 1981) and the limitations on it imposed by the representational abilities of humans (Davies and Stone 1995a, 1995b; Dienes and Perner 1999; Perner 1991; Sperber 2000b, 2000c; Wellman 1990) have been much discussed.

Finally, it should be noted that Velleman (1992) has discussed how there must be a motive that drives the process of rational integration and has proposed that that motive is the desire to act in accordance with reasons.

17. The use of the term ratified here is not to be confused with Jeffrey's (1983) concept of ratifiability in decision theory.

18. In many philosophical analyses, the so-called phenomenon of weakness of the will is defined as the person choosing based on a first-order preference that is inconsistent with a second-order preference. In such an analysis, the so-called will is identified with the second-order preference and the fact that the first-order preference dictates an alternative response is interpreted as the thwarting of this will. The present analysis rejects not only the homuncular and Cartesian-sounding concept of will, but also rejects the automatic identification of the second-order level with the true self (for discussions of this tradition and criticisms of it, see Bennett 1974; Dworkin 1988, Elster 1983; Frankfurt 1971, 1982; Hurley 1989; Lehrer 1997; MacIntyre 1990; Schick 1984; Zimmerman 1981). The possibility of a third-order judgment is not the only problem here. It is easy to concoct examples where, under the present model, the second-order judgment should have no guaranteed precedence. Suppose one developed a taste for fine wine years ago through an analytic judgment that that was a good thing, and then used explicit rules ("I'll drink three new wines each week") to acquire the taste. After twenty years, you now have a first-order preference for fine wine lodged in TASS. You have recently joined a cultish religion that as part of its memeplex has given you the second order belief: I would prefer to prefer not to drink fine wines because it is decadent. The traditional analysis gives precedence (in terms of your identity and personhood) to the meme that is the second-order preference. On the present view, there is no such privileging—both are planks in a Neurathian structure in equal need of testing. Recall again the Huckleberry Finn example mentioned in chapters 2 and 7.

Likewise, much of our enjoyment in life results from the satisfaction of first-order TASS-based tendencies. We should expect that even conscientious cognitive reform will not, on a token basis, markedly disrupt this fundamental feature of life.

19. We would also be suspicious about a strong evaluator who lacked rational integration but who never succeeded in reversing any of his/her first-order preferences. We

would question the sincerity of second-order values if they never succeeded in reversing a conflicting first-order preference.

20. Some errors can also result from the analytic system learning rules to the extent that they become instantiated as TASS defaults but that then are overgeneralized when they are inflexibly triggered in inappropriate situations. Arkes and Ayton (1999) describe something like this happening when the sunk cost fallacy is committed. People seem to be overgeneralizing a rule of "do not waste"—a rule which might be triggered automatically as a TASS default in subjects who have practiced the rule (which is why sunk costs are sometimes mistakenly honored more by adults than children).

21. Note that two-tiered rationality evaluation should not be viewed as an escape hatch for the Panglossian theorist criticized in earlier chapters. The Panglossian position (of economists, for example) is not that seeming violations of rational strictures occur because individuals are sacrificing instrumental rationality in order to engage in a program of cognitive critique based on higher-order preferences. Instead, they posit that perfect instrumental rationality is attained *whatever* the internal and external perturbations impinge on the agent.

In the course of this two-tier evaluation, we will no doubt find that some irrationality at the instrumental level arises because an individual is aspiring to something more than satisfying first-order desires (i.e., that the very critique of those desires sometimes disrupts the achievement of instrumental rationality by axiomatic criteria). However, not all thin-theory irrationality will derive from such factors, as the review in chapter 4 made clear. How can we tell how much deviation from instrumental rationality has the strong evaluator struggle as its cause? When the violations concern differential contextualization in the way precluded by the axioms of choice, we can garner critical information from the decision maker and apply Wilson and Brekke's (1994) concept of mental contamination. Mental contamination occurs when behavior is affected by factors that you wish were not implicated in your decisions. Recall the experiments on framing effects described in chapter 4 (for example, the disease problem with the 200-will-be-saved and 400-will-die-versions). In such problems, subjects agree in post-experimental interviews that the two versions are identical and that they should not be affected by the wording. In short, these are not alternative contextualizations that subjects *want* to have. Violations of descriptive invariance represent true failures of instrumental rationality. They are not alternative contextualizations caused by differential symbolic utility or by instability deriving from strong evaluation at higher levels of preference. Two-tier rationality evaluation complicates assessment at the instrumental level, but it is still possible to identify thin-theory violation.

22. The complication, and the exception, being TASS processes that have become automatized through explicit analytic system decisions to practice a certain reaction.

23. Religious views that put the focus on God might be thought to be the exception here, but upon closer inspection in the majority of cases, God is looking down on humans with a unique focus on their affairs. He is certainly not directing his eye on their genes. Nor is his attention focused on automata-like animals. God's eyes are squarely focused on us. Indeed, he is in direct contact with our conscious selves, in many religious views. If one looks closely, many religions indirectly put humans—and human consciousness—center-stage. Radcliffe Richards (2000) and Dennett (1995) discuss the tendency to feel that any complexity observed in the world needs to be explained by imputing a conscious intention—either to ourselves or God. Dennett (1995) calls this the "mind comes first" view, and it is a strong default assumption of human thinking.

24. In the classic Prisoner's Dilemma game (Sen 1982), two criminals commit a crime together and are segregated in separate cells. The prosecutor has proof of a joint

minor crime but does not have enough evidence to prove a joint major crime. Each prisoner is asked separately to confess to the major crime. If either confesses and the other does not, the confessor goes free and the other gets the full penalty of twenty years for the major crime. If both confess, both are convicted and get ten year sentences. If neither confesses, they are both convicted of the minor crime and are both sentenced to two years. Each sees the dominance logic that dictates that no matter what the other person does, they are better off confessing. Thus, each does the narrowly rational thing and confesses (NR) and are both convicted of the major crime and given ten year sentences—a much worse outcome than if they had not confessed (C, the cooperative response) and both received two year sentences.

25. This is not to diminish the importance of the problems experienced by the destitute of these societies. It is just that these social trends are more starkly apparent when the middle class is the focus—e.g., de Zengotita 2002; Frank 1999; Frank and Cook 1995; Kuttner 1998; Lasn 1999; Myers 2000; Schlosser 2001.

26. Economists use technical terms like Pareto efficiency and Pareto improvement to characterize the type of rational outcome that a market achieves. The technical vocabulary hides from the public how weak are the conditions that these versions of efficiency achieve. Imagine three hypothetical societies whose economic systems allocate economic power (money) to John and Phil. In Society A, John achieves $5,000 of wealth and Phil achieves $8,000. In two other imaginary societies, the outcome is:

Society B: John, $9,000; Phil, $55,000
Society C: John, $53,000; Phil, $54,000

A move from Society A to Society B would be technically termed a Pareto improvement, and this seems to make sense. However, we can see how weak is the efficiency that a perfect market achieves by noting that a move from Society B to Society C would *not* be deemed a Pareto improvement. I kid you not (see Hausman and McPherson 1994; Pearce 1992; Quirk 1987).

27. Markets let people reshuffle their desires and dollars in ways that optimally satisfy the desires that are connected (the desires, not the people). Market efficiency is thus a very special type of efficiency that does not map into the layperson's notions of efficiency or optimality (which almost always include distributional notions ignored in many economic analyses). Markets let everyone improve their positions by trading and they guarantee that these improvements will be efficient in this special sense, given the prior distribution of skills and assets. This reallocation is again different in a very special way—it occurs after what economists call prior determinations. Prior determinations are the skills and financial assets that people bring to the market and that enable them to trade (Lindblom 2001). As Lindblom argues, markets "do not permit people to erase or escape from the inefficiencies of prior determinations. In that sense, market efficiency is too little too late. It kicks in when earlier decisions not made with an eye to efficiency have already largely determined the results" (2001, 172). In short, the allocations of assets prior to the latest round of market transactions, the very allocations that overwhelmingly determine who gets what in the next round of trading, are totally unaddressed by notions of "market efficiency."

28. The subject of interpersonal utility comparisons is a topic of much contention and complexity; see Baron 1993a; Brink 1993; Elster and Roemer 1991; Hurley 1989, 360; Nagel 1997, 123; Parfit 1984, 330; Rawls 1971, 27–29, 187; Sen and Williams 1982.

29. The inexorable logic of this market reinforcement of first-order preferences and the erosion of our values that it encourages is at least becoming more widely commented upon and debated. For example, a letter written to the *New York Times* (Simms

2002) argues that "nothing has so harmed this country as this fixation on price and price alone, with no regard for the collateral damage." Citing ugly big-box architecture, the destruction of the viable economies of small towns, the endless sprawl that is increasing every worker's commuting time, and "a fatally high-energy lifestyle" the letter writer says that "one can only conclude that price has become a drug and that we are an addict nation."

30. Just like the market, the genes don't distinguish between a wanton and a strong evaluator as long as the first-order preference that is determining choice is the same. If the genes care that you do A rather than B and have structured your preferences accordingly, they don't give a damn what your second order preferences are (that you [(B pref A) pref (A pref B)]) as long as you don't actually switch your first-order preference to B over A.

31. The literature on different aspects of consciousness has been surveyed many times—e.g., Atkinson, Thomas, and Cleeremans, 2000; Block 1995; Carruthers 2000; P. S. Churchland 2002; Flanagan 1992; Rosenthal 1986; Wegner and Wheatley 1999; Wilkes 1988. On access consciousness, see Baars 1997, 2002; Damasio 1999; Dehaene and Naccache 2001; Dennett 2001; Nozick 2001; and Pinker 1997.

32. See chapter 4 and a host of other research programs—e.g., Ainslie 2001; Gigerenzer and Todd 1999; Gilovich, Griffin, and Kahneman 2002; Goel and Dolan 2003; Hastie and Dawes 2001; Hogarth 2001; Kahneman and Tversky 2000; Stanovich 1999; Todd and Gigerenzer 2000.

33. See Flanagan 2002; Millgram 1997; Nathanson 1994; Nozick 1993; Rawls 2001; and Scanlon 1998.

34. The empirical and philosophical literature on the representation of mental states is large—see Carruthers 1998, 2002; Clark 2001; Clark and Karmiloff-Smith 1993; Dienes and Perner 1999; Evans and Over 1999; Leslie 2000; Lillard 2001; Perner 1991, 1998; Sperber 2000b; Sterelny 2001b; Tomasello 1999; Wellman 1990; Whiten 2001; Zelazo, Astington, and Olson 1999.

35. For arguments that the causal chain might run the other way, see Baumeister, Boden, and Smart 1996; Dawes 1994; Kahne 1996; Ruggiero 2000; and Stout 2000.

REFERENCES

Abelson, R. P. 1996. The secret existence of expressive behavior. In *The rational choice controversy*, ed. J. Friedman, 25–36. New Haven, Conn.: Yale University Press.

Ackerman, P., P. Kyllonen, and R. Richards, eds. 1999. *Learning and individual differences: Process, trait, and content determinants.* Washington, D.C.: American Psychological Association.

Adams, M. J. 1989. Thinking skills curricula: Their promise and progress. *Educational Psychologist* 24:25–77.

Adams, M. J., and C. E. Feehrer. 1991. Thinking and decision making. In *Teaching decision making to adolescents*, ed. J. Baron and R. V. Brown, 79–94. Hillsdale, N.J.: Lawrence Erlbaum Associates.

Adler, J. E. 1984. Abstraction is uncooperative. *Journal for the Theory of Social Behaviour* 14:165–81.

———. 1991. An optimist's pessimism: Conversation and conjunctions. In *Probability and rationality: Studies on L. Jonathan Cohen's philosophy of science,* ed. E. Eells and T. Maruszewski, 251–82. Amsterdam: Editions Rodopi.

Ahluwalia, R., and Z. Gurhan-Canli. 2000. The effects of extensions on the family brand name: An accessibility-diagnosticity perspective. *Journal of Consumer Research* 27:371–81.

Ainslie, G. 1982. A behavioral economic approach to the defence mechanisms: Freud's energy theory revisited. *Social Science Information* 21:735–80.

———. 1984. Behavioral economics II: Motivated involuntary behavior. *Social Science Information* 23:247–74.

———. 1992. *Picoeconomics.* Cambridge: Cambridge University Press.

———. 2001. *Breakdown of will.* Cambridge: Cambridge University Press.

Allais, M. 1953. Le comportement de l'homme rationnel devant le risque: Critique des postulats et axioms de l'scole americaine. *Econometrica* 21:503–46.

Allan, L. G. 1980. A note on measurement of contingency between two binary variables in judgment tasks. *Bulletin of the Psychonomic Society* 15:147–49.

Alloy, L. B., and N. Tabachnik. 1984. Assessment of covariation by humans and animals: The joint influence of prior expectations and current situational information. *Psychological Review* 91:112–49.

Anderson, E. 1993. *Value in ethics and economics.* Cambridge, Mass.: Harvard University Press.

Anderson, J. R. 1983. *The architecture of cognition.* Cambridge, Mass.: Harvard University Press.

———. 1990. *The adaptive character of thought.* Hillsdale, N.J.: Lawrence Erlbaum Associates.

———. 1991. Is human cognition adaptive? *Behavioral and Brain Sciences* 14:471–517.

Anderson, J. R., L. M. Reder, and H. A. Simon. 1996. Situated learning and education. *Educational Researcher* 25, no. 4: 5–11.

Anderson, M. 1992. *Intelligence and development: A cognitive theory.* Oxford: Blackwell.

———. 1998. Mental retardation, general intelligence, and modularity. *Learning and Individual Differences* 10:159–78.

Arkes, H. R., and P. Ayton. 1999. The sunk cost and Concorde effects: Are humans less rational than lower animals? *Psychological Bulletin* 125:591–600.

Arkes, H. R., and A. R. Harkness. 1983. Estimates of contingency between two dichotomous variables. *Journal of Experimental Psychology: General* 112:117–35.

Atance, C. M., and D. K. O'Neill. 2001. Episodic future thinking. *Trends in Cognitive Sciences* 5:533–39.

Atkinson, A. P., M. Thomas, and A. Cleeremans. 2000. Consciousness: Mapping the theoretical landscape. *Trends in Cognitive Sciences* 4:372–82.

Atran, S. 1998. Folk biology and the anthropology of science: Cognitive universals and cultural particulars. *Behavioral and Brain Sciences* 21:547–609.

Audi, R. 1993a. *Action, intention, and reason.* Ithaca, N.Y.: Cornell University Press.

———. 1993b. *The structure of justification.* Cambridge: Cambridge University Press.

———. 2001. *The architecture of reason: The structure and substance of rationality.* Oxford: Oxford University Press.

Aunger, R. 2000a. *Darwinizing culture: The status of memetics as a science.* Oxford: Oxford University Press.

———. 2000b. Introduction. In *Darwinizing culture: The status of memetics as a science,* ed. R. Aunger, 1–23. Oxford: Oxford University Press.

———. 2002. *The electric meme: A new theory of how we think.* New York: Free Press.

Austin, E. J., and I. J. Deary. 2002. Personality dispositions. In *Why smart people can be so stupid,* ed. R. J. Sternberg, 187–211. New Haven, Conn.: Yale University Press.

Baars, B. J. 1997. *In the theater of consciousness: The workspace of the mind.* Oxford: Oxford University Press.

———. 2002. The cognitive access hypothesis: Origins and recent evidence. *Trends in Cognitive Sciences* 6:47–52.

Babad, E., and Y. Katz. 1991. Wishful thinking—Against all odds. *Journal of Applied Social Psychology* 21:1921–38.

Badcock, C. 2000. *Evolutionary psychology: A critical introduction.* Cambridge, England: Polity Press.

Baddeley, A. 1996. Exploring the central executive. *Quarterly Journal of Experimental Psychology* 49A:5–28.

Baddeley, A., D. Chincotta, and A. Adlam. 2001. Working memory and the control of action: Evidence from task switching. *Journal of Experimental Psychology: General* 130:641–57.

Baldwin, D. A. 2000. Interpersonal understanding fuels knowledge acquisition. *Current Directions in Psychological Science* 9:40–45.

Ball, J. A. 1984. Memes as replicators. *Ethology and Sociobiology* 5:145–61.

Baltes, P. B. 1987. Theoretical propositions of life-span developmental psychology: On the dynamics between growth and decline. *Developmental Psychology* 23:611–26.

Barash, D. 2001. *Revolutionary biology: The new gene-centered view of life.* New Brunswick, N.J.: Transaction.

Bargh, J. A., and T. L. Chartrand. 1999. The unbearable automaticity of being. *American Psychologist* 54: 462–79.

Bar-Hillel, M., and D. Budescu. 1995. The elusive wishful thinking effect. *Thinking and Reasoning* 1:71–103.

Barkley, R. A. 1998. Behavioral inhibition, sustained attention, and executive functions: Constructing a unifying theory of ADHD. *Psychological Bulletin* 121:65–94.

Barkow, J. H. 1989. *Darwin, sex, and status: Biological approaches to mind and culture.* Toronto: University of Toronto Press.

Barkow, J., L. Cosmides, and J. Tooby, eds. 1992. *The adapted mind.* New York: Oxford University Press.

Baron, J. 1985. *Rationality and intelligence.* Cambridge: Cambridge University Press.

———. 1993. *Morality and rational choice.* Dordrecht: Kluwer.

———. 1998. *Judgment misguided: Intuition and error in public decision making.* New York: Oxford University Press.

———. 1999. Utility maximization as a solution: Promise, difficulties, and impediments. *American Behavioral Scientist* 42:1301–21.

———. 2000. *Thinking and deciding.* 3d ed. Cambridge: Cambridge University Press.

Baron, J., and R. V. Brown, eds. 1991. *Teaching decision making to adolescents.* Hillsdale, N.J.: Lawrence Erlbaum Associates.

Baron, J., and S. Leshner. 2000. How serious are expressions of protected values? *Journal of Experimental Psychology: Applied* 6:183–94.

Baron, J., and M. Spranca. 1997. Protected values. *Organizational Behavior and Human Decision Processes* 70:1–16.

Baron-Cohen, S. 1995. *Mindblindness: An essay on autism and theory of mind.* Cambridge, Mass.: MIT Press.

———. 1998. Does the study of autism justify minimalist innate modularity? *Learning and Individual Differences* 10:179–92.

Baron-Cohen, S., H. Tager-Flusberg, and D. J. Cohen, eds. 2000. Palaeoanthropological perspectives on the theory of mind. *Understanding other minds: Perspectives from developmental cognitive neuroscience.* 2d ed. Oxford: Oxford University Press.

Barrett, L., R. Dunbar, and J. Lycett, J. 2002. *Human evolutionary psychology.* Princeton, N.J.: Princeton University Press.

Barton, R. A., and R. Dunbar. 1997. Evolution of the social brain. In *Machiavellian intelligence II: Extensions and evaluations,* ed. A. Whiten and R. W. Byrne, 240–63. Cambridge: Cambridge University Press.

Baumeister, R. F., J. M. Boden, and L. Smart. 1996. Relation of threatened egotism to violence and aggression: The dark side of high self-esteem. *Psychological Review* 103:5–33.

Bazerman, M. 1999. *Smart money decisions.* New York: Wiley.

———. 2001. Consumer research for consumers. *Journal of Consumer Research* 27: 499–504.

Bazerman, M., J. Baron, and K. Shonk. 2001. *"You can't enlarge the pie": Six barriers to effective government.* New York: Basic Books.

Bazerman, M., A. Tenbrunsel, and K. Wade-Benzoni. 1998. Negotiating with yourself and losing: Understanding and managing conflicting internal preferences. *Academy of Management Review* 23:225–41.

Beach, L. R. 1990. *Image theory: Decision making in personal and organizational contexts.* Chichester, England: Wiley.

Bechara, A., A. R. Damasio, H. Damasio, and S. Anderson. 1994. Insensitivity to future consequences following damage to human prefrontal cortex. *Cognition* 50:7–15.

Bechara, A., H. Damasio, D. Tranel, and A. R. Damasio. 1997. Deciding advantageously before knowing the advantageous strategy. *Science* 275 (Feb. 26): 1293–95.

Bechara, A., D. Tranel, and H. Damasio. 2000. Characterization of the decision-making deficit of patients with ventromedial prefrontal cortex lesions. *Brain* 123:2189–2202.

Bell, D. E. 1982. Regret in decision making under uncertainty. *Operations Research* 30:961–81.

Belsky, G., and T. Gilovich. 1999. *Why smart people make big money mistakes—And how to correct them: Lessons from the new science of behavioral economics.* New York: Simon and Schuster.

Bennett, J. 1974. The conscience of Huckleberry Finn. *Philosophy* 49:123–34.

Bereiter, C. 1997. Situated cognition and how to overcome it. In *Situated cognition: Social, semiotic, and psychological perspectives,* ed. D. Kirshner and J. A. Whitson, 3–12. Mahwah, N.J.: Lawrence Erlbaum Associates.

———. 2002. *Education and mind in the knowledge age.* Mahwah, N.J.: Lawrence Erlbaum Associates.

Beyth-Marom, R., B. Fischhoff, M. Quadrel, and L. Furby. 1991. Teaching decision making to adolescents: A critical review. In *Teaching decision making to adolescents,* ed. J. Baron and R. V. Brown, 19–59. Hillsdale, N.J.: Lawrence Erlbaum Associates.

Bickerton, D. 1995. *Language and human behavior.* Seattle: University of Washington Press.

Birnbaum, M. H. 1983. Base rates in Bayesian inference: Signal detection analysis of the cab problem. *American Journal of Psychology* 96:85–94.

Bjorklund, D. F., and A. D. Pellegrini. 2000. Child development and evolutionary psychology. *Child Development* 71:1687–1708.

———. 2002. *The origins of human nature: Evolutionary developmental psychology.* Washington D.C.: American Psychological Association.

Blackmore, S. 1999. *The meme machine.* New York: Oxford University Press.

———. 2000a. The memes' eye view. In *Darwinizing culture: The status of memetics as a science,* ed. R. Aunger, 25–42. Oxford: Oxford University Press.

———. 2000b. The power of memes. *Scientific American* 283, no. 4 (October): 64–73.

Block, N. 1995. On a confusion about a function of consciousness. *Behavioral and Brain Sciences* 18:227–87.

Bogdan, R. J. 2000. *Minding minds: Evolving a reflexive mind by interpreting others.* Cambridge, Mass.: MIT Press.

Borges, B., D. G. Goldstein, A. Ortmann, and G. Gigerenzer. 1999. Can ignorance beat the stock market? In *Simple heuristics that make us smart,* ed. G. Gigerenzer and P. M. Todd, 59–72. New York: Oxford University Press.

Boyd, R., P. Gasper, and J. D. Trout, eds. 1991. *The philosophy of science.* Cambridge, Mass.: MIT Press.

Boyd, R., and P. J. Richerson. 2000. Memes: Universal acid or a better mousetrap? In *Darwinizing culture: The status of memetics as a science,* ed. R. Aunger, 143–62. Oxford: Oxford University Press.

Boyer, P. 1994. *The naturalness of religious ideas: A cognitive theory of religion.* Berkeley and Los Angeles: University of California Press.

———. 2001. *Religion explained: The evolutionary origins of religious thought.* New York: Basic Books.

Braid, M. 2001. "All I have to do is believe." *Independent Magazine* (London). October 6, 10–17.

Brainerd, C. J., and V. F. Reyna. 2001. Fuzzy-trace theory: Dual processes in memory, reasoning, and cognitive neuroscience. In *Advances in child development and behavior,* vol. 28, ed. H. W. Reese and R. Kail, 41–100. San Diego: Academic Press.

Brase, G. L., L. Cosmides, and J. Tooby. 1998. Individuation, counting, and statistical inference: The role of frequency and whole-object representations in judgment under uncertainty. *Journal of Experimental Psychology: General* 127:3–21.

Bratman, M. E. 1987. *Intention, plans, and practical reason.* Cambridge, Mass.: Harvard University Press.

Bratman, M. E., D. J. Israel, and M. E. Pollack. 1991. Plans and resource-bounded practical reasoning. In *Philosophy and AI: Essays at the interface,* ed. J. Cummins and J. Pollock, 7–22. Cambridge, Mass.: MIT Press.

Brauer, M., W. Wasel, and P. Niedenthal. 2000. Implicit and explicit components of prejudice. *Review of General Psychology* 4:79–101.

Brenner, L. A., D. J. Koehler, V. Liberman, and A. Tversky. 1996. Overconfidence in probability and frequency judgments: A critical examination. *Organizational Behavior and Human Decision Processes* 65:212–19.

Brewer, N. T., and G. Chapman. 2002. The fragile basic anchoring effect. *Journal of Behavioral Decision Making* 15:65–77.

Brink, D. 1993. The separateness of persons, distributive norms, and moral theory. In *Value, welfare, and morality,* ed. R. G. Frey and C. W. Morris, 252–89. Cambridge: Cambridge University Press.

Brody, N. 1997. Intelligence, schooling, and society. *American Psychologist* 52:1046–50.

Bronfenbrenner, U., P. McClelland, E. Wethington, P. Moen, and S. J. Ceci. 1996. *The state of Americans.* New York: Free Press.

Broome, J. 1990. Should a rational agent maximize expected utility? In *The limits of rationality,* ed. K. S. Cook and M. Levi, 132–45. Chicago: University of Chicago Press.

———. 1991. *Weighing goods: Equality, uncertainty, and time.* Oxford: Blackwell.

Brothers, L. 1990. The social brain: A project for integrating primate behaviour and neuropsychology in a new domain. *Concepts in Neuroscience* 1:27–51.

Brown, H. I. 1977. *Perception, theory and commitment: The new philosophy of science.* Chicago: University of Chicago Press.

Bugental, D. B. 2000. Acquisitions of the algorithms of social life: A domain-based approach. *Psychological Bulletin* 126:187–219.

Buss, D. M. 1989. Sex differences in human mate preferences: Evolutionary hypotheses tested in 37 cultures. *Behavioural and Brain Sciences* 12:1–49.

———. 1999. *Evolutionary psychology: The new science of the mind.* Boston: Allyn and Bacon.

———. 2000. The evolution of happiness. *American Psychologist* 55:15–23.

Buss, D. M., M. G. Haselton, T. Shackelford, A. Beske, and J. Wakefield. 1998. Adaptations, exaptations, and spandrels. *American Psychologist* 53:533–48.

Byrne, R. W., and A. Whiten, eds. 1988. *Machiavellian intelligence: Social expertise and the evolution of intellect in monkeys, apes, and humans.* Oxford: Oxford University Press.

Cacioppo, J. T., and G. G. Berntson. 1999. The affect system: Architecture and operating characteristics. *Current Directions in Psychological Science* 8:133–37.

Cacioppo, J. T., R. E. Petty, J. Feinstein, and W. Jarvis. 1996. Dispositional differences in cognitive motivation: The life and times of individuals varying in need for cognition. *Psychological Bulletin* 119:197–253.

Cahan, S., and L. Artman. 1997. Is everyday experience dysfunctional for the development of conditional reasoning? *Cognitive Development* 12:261–79.

Cairns-Smith, A. G. 1996. *Evolving the mind: On the nature of matter and the origin of consciousness.* Cambridge: Cambridge University Press.

Calvin, W. 1990. *The cerebral symphony.* New York: Bantam.

Camerer, C. 1995. Individual decision making. In *The handbook of experimental economics,* ed. J. H. Kagel and A. E. Roth, 587–703. Princeton, N.J.: Princeton University Press.

Caporael, L. R. 1997. The evolution of truly social cognition: The core configurations model. *Personality and Social Psychology Review* 1:276–98.

Carey, S. 1985. *Conceptual change in childhood.* Cambridge, Mass.: MIT Press.

Carpenter, P. A., M. A. Just, and P. Shell. 1990. What one intelligence test measures: A theoretical account of the processing in the Raven Progressive Matrices Test. *Psychological Review* 97:404–31.

Carr, T. H. 1992. Automaticity and cognitive anatomy: Is word recognition "automatic"? *American Journal of Psychology* 105:201–37.

Carroll, J. B. 1993. *Human cognitive abilities: A survey of factor-analytic studies.* Cambridge: Cambridge University Press.

———. 1997. Psychometrics, intelligence, and public perception. *Intelligence* 24:25–52.

Carruthers, P. 1998. Thinking in language? Evolution and a modularist possibility. In *Language and thought: Interdisciplinary themes,* ed. P. Carruthers and J. Boucher, 94–119. Cambridge: Cambridge University Press.

———. 2000. The evolution of consciousness. In *Evolution and the human mind: Modularity, language and meta-cognition,* ed. P. Carruthers and A. Chamberlain, 254–75. Cambridge: Cambridge University Press.

———. 2002. The cognitive functions of language. *Behavioral and Brain Sciences* 25: 657–726.

Carruthers, P., and A. Chamberlain, eds. 2000. *Evolution and the human mind: Modularity, language and meta-cognition.* Cambridge: Cambridge University Press.

Cartwright, J. 2000. *Evolution and human behavior.* Cambridge, Mass.: MIT Press.

Case, R. 1992. The role of the frontal lobes in the regulation of cognitive development. *Brain and Cognition* 20:51–73.

Cavalli-Sforza, L. L., and M. W. Feldman. 1981. *Cultural transmission and evolution: A quantitative approach.* Princeton, N.J.: Princeton University Press.

Chaiken, S., A. Liberman, and A. H. Eagly. 1989. Heuristic and systematic information within and beyond the persuasion context. In *Unintended thought,* ed. J. S. Uleman and J. A. Bargh, 212–52. New York: Guilford Press.

Chalmers, D. J. 1996. *The conscious mind: In search of a fundamental theory.* Oxford: Oxford University Press.

Chapman, M. 1993. Everyday reasoning and the revision of belief. In *Mechanisms of everyday cognition,* ed. J. M. Puckett and H. W. Reese, 95–113. Hillsdale, N.J.: Lawrence Erlbaum Associates.

Cheesman, J., and P. M. Merikle. 1984. Priming with and without awareness. *Perception and Psychophysics* 36:387–95.

———. 1986. Distinguishing conscious from unconscious perceptual processes. *Canadian Journal of Psychology* 40:343–67.

Cheng, P. W., and K. J. Holyoak. 1989. On the natural selection of reasoning theories. *Cognition* 33:285–313.

Cherniak, C. 1986. *Minimal rationality.* Cambridge, Mass.: MIT Press.

Churchland, P. M. 1989. *A neurocomputational perspective: The nature of mind and the structure of science.* Cambridge: MIT Press.

———. 1995. *The engine of reason, the seat of the soul.* Cambridge, Mass.: MIT Press.

Churchland, P. M., and P. S. Churchland. 1998. *On the contrary: Critical essays, 1987–1997.* Cambridge, Mass.: MIT Press.

Churchland, P. S. 1986. *Neurophilosophy: Toward a unified science of the mind/brain.* Cambridge, Mass.: MIT Press.

———. 2002. *Brain-wise: Studies in neurophilosophy.* Cambridge, Mass.: MIT Press.

Clark, A. 1996. Connectionism, moral cognition, and collaborative problem solving. In *Mind and morals,* ed. A. May, M. Friedman, and A. Clark, 109–27. Cambridge, Mass.: MIT Press.

———. 1997. *Being there: Putting brain, body, and world together again.* Cambridge, Mass.: MIT Press.

———. 2001. *Mindware: An introduction to the philosophy of cognitive science.* New York: Oxford University Press.

Clark, A., and A. Karmiloff-Smith. 1993. The cognizer's innards: A psychological and philosophical perspective on the development of thought. *Mind and Language* 8:487–519.

Code, L. 1987. *Epistemic responsibility.* Hanover, N.H.: University Press of New England.

Cohen, L. J. 1981. Can human irrationality be experimentally demonstrated? *Behavioral and Brain Sciences* 4:317–70.

———. 1983. The controversy about irrationality. *Behavioral and Brain Sciences* 6:510–17.

———. 1986. *The dialogue of reason.* Oxford: Oxford University Press.

Consumer Reports. 1998. Buying or leasing a car. Vol. 63, no. 4 (April): 16–22.

Colman, A. M. 1995. *Game theory and its applications.* Oxford: Butterworth-Heinemann.

Coltheart, M. 1999. Modularity and cognition. *Trends in Cognitive Sciences* 3:115–20.

Coltheart, M., and M. Davies, eds. 2000. *Pathologies of belief.* Oxford: Blackwell.

Connolly, T., H. R. Arkes, and K. R. Hammond, eds. 2000. *Judgment and decision making: An interdisciplinary reader.* 2d ed. Cambridge: Cambridge University Press.

Cooper, W. S. 1989. How evolutionary biology challenges the classical theory of rational choice. *Biology and Philosophy* 4:457–81.

Cosmides, L. 1989. The logic of social exchange: Has natural selection shaped how humans reason? Studies with the Wason selection task. *Cognition* 31:187–276.

Cosmides, L., and J. Tooby. 1992. Cognitive adaptations for social exchange. In *The adapted mind,* ed. J. Barkow, L. Cosmides, and J. Tooby, 163–205. New York: Oxford University Press.

———. 1994a. Better than rational: Evolutionary psychology and the invisible hand. *American Economic Review* 84:327–32.

———. 1994b. Beyond intuition and instinct blindness: Toward an evolutionarily rigorous cognitive science. *Cognition* 50:41–77.

———. 1996. Are humans good intuitive statisticians after all? Rethinking some conclusions from the literature on judgment under uncertainty. *Cognition* 58:1–73.

———. 2000a. Consider the source: The evolution of adaptations for decoupling and metarepresentation. In *Metarepresentations: A multidisciplinary perspective,* ed. D. Sperber, 53–115. Oxford: Oxford University Press.

———. 2000b. Evolutionary psychology and the emotions. In *Handbook of emotions,* 2d ed., ed. M. Lewis and J. M. Haviland-Jones, 91–115. New York: Guilford Press.

Crick, F. 1994. *The astonishing hypothesis.* New York: Simon and Schuster.

Cronin, H. 1991. *The ant and the peacock.* Cambridge: Cambridge University Press.

Cummins, D. D. 1996. Evidence for the innateness of deontic reasoning. *Mind and Language* 11:160–90.

———. 2002. The evolutionary roots of intelligence and rationality. In *Common sense, reasoning, and rationality,* ed. R. Elio, 132–47. Oxford: Oxford University Press.

Currie, G., and I. Ravenscroft. 2002. *Recreative minds.* Oxford: Oxford University Press.

Cziko, G. 1995. *Without miracles: Universal selection theory and the second Darwinian revolution.* Cambridge, Mass.: MIT Press.

Daly, M., and M. Wilson. 1983. *Sex, evolution, and behavior.* 2d ed. Belmont, Calif.: Wadsworth.

Damasio, A. R. 1994. *Descartes' error.* New York: Putnam.

———. 1999. *The feeling of what happens.* New York: Harcourt Brace.

D'Andrade, R. 1987. A folk model of the mind. In *Cultural models in language and thought,* ed. D. Holland and N. Quinn, 112–48. Cambridge: Cambridge University Press.

Davenport, T., and J. Beck. 2001. *The attention economy.* Cambridge, Mass.: Harvard Business School Press.

Davies, M., and M. Coltheart. 2000. Introduction: Pathologies of belief. In *Pathologies of belief,* ed. M. Coltheart and M. Davies, 1–46. Oxford: Blackwell.

Davies, M., and T. Stone, eds. 1995a. *Folk psychology.* Oxford: Blackwell.

———, eds. 1995b. *Mental simulation.* Oxford: Blackwell.

Davis, D., and C. Holt. 1993. *Experimental economics.* Princeton, N.J.: Princeton University Press.

Dawes, R. M. 1983. Is irrationality systematic? *Behavioral and Brain Sciences* 6:491–92.

———. 1988. *Rational choice in an uncertain world.* San Diego, Calif.: Harcourt Brace Jovanovich.

———. 1989. Statistical criteria for establishing a truly false consensus effect. *Journal of Experimental Social Psychology* 25:1–17.

———. 1990. The potential nonfalsity of the false consensus effect. In *Insights into decision making,* ed. R. M. Hogarth, 179–99. Chicago: University of Chicago Press.

———. 1994. *House of cards: Psychology and psychotherapy based on myth.* New York: Free Press.

———. 1998. Behavioral decision making and judgment. In *The handbook of social psychology,* ed. D. T. Gilbert, S. T. Fiske, and G. Lindzey, 1:497–548. Boston: McGraw-Hill.

———. 2001. *Everyday irrationality.* Boulder, Colo.: Westview Press.

Dawkins, R. 1976. *The selfish gene.* New ed., 1989. New York: Oxford University Press.

———. 1982. *The extended phenotype.* New York: Oxford University Press.

———. 1983. Universal Darwinism. In *Evolution from molecules to men,* ed. D. S. Bendall, 403–25. Cambridge: Cambridge University Press.

———. 1986. *The blind watchmaker.* New York: Norton.

———. 1993. Viruses of the mind. In *Dennett and his critics,* ed. B. Dahlbom, 13–27. Cambridge, Mass.: Blackwell.

———. 1995. Putting away childish things. *Skeptical Inquirer,* January, 139.

———. 1996. *Climbing mount improbable.* New York: Norton.

———. 1999. Foreword. In S. Blackmore, *The meme machine,* vii–xvii. New York: Oxford University Press.

Dawson, E., T. Gilovich, and D. T. Regan. 2002. Motivated reasoning and performance on the Wason selection task. *Personality and Social Psychology Bulletin* 28:1379–87.

Deacon, T. 1997. *The symbolic species: The co-evolution of language and brain.* New York: Norton.

Deary, I. J. 2000. *Looking down on human intelligence: From psychometrics to the brain.* Oxford: Oxford University Press.

Dehaene, S., and L. Naccache. 2001. Towards a cognitive neuroscience of consciousness: Basic evidence and a workspace framework. *Cognition* 79:1–37.

Dempster, F. N. 1992. The rise and fall of the inhibitory mechanism: Toward a unified theory of cognitive development and aging. *Developmental Review* 12:45–75.

Dempster, F. N., and A. J. Corkill. 1999. Interference and inhibition in cognition and behavior: Unifying themes for educational psychology. *Educational Psychology Review* 11:1–88.

Denes-Raj, V., and S. Epstein. 1994. Conflict between intuitive and rational processing: When people behave against their better judgment. *Journal of Personality and Social Psychology* 66:819–29.

Dennett, D.C. 1975. Why the law of effect will not go away. *Journal of the Theory of Social Behavior* 2:169–87.

———. 1978. *Brainstorms: Philosophical essays on mind and psychology.* Cambridge, Mass.: MIT Press.

———. 1984. *Elbow room: The varieties of free will worth wanting.* Cambridge, Mass.: MIT Press.

———. 1987. *The intentional stance.* Cambridge, Mass.: MIT Press.

———. 1988. Precis of *The intentional stance. Behavioral and Brain Sciences* 11:493–544.

———. 1991. *Consciousness explained.* Boston: Little, Brown.

———. 1993. Back from the drawing board. In *Dennett and his critics,* ed. B. Dahlbom, 203–35. Cambridge, Mass.: Blackwell.

———. 1995. *Darwin's dangerous idea: Evolution and the meanings of life.* New York: Simon and Schuster.

———. 1996. *Kinds of minds: Toward an understanding of consciousness.* New York: Basic Books.

———. 2001. Are we explaining consciousness yet? *Cognition* 79:221–37.

Denzau, A. T., and D.C. North. 1994. Shared mental models: Ideologies and institutions. *Kyklos* 47:3–31.

de Sousa, R. 1987. *The rationality of emotion.* Cambridge, Mass.: MIT Press.

de Waal, F. 2002. Evolutionary psychology: The wheat and the chaff. *Current Directions in Psychological Science* 11, no. 6: 187–91.

de Zengotita, T. 2002. The numbing of the American mind: Culture as anesthetic. *Harper's Magazine,* April, 33–40.

Diener, E., E. M. Suh, R. E. Lucas, and H. L. Smith. 1999. Subjective well-being: Three decades of progress. *Psychological Bulletin* 125:276–302.

Dienes, Z., and J. Perner, J. 1999. A theory of implicit and explicit knowledge. *Behavioral and Brain Sciences* 22:735–808.

Donald, M. 1991. *Origins of the modern mind: Three stages in the evolution of culture and cognition.* Cambridge, Mass.: Harvard University Press.

———. 2001. *A mind so rare: The evolution of human consciousness.* New York: Norton.

Donaldson, M. 1978. *Children's minds.* London: Fontana Paperbacks.

———. 1993. *Human minds: An exploration.* New York: Viking Penguin.

Drabble, M. 2001. *The peppered moth.* Toronto: McClelland and Stewart.

Dube, F. 2001. Quebec man kills self in act of road rage. *National Post* (Toronto), Nov. 9, A3.

Dugatkin, L. A. 2000. *The imitation factor: Evolution beyond the gene.* New York: Free Press.

Dulany, D. E., and D. J. Hilton. 1991. Conversational implicature, conscious representation, and the conjunction fallacy. *Social Cognition* 9:85–110.

Dunbar, R. 1998. Theory of mind and the evolution of language. In *Approaches to the evolution of language,* ed. J. R. Hurford, M. Studdert-Kennedy, and C. Knight, 92–110. Cambridge: Cambridge University Press.

Duncan, J., H. Emslie, P. Williams, R. Johnson, and C. Freer. 1996. Intelligence and the frontal lobe: The organization of goal-directed behavior. *Cognitive Psychology* 30:257–303.

Durham, W. 1991. *Coevolution: Genes, culture, and human diversity.* Stanford, Calif.: Stanford University Press.

Dworkin, G. 1988. *The theory and practice of autonomy.* Cambridge: Cambridge University Press.

Dyer, F. N. 1973. The Stroop phenomenon and its use in the study of perceptual, cognitive, and response processes. *Memory & Cognition* 1:106–20.

Earman, J. 1992. *Bayes or bust*. Cambridge, Mass.: MIT Press.

Edelman, G. M. 1992. *Bright air, brilliant fire: On the matter of the mind*. New York: Basic Books.

Edwards, K., and E. E. Smith. 1996. A disconfirmation bias in the evaluation of arguments. *Journal of Personality and Social Psychology* 71:5–24.

Edwards, W. 1954. The theory of decision making. *Psychological Bulletin* 51:380–417.

Eigen, M. 1992. *Steps towards life*. Oxford: Oxford University Press.

Einhorn, H. J., and R. M. Hogarth. 1981. Behavioral decision theory: Processes of judgment and choice. *Annual Review of Psychology* 32:53–88.

Elster, J. 1983. *Sour grapes: Studies in the subversion of rationality*. Cambridge: Cambridge University Press.

———. 1989. *Solomonic judgements*. Cambridge: Cambridge University Press.

Elster, J., and J. E. Roemer, eds. 1991. *Interpersonal comparisons of well-being*. Cambridge: Cambridge University Press.

Engle, R. W., S. W. Tuholski, J. E. Laughlin, and A. R. A. Conway. 1999. Working memory, short-term memory, and general fluid intelligence: A latent-variable approach. *Journal of Experimental Psychology: General* 128:309–31.

Ennis, R. H. 1987. A taxonomy of critical thinking dispositions and abilities. In *Teaching thinking skills: Theory and practice*, ed. J. Baron and R. Sternberg, 9–26. New York: W. H. Freeman.

Epstein, S. 1994. Integration of the cognitive and the psychodynamic unconscious. *American Psychologist* 49:709–24.

Estes, W. K. 1964. Probability learning. In *Categories of human learning*, ed. A. W. Melton, 89–128. New York: Academic Press.

———. 1976. The cognitive side of probability learning. *Psychological Review* 83:37–64.

———. 1984. Global and local control of choice behavior by cyclically varying outcome probabilities. *Journal of Experimental Psychology: Learning, Memory, and Cognition* 10, no. 2: 258–70.

Evans, J. St. B. T. 1984. Heuristic and analytic processes in reasoning. *British Journal of Psychology* 75:451–68.

———. 1989. *Bias in human reasoning: Causes and consequences*. London: Lawrence Erlbaum Associates.

———. 1995. Relevance and reasoning. In *Perspectives on thinking and reasoning*, ed. S. E. Newstead and J. St. B. T. Evans, 147–71. Hove, England: Lawrence Erlbaum Associates.

———. 1996. Deciding before you think: Relevance and reasoning in the selection task. *British Journal of Psychology* 87:223–40.

———. 1998. Matching bias in conditional reasoning: Do we understand it after 25 years? *Thinking and Reasoning* 4:45–82.

———. 2002a. Logic and human reasoning: An assessment of the deduction paradigm. *Psychological Bulletin* 128:978–96.

———. 2002b. Matching bias and set sizes: A discussion of Yama (2001). *Thinking and Reasoning* 8:153–63.

———. 2002c. The influence of prior belief on scientific thinking. In *The cognitive basis of science*, ed. P. Carruthers, S. Stich, and M. Siegal, 193–210. Cambridge: Cambridge University Press.

Evans, J. St. B. T., J. Barston, and P. Pollard. 1983. On the conflict between logic and belief in syllogistic reasoning. *Memory & Cognition* 11:295–306.

Evans, J. St. B. T., S. E. Newstead, and R. M. J. Byrne. 1993. *Human reasoning: The psychology of deduction.* Hove, England: Lawrence Erlbaum Associates.

Evans, J. St. B. T., and D. E. Over. 1996. *Rationality and reasoning.* Hove, England: Psychology Press.

———. 1999. Explicit representations in hypothetical thinking. *Behavioral and Brain Sciences* 22:763–64.

Evans, J. St. B. T., D. E. Over, and K. Manktelow. 1993. Reasoning, decision making and rationality. *Cognition* 49:165–87.

Evans, J. St. B. T., J. H. Simon, N. Perham, D. E. Over, and V. A. Thompson. 2000. Frequency versus probability formats in statistical word problems. *Cognition* 77:197–213.

Evans, J. St. B. T., and P. C. Wason. 1976. Rationalization in a reasoning task. *British Journal of Psychology* 67:479–86.

Fantino, E., and A. Esfandiari. 2002. Probability matching: Encouraging optimal responding in humans. *Canadian Journal of Experimental Psychology* 56:58–63.

Fiddick, L., L. Cosmides, and J. Tooby. 2000. No interpretation without representation: The role of domain-specific representations and inferences in the Wason selection task. *Cognition* 77:1–79.

Fischhoff, B. 1988. Judgment and decision making. In *The psychology of human thought,* ed. R. J. Sternberg and E. E. Smith, 153–87. Cambridge: Cambridge University Press.

———. 1991. Value elicitation: Is there anything there? *American Psychologist* 46:835–47.

Fischler, I. 1977. Semantic facilitation without association in a lexical decision task. *Memory & Cognition* 5:335–39.

Fishburn, P. C. 1981. Subjective expected utility: A review of normative theories. *Theory and Decision* 13:139–99.

———. 1999. The making of decision theory. In *Decision science and technology: Reflections on the contributions of Ward Edwards,* ed. J. Shanteau, B. A. Mellers, and D. A. Schum, 369–88. Boston: Kluwer Academic Publishers.

Flanagan, O. 1992. *Consciousness reconsidered.* Cambridge, Mass.: MIT Press.

———. 1996. *Self expressions: Mind, morals, and the meaning of life.* New York: Oxford University Press.

———. 2002. *The problem of the soul.* New York: Basic Books.

Floden, R. E., M. Buchmann, and J. R. Schwille. 1987. Breaking with everyday experience. *Teacher's College Record* 88:485–506.

Fodor, J. 1983. *The modularity of mind.* Cambridge, Mass.: MIT Press.

———. 1985. Precis of *The modularity of mind. Behavioral and Brain Sciences* 8:1–42.

Fodor, J. A., and Z. W. Pylyshyn. 1988. Connectionism and cognitive architecture: A critical analysis. *Cognition* 28:3–71.

Foley, Richard. 1987. *The theory of epistemic rationality.* Cambridge, Mass.: Harvard University Press.

———. 1991. Rationality, belief, and commitment. *Synthese* 89:365–92.

Foley, Robert. 1996. The adaptive legacy of human evolution: A search for the EEA. *Evolutionary Anthropology* 4:194–203.

Frank, M. G., and T. Gilovich. 1988. The dark side of self- and social perception: Black uniforms and aggression in professional sports. *Journal of Personality and Social Psychology* 54:74–85.

Frank, R. H. 1988. *Passions within reason: The strategic role of the emotions.* New York: Norton.

———. 1999. *Luxury fever: Why money fails to satisfy in an era of excess.* New York: Free Press.

Frank, R. H., and P. J. Cook. 1995. *The winner-take-all society.* New York: Free Press.

Frankfurt, H. 1971. Freedom of the will and the concept of a person. *Journal of Philosophy* 68:5–20.

———. 1982. The importance of what we care about. *Synthese* 53:257–72.

Franzen, J. 2001. My father's brain: What Alzheimer's takes away. *The New Yorker,* September 10, 81–91.

Fridson, M. S. 1993. *Investment illusions.* New York: Wiley.

Friedrich, J. 1993. Primary error detection and minimization (PEDMIN) strategies in social cognition: A reinterpretation of confirmation bias phenomena. *Psychological Review* 100:298–319.

Frisch, D., and J. K. Jones. 1993. Assessing the accuracy of decisions. *Theory & Psychology,* 3:115–35.

Fry, A. F., and S. Hale. 1996. Processing speed, working memory, and fluid intelligence. *Psychological Science* 7:237–41.

Funder, D.C. 1987. Errors and mistakes: Evaluating the accuracy of social judgment. *Psychological Bulletin* 101:75–90.

Gal, I., and J. Baron. 1996. Understanding repeated simple choices. *Thinking and Reasoning* 2:81–98.

Gardenfors, P., and N. Sahlin, eds. 1988. *Decision, probability, and utility: Selected readings.* Cambridge: Cambridge University Press.

Gardner, H. 1983. *Frames of mind.* New York: Basic Books.

Garry, M., S. Frame, and E. F. Loftus. 1999. Lie down and tell me about your childhood. In *Mind myths: Exploring popular assumptions about the mind and brain,* ed. S. Della Sala, 113–24. Chichester, England: Wiley.

Gauthier, D. 1975. Reason and maximization. *Canadian Journal of Philosophy* 4:411–33.

———. 1986. *Morals by agreement.* Oxford: Oxford University Press.

———, ed. 1990. *Moral dealing: Contract, ethics, and reason.* Ithaca, N.Y.: Cornell University Press.

Gazzaniga, M. S. 1989. Organization of the human brain. *Science* 245:947–52.

———. 1997. Why can't I control my brain? Aspects of conscious experience. In *Cognition, computation, and consciousness,* ed. M. Ito, Y. Miyashita, and E. T. Rolls, 69–80. Oxford: Oxford University Press.

———. 1998a. *The mind's past.* Berkeley and Los Angeles: University of California Press.

———. 1998b. The split brain revisited. *Scientific American,* July, 51–55.

Gazzaniga, M. S., and J. E. LeDoux. 1978. *The integrated mind.* New York: Plenum.

Geary, D.C., and D. F. Bjorklund. 2000. Evolutionary developmental psychology. *Child Development* 71:57–65.

Geary, D.C., and K. J. Huffman. 2002. Brain and cognitive evolution: Forms of modularity and functions of mind. *Psychological Bulletin* 128:667–98.

Gebauer, G., and D. Laming. 1997. Rational choices in Wason's selection task. *Psychological Research* 60:284–93.

Gewirth, A. 1998. *Self-fulfillment.* Princeton, N.J.: Princeton University Press.

Gibbard, A. 1990. *Wise choices, apt feelings: A theory of normative judgment.* Cambridge, Mass.: Harvard University Press.

Gigerenzer, G. 1991. How to make cognitive illusions disappear: Beyond "heuristics and biases." *European Review of Social Psychology* 2:83–115.

———. 1993. The bounded rationality of probabilistic mental models. In *Rationality: Psychological and philosophical perspectives,* ed. K. Manktelow and D. Over, 284–313. London: Routledge.

———. 1994. Why the distinction between single-event probabilities and frequencies is important for psychology (and vice versa). In *Subjective probability,* ed. G. Wright and P. Ayton, 129–61. Chichester, England: Wiley.

———. 1996a. On narrow norms and vague heuristics: A reply to Kahneman and Tversky (1996). *Psychological Review* 103:592–96.

———. 1996b. Rationality: Why social context matters. In *Interactive minds: Life-span perspectives on the social foundation of cognition,* ed. P. B. Baltes and U. Staudinger, 319–46. Cambridge: Cambridge University Press.

———. 1998. Ecological intelligence: An adaptation for frequencies. In *The evolution of mind,* ed. D. D. Cummins and C. Allen, 9–29. New York: Oxford University Press.

———. 2002. *Calculated risks: How to know when numbers deceive you.* New York: Simon and Schuster.

Gigerenzer, G., and D. G. Goldstein. 1996. Reasoning the fast and frugal way: Models of bounded rationality. *Psychological Review* 103:650–69.

Gigerenzer, G., and U. Hoffrage. 1995. How to improve Bayesian reasoning without instruction: Frequency formats. *Psychological Review* 102:684–704.

Gigerenzer, G., U. Hoffrage, and A. Ebert. 1998. AIDS counselling for low-risk clients. *AIDS Care* 10:197–211.

Gigerenzer, G., U. Hoffrage, and H. Kleinbolting. 1991. Probabilistic mental models: A Brunswikian theory of confidence. *Psychological Review* 98:506–28.

Gigerenzer, G., Z. Swijtink, T. Porter, L. Daston, J. Beatty, and L. Kruger. 1989. *The empire of chance.* Cambridge: Cambridge University Press.

Gigerenzer, G., and P. M. Todd. 1999. *Simple heuristics that make us smart.* New York: Oxford University Press.

Gilbert, D. 1991. How mental systems believe. *American Psychologist* 46:107–19.

Gilovich, T. 1991. *How we know what isn't so.* New York: Free Press.

Gilovich, T., D. W. Griffin, and D. Kahneman, eds. 2002. *Heuristics and biases: The psychology of intuitive judgment.* New York: Cambridge University Press.

Girotto, V., and M. Gonzalez. 2001. Solving probabilistic and statistical problems: A matter of information structure and question form. *Cognition* 78:247–76.

Glenberg, A. M. 1997. What memory is for. *Behavioral and Brain Sciences* 20:1–55.

Godfrey-Smith, P. 1996. *Complexity and the function of mind in nature.* Cambridge: Cambridge University Press.

Goel, V., and R. J. Dolan. 2003. Explaining modulation of reasoning by belief. *Cognition* 87:B11–B22.

Goldman, A. I. 1978. Epistemics: The regulative theory of cognition. *Journal of Philosophy* 75:509–23.

———. 1986. *Epistemology and cognition.* Cambridge, Mass.: Harvard University Press.

Goldstein, A. 1994. *Addiction: From biology to drug policy.* New York: W. H. Freeman.

Goldstein, D. G., and G. Gigerenzer. 1999. The recognition heuristic: How ignorance makes us smart. In *Simple heuristics that make us smart,* ed. G. Gigerenzer and P. M. Todd, 37–58. New York: Oxford University Press.

———. 2002. Models of ecological rationality: The recognition heuristic. *Psychological Review* 109:75–90.

Goody, E. N., ed. 1995. *Social intelligence and interaction: Expressions and implications of the social bias in human intelligence.* Cambridge: Cambridge University Press.

Gopnik, A. 1993. How we know our minds: The illusion of first-person knowledge of intentionality. *Behavioral and Brain Sciences* 16:1–14.

Gordon, R. A. 1997. Everyday life as an intelligence test: Effects of intelligence and intelligence context. *Intelligence* 24:203–320.

Gottfredson, L. S. 1997. Why g matters: The complexity of everyday life. *Intelligence* 24:79–132.

Gould, S. J. 1989. *Wonderful life: The Burgess shale and the nature of history.* New York: Norton.

———. 1996. *Full house: The spread of excellence from Plato to Darwin.* New York: Harmony Books.

———. 2002. *The structure of evolutionary theory.* Cambridge, Mass.: Harvard University Press.

Grealy, L. 1995. *Autobiography of a face.* New York: HarperPerennial.

Greenwald, A. G. 1992. New look 3: Unconscious cognition reclaimed. *American Psychologist* 47:766–79.

Greenwald, A. G., and M. R. Banaji. 1995. Implicit social cognition: Attitudes, self-esteem, and stereotypes. *Psychological Review* 102:4–27.

Greenwald, A. G., M. R. Banaji, L. A. Rudman, S. D. Farnham, B. A. Nosek, and D. S. Mellott. 2002. A unified theory of implicit attitudes, stereotypes, self-esteem, and self-concept. *Psychological Review* 109:3–25.

Greider, W. 1992. *Who will tell the people?* New York: Simon and Schuster.

Grether, D. M., and C. R. Plott. 1979. Economic theory of choice and the preference reversal phenomenon. *American Economic Review* 69:623–38.

Grice, H. P. 1975. Logic and conversation. In *Syntax and semantics,* vol. 3, *Speech acts,* ed. P. Cole and J. Morgan, 41–58. New York: Academic Press.

Griffin, D. W., and C. A. Varey. 1996. Toward a consensus on overconfidence. *Organizational Behavior and Human Decision Processes* 65:227–31.

Griffiths, P. E. 1997. *What emotions really are.* Chicago: University of Chicago Press.

Grigorenko, E. L. 1999. Heredity versus environment as the basis of cognitive ability. In *The nature of cognition,* ed. R. J. Sternberg, 665–96. Cambridge, Mass.: MIT Press.

Gur, R., and H. Sackheim. 1979. Self-deception: A concept in search of a phenomenon. *Journal of Personality and Social Psychology* 37:147–69.

Hacking, I. 1975. *The emergence of probability.* Cambridge: Cambridge University Press.

———. 1990. *The taming of chance.* Cambridge: Cambridge University Press.

Haidt, J. 2001. The emotional dog and its rational tail: A social intuitionist approach to moral judgment. *Psychological Review* 108:814–34.

Hallman, T. 2002. *Sam: The boy behind the mask.* New York: Putnam.

Hamilton, W. D. 1964. The genetical evolution of social behaviour (I and II). *Journal of Theoretical Biology* 7:1–52.

———. 1966. The moulding of senescence by natural selection. *Journal of Theoretical Biology* 12:12–45.

———. 1996. *Narrow roads of gene land.* Oxford: W. H. Freeman.

Hardin, G. 1968. The tragedy of the commons. *Science* 162:1243–48.

Hardman, D. 1998. Does reasoning occur on the selection task? A comparison of relevance-based theories. *Thinking and Reasoning* 4:353–76.

Hargreaves Heap, S. P. 1992. *Rationality and economics.* Cambridge: Cambridge University Press.

Hargreaves Heap, S. P., and Y. Varoufakis. 1995. *Game theory: A critical introduction.* London: Routledge.

Harman, G. 1993. Desired desires. In *Value, welfare, and morality,* ed. R. G. Frey and C. W. Morris, 138–57. Cambridge: Cambridge University Press.

———. 1995. Rationality. In *Thinking,* ed. E. E. Smith and D. N. Osherson, 3:175–211. Cambridge, Mass.: MIT Press.

Harnish, R. 2002. *Minds, brains, and computers.* Oxford: Blackwell.

Harnishfeger, K. K., and D. F. Bjorklund. 1994. A developmental perspective on individual differences in inhibition. *Learning and Individual Differences* 6:331–56.

Harries, C., and N. Harvey. 2000. Are absolute frequencies, relative frequencies, or both effective in reducing cognitive biases? *Journal of Behavioral Decision Making* 13:431–44.

Harris, P. L. 2001. The veridicality assumption. *Mind and Language* 16:247–62.

Harvey, N. 1992. Wishful thinking impairs belief-desire reasoning: A case of decoupling failure in adults? *Cognition* 45:141–62.

Hasher, L., and R. T. Zacks. 1979. Automatic processing of fundamental information: The case of frequency of occurrence. *Journal of Experimental Psychology: General* 39:1372–88.

Haslam, N., and J. Baron. 1994. Intelligence, personality, and prudence. In *Personality and intelligence,* ed. R. J. Sternberg and P. Ruzgis, 32–58. Cambridge: Cambridge University Press.

Hastie, R., and R. M. Dawes. 2001. *Rational choice in an uncertain world.* Thousand Oaks, Calif.: Sage.

Hastie, R., and K. A. Rasinski. 1988. The concept of accuracy in social judgment. In *The social psychology of knowledge,* ed. D. Bar-Tal and A. Kruglanski, 193–208. Cambridge: Cambridge University Press.

Hausman, D. M. 1991. On dogmatism in economics: The case of preference reversals. *Journal of Socio-Economics* 20:205–25.

Hausman, D. M., and M. McPherson. 1994. Economics, rationality, and ethics. In *The philosophy of economics: An anthology,* 2d ed., ed. D. M. Hausman, 252–77. Cambridge: Cambridge University Press.

Hawking, S. 1988. *A brief history of time.* New York: Bantam Books.

Henle, M. 1962. On the relation between logic and thinking. *Psychological Review* 69:366–78.

Hentoff, N. 1980. *The first freedom: The tumultuous history of free speech in America.* New York: Delacorte Press.

———. 1992. *Free speech for me—But not for thee.* New York: HarperCollins.

Hertwig, R., and G. Gigerenzer. 1999. The 'conjunction fallacy' revisited: How intelligent inferences look like reasoning errors. *Journal of Behavioral Decision Making* 12:275–305.

Hilton, D. J. 1995. The social context of reasoning: Conversational inference and rational judgment. *Psychological Bulletin* 118:248–71.

Hilton, D. J., and B. R. Slugoski. 2000. Judgment and decision making in social context: Discourse processes and rational inference. In *Judgment and decision making: An interdisciplinary reader,* 2d ed., ed. T. Connolly, H. R. Arkes, and K. R. Hammond, 651–76. Cambridge, Mass.: Cambridge University Press.

Hirsch, E. D. 1996. *The schools we need: And why we don't have them.* New York: Doubleday.

Hirschfeld, L. A., and S. A. Gelman, eds. 1994. *Mapping the mind: Domain specificity in cognition and culture.* Cambridge: Cambridge University Press.

Hirschman, A. O. 1986. *Rival views of market society and other recent essays.* New York: Viking.

Hoch, S. J. 1987. Perceived consensus and predictive accuracy: The pros and cons of projection. *Journal of Personality and Social Psychology* 53:221–34.

Hofstadter, D. R. 1982. Can creativity be mechanized? *Scientific American,* October, 18–34.

Hogarth, R. M. 2001. *Educating intuition.* Chicago: University of Chicago Press.

Holender, D. 1986. Semantic activation without conscious identification in dichotic listening, parafoveal vision, and visual masking: A survey and appraisal. *Behavioral and Brain Sciences* 9:1–66.

Hollis, M. 1992. Ethical preferences. In *The theory of choice: A critical guide,* ed. S. Hargreaves Heap, M. Hollis, B. Lyons, R. Sugden, and A. Weale, 308–10. Oxford: Blackwell.

Horgan, T., and J. Tienson. 1993. Levels of description in nonclassical cognitive science. In *Philosophy and cognitive science,* ed. C. Hookway and D. Peterson, 159–88. Cambridge: Cambridge University Press.

Horn, J. L. 1982. The theory of fluid and crystallized intelligence in relation to concepts of cognitive psychology and aging in adulthood. In *Aging and cognitive processes,* ed. F. I. M. Craik and S. Trehub, 847–70. New York: Plenum.

Horn, J. L., and R. B. Cattell. 1967. Age differences in fluid and crystallized intelligence. *Acta Psychologica* 26:1–23.

Hsee, C. K., G. F. Loewenstein, S. Blount, and M. H. Bazerman. 1999. Preference reversals between joint and separate evaluations of options: A review and theoretical analysis. *Psychological Bulletin* 125:576–90.

Hull, D. L. 1982. The naked meme. In *Learning, development, and culture: Essays in evolutionary epistemology,* ed. H. C. Plotkin, 273–327. Chichester, England: Wiley.

———. 1988. *Science as a process: An evolutionary account of the social and conceptual development of science.* Chicago: University of Chicago Press.

———. 2000. Taking memetics seriously: Memetics will be what we make it. In *Darwinizing culture: The status of memetics as a science*, ed. R. Aunger, 43–67. Oxford: Oxford University Press.

———. 2001. *Science and selection: Essays on biological evolution and the philosophy of science*. Cambridge: Cambridge University Press.

Hume, D. [1740] 1888. *A treatise of human nature*. Edited by L. A. Selby-Bigge. London: Oxford University Press.

Humphrey, N. 1976. The social function of intellect. In *Growing points in ethology*, ed. P. P. G. Bateson and R. A. Hinde, 303–17. London: Faber and Faber.

———. 1986. *The inner eye*. London: Faber and Faber.

———. 1993. *A history of mind: Evolution and the birth of consciousness*. New York: HarperPerennial.

Hunt, E. 1978. Mechanics of verbal ability. *Psychological Review* 85:109–30.

———. 1987. The next word on verbal ability. In *Speed of information-processing and intelligence*, ed. P. A. Vernon, 347–92. Norwood, N.J.: Ablex.

———. 1995. *Will we be smart enough? A cognitive analysis of the coming workforce*. New York: Russell Sage Foundation.

———. 1999. Intelligence and human resources: Past, present, and future. In *The future of learning and individual differences research: Processes, traits, and content*, ed. P. Ackerman and P. Kyllonen, 3–28. Washington, D.C.: American Psychological Association.

Hurley, S. L. 1989. *Natural reasons: Personality and polity*. New York: Oxford University Press.

Hurst, L. 2001. Suicide warriors: Deactivating extremists not easy, experts say. *Toronto Star*. November 3, A1–A4.

Huxley, T. H. [1894] 1989. *Evolution and ethics*. Princeton, N.J.: Princeton University Press.

Jackendoff, R. 1996. How language helps us think. *Pragmatics and Cognition* 4:1–34.

James, L., and D. Nahl. 2000. *Road rage and aggressive driving: Steering clear of highway warfare*. Amherst, N.Y.: Prometheus Books.

Jeffrey, R. C. 1974. Preferences among preferences. *Journal of Philosophy* 71:377–91.

———. 1983. *The logic of decision*. 2d ed. Chicago: University of Chicago Press.

Jepson, C., D. Krantz, and R. Nisbett. 1983. Inductive reasoning: Competence or skill? *Behavioral and Brain Sciences* 6:494–501.

Johnson, E. J., J. Hershey, J. Meszaros, and H. Kunreuther. 2000. Framing, probability distortions, and insurance decisions. In *Choices, values, and frames*, ed. D. Kahneman and A. Tversky, 224–40. Cambridge: Cambridge University Press.

Johnson, H., and D. S. Broder. 1996. *The system: The American way of politics at the breaking point*. Boston: Little, Brown.

Johnson, M. K. 1991. Reality monitoring: Evidence from confabulation in organic brain disease patients. In *Awareness of deficit after brain injury*, ed. G. Prigantano and D. Schacter, 121–40. New York: Oxford University Press.

Johnson-Laird, P. N. 1983. *Mental models*. Cambridge, Mass.: Harvard University Press.

———. 1988. *The computer and the mind: An introduction to cognitive science*. Cambridge, Mass.: Harvard University Press.

———. 1999. Deductive reasoning. *Annual Review of Psychology* 50:109–35.

———. 2001. Mental models and deduction. *Trends in Cognitive Sciences* 5:434–42.

Johnson-Laird, P. N., and R. M. J. Byrne. 1991. *Deduction*. Hillsdale, N.J.: Lawrence Erlbaum Associates.

———. 1993. Models and deductive rationality. In *Rationality: Psychological and philosophical perspectives*, ed. K. Manktelow and D. Over, 177–210. London: Routledge.

Johnson-Laird, P., and K. Oatley. 1992. Basic emotions, rationality, and folk theory. *Cognition and Emotion* 6:201–23.

———. 2000. Cognitive and social construction in emotions. In *Handbook of emotions*, 2d ed., ed. M. Lewis and J. Haviland-Jones, 458–75. New York: Guilford Press.

Johnson-Laird, P. N., and F. Savary. 1996. Illusory inferences about probabilities. *Acta Psychologica* 93:69–90.

Jolly, A. 1966. Lemur social behaviour and primate intelligence. *Science* 153:501–6.

Kagel, C. J. 1987. Economics according to the rats (and pigeons too): What have we learned and what we hope to learn. In *Laboratory experimentation in economics: Six points of view*, ed. A. Roth, 587–703. New York: Cambridge University Press.

Kahne, J. 1996. The politics of self-esteem. *American Educational Research Journal* 33: 3–22.

Kahneman, D. 1981. Who shall be the arbiter of our intuitions? *Behavioral and Brain Sciences* 4:339–40.

———. 1991. Judgment and decision making: A personal view. *Psychological Science* 2:142–45.

———. 1994. New challenges to the rationality assumption. *Journal of Institutional and Theoretical Economics* 150:18–36.

———. 2000. A psychological point of view: Violations of rational rules as a diagnostic of mental processes. *Behavioral and Brain Sciences* 23:681–83.

Kahneman, D., E. Diener, and N. Schwarz, eds. 1999. *Well-being: The foundations of hedonic psychology.* Thousand Oaks, Calif.: Sage.

Kahneman, D., and S. Frederick. 2002. Representativeness revisited: Attribute substitution in intuitive judgment. In *Heuristics and biases: The psychology of intuitive judgment*, ed. T. Gilovich, D. Griffin, and D. Kahneman, 49–81. New York: Cambridge University Press.

Kahneman, D., and J. Snell. 1990. Predicting utility. In *Insights into decision making*, ed. R. M. Hogarth, 295–310. Chicago: University of Chicago Press.

Kahneman, D., and A. Tversky. 1973. On the psychology of prediction. *Psychological Review* 80:237–51.

———. 1979. Prospect theory: An analysis of decision under risk. *Econometrica* 47: 263–91.

———. 1982. On the study of statistical intuitions. *Cognition* 11:123–41.

———. 1983. Can irrationality be intelligently discussed? *Behavioral and Brain Sciences* 6:509–10.

———. 1984. Choices, values, and frames. *American Psychologist* 39:341–50.

———. 1996. On the reality of cognitive illusions. *Psychological Review* 103:582–91.

———, eds. 2000. *Choices, values, and frames.* Cambridge: Cambridge University Press.

Kahneman, D., P. P. Wakker, and R. Sarin. 1997. Back to Bentham? Explorations of experienced utility. *The Quarterly Journal of Economics* 112, no. 2: 375–405.

Kane, M. J., and R. W. Engle. 2002. The role of prefrontal cortex working-memory capacity, executive attention, and general fluid intelligence: An individual-differences perspective. *Psychonomic Bulletin and Review* 9:637–71.

Kao, S. F., and E. A. Wasserman. 1993. Assessment of an information integration account of contingency judgment with examination of subjective cell importance and method of information presentation. *Journal of Experimental Psychology: Learning, Memory, and Cognition* 19:1363–86.

Kardash, C. M., and R. J. Scholes. 1996. Effects of pre-existing beliefs, epistemological beliefs, and need for cognition on interpretation of controversial issues. *Journal of Educational Psychology* 88:260–71.

Karmiloff-Smith, A. 1992. *Beyond modularity: A developmental perspective on cognitive science.* Cambridge, Mass.: MIT Press.

Keating, D. P. 1990. Charting pathways to the development of expertise. *Educational Psychologist* 25:243–67.

Keil, F. C., and R. A. Wilson, eds. 2000. *Explanation and cognition.* Cambridge, Mass.: MIT Press.

Kekes, J. 1990. *Facing evil.* Princeton, N.J.: Princeton University Press.

Kenrick, D. T. 2001. Evolutionary psychology, cognitive science, and dynamical systems: Building an integrative paradigm. *Current Directions in Psychological Science* 10:13–17.

Kettlewell, H. 1973. *The evolution of melanism.* Oxford: Oxford University Press.

Kimberg, D. Y., M. D'Esposito, and M. J. Farah. 1998. Cognitive functions in the prefrontal cortex—working memory and executive control. *Current Directions in Psychological Science* 6:185–92.

Kimberg, D. Y., and M. J. Farah. 1993. A unified account of cognitive impairments following frontal lobe damage: The role of working memory in complex, organized behavior. *Journal of Experimental Psychology: General* 122:411–28.

Kirkpatrick, L., and S. Epstein. 1992. Cognitive-experiential self-theory and subjective probability: Evidence for two conceptual systems. *Journal of Personality and Social Psychology* 63:534–44.

Kirkwood, T., and R. Holliday. 1979. The evolution of ageing and longevity. *Proceedings of the Royal Society of London B,* 205:531–46.

Kitcher, P. 1993. *The advancement of science.* New York: Oxford University Press.

Klaczynski, P. A. 2001. Analytic and heuristic processing influences on adolescent reasoning and decision making. *Child Development* 72:844–61.

Klaczynski, P. A., D. H. Gordon, and J. Fauth. 1997. Goal-oriented critical reasoning and individual differences in critical reasoning biases. *Journal of Educational Psychology* 89:470–85.

Klayman, J., and Y. Ha. 1987. Confirmation, disconfirmation, and information in hypothesis testing. *Psychological Review* 94:211–28.

Klein, G. 1998. *Sources of power: How people make decisions.* Cambridge, Mass.: MIT Press.

Klein, G. S. 1964. Semantic power measured through the interference of words with color-naming. *American Journal of Psychology* 77:576–88.

Kleindorfer, P. R., H. C. Kunreuther, and P. J. H. Schoemaker. 1993. *Decision sciences: An integrative perspective.* Cambridge: Cambridge University Press.

Koehler, J. J. 1993. The influence of prior beliefs on scientific judgments of evidence quality. *Organizational Behavior and Human Decision Processes* 56:28–55.

———. 1996. The base rate fallacy reconsidered: Descriptive, normative and method-ological challenges. *Behavioral and Brain Sciences* 19:1–53.

Kohler, W. 1927. *The mentality of apes.* 2d ed. London: Routledge and Kegan Paul.

Kokis, J., R. Macpherson, M. Toplak, R. F. West, and K. E. Stanovich. 2002. Heuristic and analytic processing: Age trends and associations with cognitive ability and cognitive styles. *Journal of Experimental Child Psychology* 83:26–52.

Komorita, S. S., and C. D. Parks. 1994. *Social dilemmas.* Boulder, Colo.: Westview Press.

Krantz, D. H. 1981. Improvements in human reasoning and an error in L. J. Cohen's. *Behavioral and Brain Sciences* 4:340–41.

Kroll, L., and L. Goldman. 2002. The global billionaires. *Forbes* 169, no. 6 (March 18): 119–32.

Krueger, J., and J. Zeiger. 1993. Social categorization and the truly false consensus effect. *Journal of Personality and Social Psychology* 65:670–80.

Kruglanski, A. W. 1990. Lay epistemics theory in social-cognitive psychology. *Psychological Inquiry* 1:181–97.

Kruglanski, A. W., and I. Ajzen. 1983. Bias and error in human judgment. *European Journal of Social Psychology* 13:1–44.

Kruglanski, A. W., and D. M. Webster. 1996. Motivated closing the mind: "Seizing" and "freezing." *Psychological Review* 103:263–83.

Kuhberger, A. 2002. The rationality of risky decisions: A changing message. *Theory & Psychology* 12:427–52.

Kuhn, D. 1991. *The skills of argument.* Cambridge: Cambridge University Press.

———. 1996. Is good thinking scientific thinking? In *Modes of thought: Explorations in culture and cognition,* ed. D. R. Olson and N. Torrance, 261–81. New York: Cambridge University Press.

———. 2001. How do people know? *Psychological Science* 12:1–8.

Kummer, H., L. Daston, G. Gigerenzer, and J. B. Silk. 1997. The social intelligence hy-pothesis. In *Human by nature: Between biology and the social sciences,* ed. P. Weingart, S. D. Mitchell, P. J. Richerson, and S. Maasen, 157 79. Mahwah, N.J.: Lawrence Erlbaum Associates.

Kunda, Z. 1990. The case for motivated reasoning. *Psychological Bulletin* 108:480–98.

———. 1999. *Social cognition: Making sense of people.* Cambridge, Mass.: MIT Press.

Kuttner, R. 1998. *Everything for sale: The virtues and limits of markets.* Chicago: University of Chicago Press.

LaBerge, D., and S. Samuels. 1974. Toward a theory of automatic information pro-cessing in reading. *Cognitive Psychology* 6:293–323.

LaCerra, P., and R. Bingham. 1998. The adaptive nature of the human neurocognitive architecture: An alternative model. *Proceeds of the National Academy of Sciences* 95:11290–94.

Langlois, J. H., L. Kalakanis, A. J. Rubenstein, A. Larson, M. Hallam, and M. Smott. 2000. Maxims or myths of beauty? A meta-analytic and theoretical review. *Psychological Bulletin* 126:390–423.

Langer, E. J. 1989. *Mindfulness.* Reading, Mass.: Addison-Wesley.

Langer, E. J., A. Blank, and B. Chanowitz. 1978. The mindlessness of ostensibly thought-ful action: The role of "placebic" information in interpersonal interaction. *Journal of Personality and Social Psychology* 36:635–42.

Lasn, K. 1999. *Culture jam: The uncooling of America*. New York: William Morrow.

Laudan, L. 1996. *Beyond positivism and relativism*. Boulder, Colo.: Westview Press.

Lefkowitz, B. 1997. *Our guys: The Glen Ridge rape and the secret life of the perfect suburb*. Berkeley and Los Angeles: University of California Press.

Lehrer, K. 1990. *Theory of knowledge*. London: Routledge.

———. 1997. *Self-trust: A study of reason, knowledge, and autonomy*. Oxford: Oxford University Press.

Leslie, A. M. 1987. Pretense and representation: The origins of "theory of mind." *Psychological Review* 94:412–26.

———. 1994. ToMM, ToBY, and agency: Core architecture and domain specificity. In *Mapping the mind: Domain specificity in cognition and culture*, ed. L. A. Hirschfeld and S. A. Gelman, 119–48. Cambridge: Cambridge University Press.

———. 2000. How to acquire a representational theory of mind. In *Metarepresentations: A multidisciplinary perspective*, ed. D. Sperber, 197–223. Oxford: Oxford University Press.

Levelt, W. 1995. Chapters of psychology. In *The science of the mind: 2001 and beyond*, ed. R. L. Solso and D. W. Massaro, 184–202. New York: Oxford University Press.

Levinson, S. C. 1995. Interactional biases in human thinking. In *Social intelligence and interaction*, ed. E. Goody, 221–260. Cambridge: Cambridge University Press.

Lewis, C., and G. Keren. 1999. On the difficulties underlying Bayesian reasoning: A comment on Gigerenzer and Hoffrage. *Psychological Review* 106:411–16.

Lewis, D. 1989. Dispositional theories of value. *Proceedings of the Aristotelian Society*, supplementary vol. 63, 113–37.

Liberman, N., and Y. Klar. 1996. Hypothesis testing in Wason's selection task: Social exchange cheating detection or task understanding. *Cognition* 58:127–56.

Lichtenstein, S., and P. Slovic. 1971. Reversal of preferences between bids and choices in gambling decisions. *Journal of Experimental Psychology* 89:46–55.

Lieberman, M. D. 2000. Intuiton: A social cognitive neuroscience approach. *Psychological Bulletin* 126:109–37.

Lillard, A. 2001. Pretend play as twin Earth: A social-cognitive analysis. *Developmental Review* 21:495–531.

Lindblom, C. E. 2001. *The market system*. New Haven, Conn.: Yale University Press.

Lindsay, P., and D. Norman. 1977. *Human information processing*. 2d ed. New York: Academic Press.

Lipman, M. 1991. *Thinking in education*. Cambridge: Cambridge University Press.

Lodge, D. 2001. *Thinks . . .* London: Secker and Warburg.

Loewenstein, G. F. 1996. Out of control: Visceral influences on behavior. *Organizational Behavior and Human Decision Processes* 65:272–92.

Loewenstein, G. F., E. U. Weber, C. K. Hsee, and N. Welch. 2001. Risk as feelings. *Psychological Bulletin* 127:267–86.

Loftus, E. F. 1997. Memory for a past that never was. *Current Directions in Psychological Science* 6:60–65.

Loftus, E. F., and K. Ketcham. 1994. *The myth of repressed memory: False memories and allegations of sexual abuse*. New York: St. Martins.

Logan, G. D. 1985. Skill and automaticity: Relations, implications, and future directions. *Canadian Journal of Psychology* 39:367–86.

Lohman, D. F. 2000. Complex information processing and intelligence. In *Handbook of intelligence,* ed. R. J. Sternberg, 285–340. Cambridge: Cambridge University Press.

Loomes, G., and R. Sugden. 1982. Regret theory: An alternative theory of rational choice under uncertainty. *The Economic Journal* 92:805–24.

Lubinski, D. 2000. Scientific and social significance of assessing individual differences: "Sinking shafts at a few critical points." *Annual Review of Psychology* 51:405–44.

Lubinski, D., and L. G. Humphreys. 1997. Incorporating general intelligence into epidemiology and the social sciences. *Intelligence* 24:159–201.

Luce, R. D., and H. Raiffa. 1957. *Games and decisions.* New York: Wiley.

Lumsden, C. J., and E. O. Wilson. 1981. *Genes, mind and culture.* Cambridge, Mass.: Harvard University Press.

Luria, A. R. 1976. *Cognitive development: Its cultural and social foundations.* Cambridge, Mass.: Harvard University Press.

Luttwak, E. 1999. *Turbo-capitalism: Winners and losers in the global economy.* New York: HarperCollins.

Lynch, A. 1996. *Thought contagion.* New York: Basic Books.

Macchi, L., and G. Mosconi. 1998. Computational features vs frequentist phrasing in the base-rate fallacy. *Swiss Journal of Psychology* 57:79–85.

Macdonald, R. R., and K. J. Gilhooly. 1990. More about Linda *or* conjunctions in context. *European Journal of Cognitive Psychology* 2:57–70.

MacIntyre, A. 1990. Is akratic action always irrational? In *Identity, character, and morality,* ed. O. Flanagan and A. O. Rorty, 379–400. Cambridge, Mass.: MIT Press.

Macintyre, B. 2001. Word bombs. *The Times* (London), October 9, sect. 2, 2–3.

MacLeod, C. M. 1991. Half a century of research on the Stroop effect: An integrative review. *Psychological Bulletin* 109:163–203.

———. 1992. The Stroop task: The "gold standard" of attentional measures. *Journal of Experimental Psychology: General* 121:12–14.

MacLeod, C. M., and P. A. MacDonald. 2000. Interdimensional interference in the Stroop effect: Uncovering the cognitive and neural anatomy of attention. *Trends in Cognitive Sciences* 4, no. 10: 383–91.

Maher, P. 1993. *Betting on theories.* Cambridge: Cambridge University Press.

Majerus, M. E. N. 1998. *Melanism: Evolution in action.* Oxford: Oxford University Press.

Makin, K. 2001. Man recants repressed "memories." *Globe and Mail* (Toronto), November 3, A12.

Malik, K. 2000. *Man, beast and zombie: What science can and cannot tell us about human nature.* London: Weidenfeld and Nicolson.

Malle, B. F., L. J. Moses, and D. A. Baldwin, eds. 2001. *Intentions and intentionality: Foundations of social cognition.* Cambridge, Mass.: MIT Press.

Manktelow, K. I. 1999. *Reasoning and Thinking.* Hove, England: Psychology Press.

Manktelow, K. I., and J. St. B. T. Evans. 1979. Facilitation of reasoning by realism: Effect or non-effect? *British Journal of Psychology* 70:477–88.

Manktelow, K. I., and D. E. Over. 1991. Social roles and utilities in reasoning with deontic conditionals. *Cognition* 39:85–105.

———. 1995. Deontic reasoning. In *Perspectives on thinking and reasoning: Essays in honour of Peter Wason,* ed. S. E. Newstead and J. St. B. T. Evans, 91–114. Hove, England: Lawrence Erlbaum Associates.

Mann, L., R. Harmoni, and C. Power. 1991. The GOFER course in decision making. In *Teaching decision making to adolescents,* ed. J. Baron and R. V. Brown, 61–78. Hillsdale, N.J.: Lawrence Erlbaum Associates.

Marcel, A. J. 1983. Conscious and unconscious perception: Experiments on visual masking and word recognition. *Cognitive Psychology* 15:197–237.

———. 1988. Phenomenal experience and functionalism. In *Consciousness in contemporary science,* ed. A. J. Marcel and E. Bisiach, 121–58. Oxford: Oxford University Press.

Margolis, H. 1996. *Dealing with risk.* Chicago: University of Chicago Press.

Markovits, H., and G. Nantel. 1989. The belief-bias effect in the production and evaluation of logical conclusions. *Memory and Cognition* 17:11–17.

Markowitz, H. M. 1952. The utility of wealth. *Journal of Political Economy* 60:151–58.

Marr, D. 1982. *Vision.* San Francisco: W. H. Freeman.

Masson, M. E. J. 1995. A distributed memory model of semantic priming. *Journal of Experimental Psychology: Learning, Memory and Cognition* 21:3–23.

Matthews, G., and I. J. Deary. 1998. *Personality traits.* Cambridge: Cambridge University Press.

Maynard Smith, J. M. 1974. The theory of games and the evolution of animal conflict. *Journal of Theoretical Biology* 47:209–21.

———. 1975. *The theory of evolution.* 3d ed. Cambridge: Cambridge University Press.

———. 1976. Evolution and the theory of games. *American Scientist* 64:41–45.

———. 1998. *Evolutionary genetics.* 2d ed. Oxford: Oxford University Press.

Maynard Smith, J. M., and E. Szathmáry. 1999. *The origins of life.* Oxford: Oxford University Press.

McCauley, R. N. 2000. The naturalness of religion and the unnaturalness of science. In *Explanation and cognition,* ed. F. C. Keil and R. A. Wilson, 61–85. Cambridge, Mass.: MIT Press.

McEwan, I. 1998. *Enduring love.* London: Vintage.

McFadden, D. 1999. Rationality for economists? *Journal of Risk and Uncertainty* 19: 73–105.

McFarland, D. 1989. Goals, no goals, and own goals. In *Goals, no goals, and own goals: A debate on goal-directed and intentional behavior,* ed. A. Montefiori and D. Noble, 39–57. London: Unwin Hyman.

McGinn, C. 1999. *The mysterious flame: Conscious minds in a material world.* New York: Basic Books.

McKenzie, C. R. M. 1994. The accuracy of intuitive judgment strategies: Covariation assessment and Bayesian inference. *Cognitive Psychology* 26:209–39.

McNeil, B., S. Pauker, H. Sox, and A. Tversky. 1982. On the elicitation of preferences for alternative therapies. *New England Journal of Medicine* 306:1259–62.

Medin, D. L., and M. H. Bazerman. 1999. Broadening behavioral decision research: Multiple levels of cognitive processing. *Psychonomic Bulletin and Review* 6:533–46.

Medin, D. L., H. C. Schwartz, S. V. Blok, and L. A. Birnbaum. 1999. The semantic side of decision making. *Psychonomic Bulletin and Review* 6:562–69.

Mele, A. R. 1987. Recent work on self-deception. *American Philosophical Quarterly* 24:1–17.

———. 1997. Real self-deception. *Behavioral and Brain Sciences* 20:91–136.

———. 2001. *Self-deception unmasked.* Princeton, N.J.: Princeton University Press.

Mellers, B. A. 2000. Choice and the relative pleasure of consequences. *Psychological Bulletin* 126:910–24.

Mellers, B. A., and K. Biagini. 1994. Similarity and choice. *Psychological Review* 101: 505–18.

Mellers, B. A., R. Hertwig, and D. Kahneman. 2001. Do frequency representations eliminate conjunction effects? An exercise in adversarial collaboration. *Psychological Science* 12:269–75.

Mellers, B. A., and A. P. McGraw. 1999. How to improve Bayesian reasoning: Comment on Gigerenzer and Hoffrage (1995). *Psychological Review* 106:417–24.

Mercer, T. 2000. Navigating the shark-eat-shark world of 'compassionate' airfares. *The Globe and Mail* (Toronto), September 16, T5.

Merikle, P. M., D. Smilek, and J. D. Eastwood. 2001. Perception without awareness: Perspectives from cognitive psychology. *Cognition* 79:115–34.

Messick, S. 1984. The nature of cognitive styles: Problems and promise in educational practice. *Educational Psychologist* 19:59–74.

———. 1994. The matter of style: Manifestations of personality in cognition, learning, and teaching. *Educational Psychologist* 29:121–36.

Metcalfe, J., and W. Mischel. 1999. A hot/cool-system analysis of delay of gratification: Dynamics of will power. *Psychological Review* 106:3–19.

Milgram, S. 1974. *Obedience to authority*. New York: Harper and Row.

Miller, G. 2001. *The mating mind*. New York: Anchor Books.

Millgram, E. 1997. *Practical induction*. Cambridge, Mass.: Harvard University Press.

Millikan, R. G. 1993. *White Queen psychology and other essays for Alice*. Cambridge, Mass.: MIT Press.

Minsky, M. 1985. *The society of mind*. New York: Simon and Schuster.

Mitchell, P., E. J. Robinson, J. E. Isaacs, and R. M. Nye. 1996. Contamination in reasoning about false belief: An instance of realist bias in adults but not children. *Cognition* 59:1–21.

Mithen, S. 1996. *The prehistory of mind: The cognitive origins of art and science*. London: Thames and Hudson.

———. Palaeoanthropological perspectives on the theory of mind. In *Understanding other minds: Perspectives from developmental cognitive neuroscience*, 2d ed., ed S. Baron-Cohen, H. Tager-Flusberg, and D. J. Cohen, 488–502. Oxford: Oxford University Press.

———. 2002. Human evolution and the cognitive basis of science. In *The cognitive basis of science*, ed. P. Carruthers, S. Stich, and M. Siegel, 23–40. Cambridge: Cambridge University Press.

Miyake, A., and P. Shah, eds. 1999. *Models of working memory: Mechanisms of active maintenance and executive control*. New York: Cambridge University Press.

Moldoveanu, M., and E. Langer. 2002. False memories of the future: A critique of the application of probabilistic reasoning to the study of cognitive processes. *Psychological Review* 109:358–75.

Morton, O. 1997. Doing what comes naturally: A new school of psychology finds reasons for your foolish heart. *The New Yorker* 73 (Nov. 3): 102–7.

Moscovitch, M. 1989. Confabulation and the frontal systems: Strategic versus associative retrieval in neuropsychological theories of memory. In *Varieties of memory and consciousness*, ed. H. L. Roediger and F. I. M. Craik, 133–60. Hillsdale, N.J.: Lawrence Erlbaum Associates.

Moshman, D. 1994. Reasoning, metareasoning, and the promotion of rationality. In *Intelligence, mind, and reasoning: Structure and development*, ed. A. Demetriou and A. Efklides, 135–50. Amsterdam: Elsevier.

Mueser, P. R., N. Cowan, and K. T. Mueser. 1999. A generalized signal detection model to predict rational variation in base rate use. *Cognition* 69:267–312.

Mussweiler, T., F. Strack, and T. Pfeiffer. 2000. Overcoming the inevitable anchoring effect: Considering the opposite compensates for selective accessibility. *Personality and Social Psychology Bulletin* 9:1142–50.

Myers, D. G. 2000. *The American paradox: Spiritual hunger in an age of plenty.* New Haven, Conn.: Yale University Press.

———. 2002. *Intuition: Its powers and perils.* New Haven, Conn.: Yale University Press.

Mynatt, C. R., M. E. Doherty, and W. Dragan. 1993. Information relevance, working memory, and the consideration of alternatives. *Quarterly Journal of Experimental Psychology* 46A:759–78.

Nagel, T. 1997. *The last word.* New York: Oxford University Press.

Nathanson, S. 1994. *The ideal of rationality.* Chicago: Open Court.

National Highway Traffic Safety Administration. 1999. *Traffic safety facts 1999: Children.* Fact sheet: DOT HS 809 087. Washington, D.C.: NHTSA. Retrieved November 24, 2000, from the World Wide Web: http://www.nhtsa.dot.gov/people/ncsa/pdf/child99.pdf.

Navon, D. 1989. The importance of being visible: On the role of attention in a mind viewed as an anarchic intelligence system. *European Journal of Cognitive Psychology* 1:191–238.

Neely, J. H., and D. E. Keefe. 1989. Semantic context effects on visual word processing: A hybrid prospective-retrospective processing theory. In *The psychology of learning and motivation*, ed. G. H. Bower, 24:207–48. San Diego: Academic Press.

Neimark, E. 1987. *Adventures in thinking.* San Diego: Harcourt Brace Jovanovich.

Neisser, U., G. Boodoo, T. Bouchard, A. W. Boykin, N. Brody, S. J. Ceci, D. Halpern, J. Loehlin, R. Perloff, R. Sternberg, and S. Urbina. 1996. Intelligence: Knowns and unknowns. *American Psychologist* 51:77–101.

Nelson, K. 1996. *Language in cognitive development: The emergence of the mediated mind.* Cambridge: Cambridge University Press.

Neumann, P. J., and P. E. Politser. 1992. Risk and optimality. In *Risk-taking behavior*, ed. J. F. Yates, 27–47. Chichester, England: Wiley.

Neurath, O. 1932–33. Protokollsatze. *Erkenntis* 3:204–14.

Newell, A. 1982. The knowledge level. *Artificial Intelligence* 18:87–127.

———. 1990. *Unified theories of cognition.* Cambridge, Mass.: Harvard University Press.

Newstead, S. E., and J. St. B. T. Evans, eds. 1995. *Perspectives on thinking and reasoning.* Hove, England: Lawrence Erlbaum Associates.

Nickerson, R. S. 1988. On improving thinking through instruction. In *Review of Research in Education*, vol. 15, ed. E. Z. Rothkopf, 3–57. Washington, D.C.: American Educational Research Association.

———. 1996. Hempel's paradox and Wason's selection task: Logical and psychological puzzles of confirmation. *Thinking and Reasoning* 2:1–31.

———. 1998. Confirmation bias: A ubiquitous phenomenon in many guises. *Review of General Psychology* 2:175–220.

Nicolson, A. 2000. *Perch Hill.* London: Penguin Books.

Nisbett, R. E. 1993. *Rules for reasoning.* Hillsdale, N.J.: Lawrence Erlbaum Associates.

Nisbett, R. E., and L. Ross, L. 1980. *Human inference: Strategies and shortcomings of social judgment.* Englewood Cliffs, N.J.: Prentice-Hall.

Nisbett, R. E., and S. Schachter. 1966. Cognitive manipulation of pain. *Journal of Experimental Social Psychology* 21:227–36.

Nisbett, R. E., and T. D. Wilson. 1977. Telling more than we can know: Verbal reports on mental processes. *Psychological Review* 84:231–59.

Norman, D. A., and T. Shallice. 1986. Attention to action: Willed and automatic control of behavior. In *Consciousness and self-regulation,* ed. R. J. Davidson, G. E. Schwartz, and D. Shapiro, 1–18. New York: Plenum.

Nozick, R. 1974. *Anarchy, state, and utopia.* New York: Basic Books.

———. 1981. *Philosophical explanations.* Cambridge, Mass.: Harvard University Press.

———. 1989. *The examined life.* New York: Simon and Schuster.

———. 1993. *The nature of rationality.* Princeton, N.J.: Princeton University Press.

———. 2001. *Invariances: The structure of the objective world.* Cambridge, Mass.: Harvard University Press.

Oaksford, M., and N. Chater. 1993. Reasoning theories and bounded rationality. In *Rationality: Psychological and philosophical perspectives,* ed. K. Manktelow and D. Over, 31–60. London: Routledge.

———. 1994. A rational analysis of the selection task as optimal data selection. *Psychological Review* 101:608–31.

———. 1995. Theories of reasoning and the computational explanation of everyday inference. *Thinking and Reasoning* 1:121–52.

———. 1996. Rational explanation of the selection task. *Psychological Review* 103:381–91.

———, eds. 1998. *Rationality in an uncertain world.* Hove, England: Psychology Press.

———. 2001. The probabilistic approach to human reasoning. *Trends in Cognitive Sciences* 5:349–57.

Oatley, K. 1992. *Best laid schemes: The psychology of emotions.* Cambridge: Cambridge University Press.

———. 1998. The structure of emotions. In *Mind readings,* ed. P. Thagard, 239–57. Cambridge, Mass.: MIT Press.

Orwell, G. 1950. *Shooting an elephant and other essays.* New York: Harcourt, Brace and World.

Osherson, D. N. 1995. Probability judgment. In *Thinking,* vol. 3, ed. E. E. Smith and D. N. Osherson, 35–75. Cambridge, Mass.: MIT Press.

Over, D. E. 2000. Ecological rationality and its heuristics. *Thinking and Reasoning* 6: 182–92.

———. 2002. The rationality of evolutionary psychology. In *Reason and nature: Essays in the theory of rationality,* ed. J. L. Bermudez and A. Millar, 187–207. Oxford: Oxford University Press.

———, ed., 2003a. *Evolution and the psychology of thinking: The debate.* Hove, England: Psychology Press.

———. 2003b. From massive modularity to metarepresentation: The evolution of higher cognition. In *Evolution and the psychology of thinking: The debate*, ed. D. Over, 121–44. Hove, England: Psychology Press.

Over, D. E., and J. St. B. T. Evans. 2000. Rational distinctions and adaptations. *Behavioral and Brain Sciences* 23:693–94.

Over, D. E., and D. W. Green. 2001. Contingency, causation, and adaptive inference. *Psychological Review* 108:682–84.

Pacini, R., and S. Epstein. 1999. The interaction of three facets of concrete thinking in a game of chance. *Thinking and Reasoning* 5:303–25.

Papineau, D. 2001. The evolution of means-ends reasoning. In *Naturalism, evolution and mind*, ed. D. M. Walsh, 145–78. Cambridge: Cambridge University Press.

Parfit, D. 1984. *Reasons and persons.* Oxford: Oxford University Press.

Parkin, A. J. 1996. *Explorations in cognitive neuropsychology.* Oxford: Blackwell.

Partridge, J. 1997. *Changing faces: The challenge of facial disfigurement.* 3d ed. East Grand Rapids, Mich.: Phoenix Society for Burn Survivors.

Paul, R. W. 1984. Critical thinking: Fundamental to education for a free society. *Educational Leadership* 42, no. 1: 4–14.

———. 1987. Critical thinking and the critical person. In *Thinking: The second international conference*, ed. D. N. Perkins, J. Lockhead, and J. Bishop, 373–403. Hillsdale, N.J.: Lawrence Erlbaum Associates.

Paul, R., A. Binker, D. Martin, and K. Adamson. 1989. *Critical thinking handbook: High school.* Rohnert Park, Calif.: Center for Critical Thinking and Moral Critique.

Payne, J. W., J. R. Bettman, and E. J. Johnson. 1992. Behavioral decision research: A constructive processing perspective. *Annual Review of Psychology* 43:87–131.

Payne, J., J. Bettman, and D. Schkade. 1999. Measuring constructed preferences: Towards a building code. *Journal of Risk and Uncertainty* 19:243–70.

Pearce, D. W. 1995. *The MIT dictionary of modern economics.* Cambridge: MIT Press.

Pennington, B. F., and S. Ozonoff. 1996. Executive functions and developmental psychopathology. *Journal of Child Psychology and Psychiatry* 37:51–87.

Perkins, D. N. 1995. *Outsmarting IQ: The emerging science of learnable intelligence.* New York: Free Press.

Perkins, D. N., and T. A. Grotzer. 1997. Teaching intelligence. *American Psychologist* 52:1125–33.

Perkins, D. N., E. Jay, and S. Tishman. 1993. Beyond abilities: A dispositional theory of thinking. *Merrill-Palmer Quarterly* 39:1–21.

Perner, J. 1991. *Understanding the representational mind.* Cambridge, Mass.: MIT Press.

———. 1998. The meta-intentional nature of executive functions and theory of mind. In *Language and thought: Interdisciplinary themes*, ed. P. Carruthers and J. Boucher, 270–83. Cambridge: Cambridge University Press.

Pezdek, K., and W. P. Banks, eds. 1996. *The recovered memory/false memory debate.* San Diego: Academic Press.

Pezdek, K., and D. Hodge. 1999. Planting false childhood memories in children: The role of event plausibility. *Child Development* 70:887–95.

Philip, I. 2001. Ban on religious hate. *The Independent* (London), October 5, Review section, 2.

Piaget, J. 1972. Intellectual evolution from adolescence to adulthood. *Human Development* 15:1–12.

Piattelli-Palmarini, M. 1994. *Inevitable illusions: How mistakes of reason rule our minds.* New York: Wiley.

Pinel, J., S. Assanand, and D. R. Lehman. 2000. Hunger, eating, and ill health. *American Psychologist* 55:1105–16.

Pinker, S. 1994. *The language instinct.* New York: William Morrow.

———. 1997. *How the mind works.* New York: Norton.

———. 2002. *The blank slate: The modern denial of human nature.* New York: Viking.

Piper, A. 1998. Multiple personality disorder: Witchcraft survives in the twentieth century. *Skeptical Inquirer* 22, no. 3: 44–50.

Plato. 1945. *The republic.* Translated by Francis MacDonald Cornford. New York: Oxford University Press.

Plomin, R., J. C. DeFries, G. E. McClearn, and P. McGuffin. 2001. *Behavior genetics.* 4th ed. New York: Worth.

Plomin, R., and S. A. Petrill. 1997. Genetics and intelligence: What's new? *Intelligence* 24:53–77.

Plotkin, H. C. 1988. Behavior and evolution. In *The role of behavior in evolution*, ed. H. C. Plotkin, 1–17. Cambridge, Mass.: MIT Press.

———. 1994. *Darwin machines and the nature of knowledge.* Cambridge, Mass.: Harvard University Press.

———. 1998. *Evolution in mind: An introduction to evolutionary psychology.* Cambridge, Mass.: Harvard University Press.

Plous, S. 1993. *The psychology of judgment and decision making.* New York: McGraw-Hill.

Politzer, G., and I. A. Noveck. 1991. Are conjunction rule violations the result of conversational rule violations? *Journal of Psycholinguistic Research* 20:83–103.

Pollock, J. L. 1991. OSCAR: A general theory of rationality. In *Philosophy and AI: Essays at the interface*, ed. J. Cummins and J. L. Pollock, 189–213. Cambridge, Mass.: MIT Press.

———. *Cognitive carpentry: A blueprint for how to build a person.* Cambridge, Mass.: MIT Press.

Popper, K. R. 1963. *Conjectures and refutations: The growth of scientific knowledge.* New York: Harper and Row.

Posner, M. I., and C. R. R. Snyder. 1975. Attention and cognitive control. In *Information processing and cognition: The Loyola Symposium*, ed. R. L. Solso, 55–85. New York: Wiley.

Pratt, J. W., H. Raiffa, and R. Schlaifer. 1995. *Introduction to statistical decision theory.* Cambridge, Mass.: MIT Press.

Purcell, D. G., A. Stewart, and K. E. Stanovich. 1983. Another look at semantic priming without awareness. *Perception & Psychophysics* 34:65–71.

Pylyshyn, Z. 1984. *Computation and cognition.* Cambridge, Mass.: MIT Press.

Quattrone, G., and A. Tversky. 1984. Causal versus diagnostic contingencies: On self-deception and on the voter's illusion. *Journal of Personality and Social Psychology* 46:237–48.

Quine, W. 1960. *Word and object.* Cambridge, Mass.: MIT Press.

Quirk, J. P. 1987. *Intermediate microeconomics.* 3d ed. Chicago: Science Research Associates.

Radcliffe Richards, J. 2000. *Human nature after Darwin: A philosophical introduction.* London: Routledge.

Radnitzky, G., and W. W. Bartley, eds. 1987. *Evolutionary epistemology, rationality, and the sociology of knowledge.* La Salle, Ill.: Open Court.

Ramachandran, V. S., and S. Blakeslee. 1998. *Phantoms in the brain.* New York: William Morrow.

Rauch, J. 1993. *Kindly inquisitors: The new attacks on free thought.* Chicago: University of Chicago Press.

Rawls, J. 1971. *A theory of justice.* Oxford: Oxford University Press.

———. 2001. *Justice as fairness: A restatement.* Cambridge, Mass.: Harvard University Press.

Raymo, C. 1999. *Skeptics and true believers.* Toronto: Doubleday Canada.

Real, L. A. 1991. Animal choice behavior and the evolution of cognitive architecture. *Science* 253:980–86.

Reber, A. S. 1992a. An evolutionary context for the cognitive unconscious. *Philosophical Psychology* 5:33–51.

———. 1992b. The cognitive unconscious: An evolutionary perspective. *Consciousness and Cognition* 1:93–133.

———. 1993. *Implicit learning and tacit knowledge.* New York: Oxford University Press.

Redelmeier, D. A., E. Shafir, and P. S. Aujla. 2001. The beguiling pursuit of more information. *Medical Decision Making* 21, no. 5: 376–81.

Redelmeier, D. A., and A. Tversky. 1990. Discrepancy between medical decisions for individual patients and for groups. *New England Journal of Medicine* 322:1162–64.

———. 1992. On the framing of multiple prospects. *Psychological Science* 3:191–93.

Resnik, M. D. 1987. *Choices: An introduction to decision theory.* Minneapolis: University of Minnesota Press.

Rhoads, S. E. 1985. *The economist's view of the world: Government, markets, and public policy.* Cambridge: Cambridge University Press.

Richardson, H. S. 1997. *Practical reasoning about final ends.* Cambridge: Cambridge University Press.

Richardson, J. H. 2001. *In the little world: A true story of dwarfs, love, and trouble.* New York: HarperCollins.

Ridley, Mark. 1996. *Evolution.* 2d ed. Cambridge, Mass.: Blackwell Science.

———. 2000. *Mendel's demon: Gene justice and the complexity of life.* London: Weidenfeld and Nicolson.

Ridley, Matt. 1999. *Genome: The autobiography of a species in 23 chapters.* New York: HarperCollins.

Rips, L. J. 1994. *The logic of proof.* Cambridge, Mass.: MIT Press.

Rips, L. J., and F. G. Conrad. 1983. Individual differences in deduction. *Cognition and Brain Theory* 6:259–85.

Roberts, M. J., and E. J. Newton. 2001. Inspection times, the change task, and the rapid-response selection task. *Quarterly Journal of Experimental Psychology* 54A:1031–48.

Rode, C., L. Cosmides, W. Hell, and J. Tooby. 1999. When and why do people avoid unknown probabilities in decisions under uncertainty? Testing some predictions from
optimal foraging theory. *Cognition* 72:269–304.

Rodkin, L. I., E. J. Hunt, and S. D. Cowan. 1982. A men's support group for significant
others of rape victims. *Journal of Marital and Family Therapy* 8:91–97.

Rorty, R. 1995. Cranes and skyhooks. *Lingua Franca*, July/August, 62–66.

Rose, M. R. 1991. *Evolutionary biology of aging*. New York: Oxford University Press.

Rose, S., L. J. Kamin, and R. C. Lewontin. 1984. *Not in our genes: Biology, ideology and
human nature*. London: Penguin Books.

Rosenthal, D. 1986. Two concepts of consciousness. *Philosophical Studies* 49:329–59.

Rozin, P. 1976. The evolution of intelligence and access to the cognitive unconscious.
Progress in Psychobiology and Physiological Psychology 6:245–80.

———. 1996. Towards a psychology of food and eating: From motivation to module to
model to marker, morality, meaning and metaphor. *Current Directions in
Psychological Science* 5, no. 1: 18–24.

Rozin, P., and A. E. Fallon. 1987. A perspective on disgust. *Psychological Review* 94:23–41.

Rozin, P., L. Millman, and C. Nemeroff. 1986. Operation of the laws of sympathetic
magic in disgust and other domains. *Journal of Personality and Social Psychology*
50:703–12.

Ruggiero, V. 2000. Bad attitude: Confronting the views that hinder students' learning.
American Educator 24, no. 2: 10–15.

Rumelhart, D. E., P. Smolensky, J. L. McClelland, and G. E. Hinton. 1986. Schemata and
sequential thought processes in PDP models. In *Parallel distributed processing:
Explorations in the microstructure of cognition*, vol. 2, ed. J. L. McClelland and D. E.
Rumelhart, 7–57. Cambridge, Mass.: MIT Press.

Runciman, W. G. 1998. The selectionist paradigm and its implications for sociology.
Sociology 32:163–88.

Ruse, M. 1998. *Taking Darwin seriously*. Amherst, N.Y.: Prometheus Books.

Russo, J. E., and P. Schoemaker. 2002. *Winning decisions: Getting it right the first time*.
New York: Doubleday.

Sá, W., R. F. West, and K. E. Stanovich. 1999. The domain specificity and generality of
belief bias: Searching for a generalizable critical thinking skill. *Journal of Educational
Psychology* 91:497–510.

Sabini, J., and M. Silver. 1998. *Emotion, character, and responsibility*. New York: Oxford
University Press.

Sahlin, N. 1981. Preference among preferences as a method for obtaining a higher-
ordered metric scale. *British Journal of Mathematical and Statistical Psychology* 34:
62–75.

Saks, M. J., and R. F. Kidd. 1980. Human information processing and adjudication: Trial
by heuristics. *Law and Society* 15:123–60.

Samuels, R. 1998. Evolutionary psychology and the massive modularity hypothesis.
British Journal for the Philosophy of Science 49:575–602.

Samuels, R., S. P. Stich, and M. Bishop. 2002. Ending the rationality wars: How to make
disputes about human rationality disappear. In *Common sense, reasoning and rationality*, ed. R. Elio, 236–68. New York: Oxford University Press.

Samuels, R., S. P. Stich, and P. D. Tremoulet. 1999. Rethinking rationality: From bleak implications to Darwinian modules. In *What is cognitive science?*, ed. E. Lepore and Z. Pylyshyn, 74–120. Oxford: Blackwell.

Satz, D., and J. Ferejohn. 1994. Rational choice and social theory. *Journal of Philosophy* 91:71–87.

Savage, L. J. 1954. *The foundations of statistics.* New York: Wiley.

Scanlon, T. M. 1998. *What we owe to each other.* Cambridge, Mass.: Harvard University Press.

Scheffler, I. 1991. *In praise of the cognitive emotions.* New York: Routledge.

Schelling, T. C. 1984. *Choice and consequence: Perspectives of an errant economist.* Cambridge, Mass.: Harvard University Press.

Schick, F. 1984. *Having reasons: An essay on rationality and sociality.* Princeton, N.J.: Princeton University Press.

———. 1987. Rationality: A third dimension. *Economics and Philosophy* 3:49–66.

———. 1997. *Making choices: A recasting of decision theory.* Cambridge: Cambridge University Press.

Schlosser, E. 2001. *Fast food nation.* Boston: Houghton Mifflin.

Schmidt, F. L., and J. E. Hunter. 1992. Development of a causal model of processes determining job performance. *Current Directions in Psychological Science* 1:89–92.

———. 1998. The validity and utility of selection methods in personnel psychology: Practical and theoretical implications of 85 years of research findings. *Psychological Bulletin* 124:262–74.

Schmidtz, D. 1995. *Rational choice and moral agency.* Princeton, N.J.: Princeton University Press.

———, ed. 2002. *Robert Nozick.* Cambridge: Cambridge University Press.

Schoenfeld, A. H. 1983. Beyond the purely cognitive: Belief systems, social cognitions, and metacognitions as driving forces in intellectual performance. *Cognitive Science* 7:329–63.

Scholl, B. J., and A. M. Leslie. 2001. Minds, modules, and meta-analysis. *Child Development* 72:696–701.

Schommer, M. 1990. Effects of beliefs about the nature of knowledge on comprehension. *Journal of Educational Psychology* 82:498–504.

———. 1994. Synthesizing epistemological belief research: Tentative understandings and provocative confusions. *Educational Psychology Review* 6:293–319.

Schueler, G. F. 1995. *Desire: Its role in practical reason and the explanation of action.* Cambridge, Mass.: MIT Press.

Schustack, M. W., and R. J. Sternberg. 1981. Evaluation of evidence in causal inference. *Journal of Experimental Psychology: General* 110:101–20.

Scitovsky, T. 1976. *The joyless economy: The psychology of human satisfaction.* Oxford: Oxford University Press.

Searle, J. R. 1992. *The rediscovery of mind.* Cambridge, Mass.: MIT Press.

———. 2001. *The rationality of action.* Cambridge, Mass.: MIT Press.

Selfridge, O., and U. Neisser. 1960. Pattern recognition by machine. *Scientific American* 203:60–68.

Sen, A. K. 1977. Rational fools: A critique of the behavioral foundations of economic theory. *Philosophy and Public Affairs* 6:317–44.

———. 1982. Choices, orderings and morality. In A. K. Sen, *Choice, welfare and measurement*, 74–83. Cambridge, Mass.: Harvard University Press.

———. 1987. *On ethics and economics.* Oxford: Blackwell.

———. 1993. Internal consistency of choice. *Econometrica* 61:495–521.

———. 1999. *Development as freedom.* New York: Knopf.

———. 2002. On the Darwinian view of progress. In A. K. Sen, *Rationality and freedom*, 484–500. Cambridge, Mass.: Harvard University Press.

Sen, A. K., and B. Williams, eds. 1982. *Utilitarianism and beyond.* Cambridge: Cambridge University Press.

Sennett, R. 1998. *The corrosion of character.* New York: Norton.

Shafer, G. 1988. Savage revisited. In *Decision making: Descriptive, normative, and prescriptive interactions*, ed. D. Bell, H. Raiffa, and A. Tversky, 193–234. Cambridge: Cambridge University Press.

Shafir, E. 1994. Uncertainty and the difficulty of thinking through disjunctions. *Cognition* 50:403–30.

Shafir, E., and R. A. LeBoeuf. 2002. Rationality. *Annual Review of Psychology* 53:491–517.

Shafir, E., and A. Tversky. 1995. Decision making. In *Thinking*, ed. E. E. Smith and D. N. Osherson, 3:77–100. Cambridge, Mass.: MIT Press.

Shallice, T. 1988. *From neuropsychology to mental structure.* Cambridge: Cambridge University Press.

———. 1991. Precis of *From neuropsychology to mental structure. Behavioral and Brain Sciences* 14:429–69.

Shanks, D. R. 1995. Is human learning rational? *Quarterly Journal of Experimental Psychology* 48A:257–79.

Shaw, G. B. 1921. *Back to Methuselah.* London: Constable and Co.

Shawn, W. 1991. *The fever.* London: Faber and Faber.

Shear, J., ed. 1998. *Explaining consciousness—The 'hard problem.'* Cambridge, Mass.: MIT Press.

Shenk, D. 2001. *The forgetting: Alzheimer's, portrait of an epidemic.* New York: Doubleday.

Shepard, R. N. 1987. Evolution of a mesh between principles of the mind and regularities of the world. In *The latest on the best: Essays on evolution and optimality*, ed. J. Dupre, 251–75. Cambridge, Mass.: MIT Press.

Shermer, M. 1997. *Why people believe weird things.* New York: W. H. Freeman.

Shiffrin, R. M., and W. Schneider. 1977. Controlled and automatic human information processing: II. Perceptual learning, automatic attending, and a general theory. *Psychological Review* 84:127–90.

Shweder, R. A. 1987. Comments on Plott and on Kahneman, Knetsch, and Thaler. In *Rational choice: The contrast between economics and psychology*, ed. R. M. Hogarth and M. W. Reder, 161–70. Chicago: University of Chicago Press.

Siegel, H. 1988. *Educating reason: Rationality, critical thinking, and education.* New York: Routledge.

Sigel, I. E. 1993. The centrality of a distancing model for the development of representational competence. In *The development and meaning of psychological distance*, ed. R. Cocking and K. Renninger, 141–58. Hillsdale, N.J.: Lawrence Erlbaum Associates.

Simms, R. 2002. Saving at Wal-Mart, but paying a price. *New York Times*, March 3.

Simon, H. A. 1956. Rational choice and the structure of the environment. *Psychological Review* 63:129–38.

———. 1957. *Models of man*. New York: Wiley.

———. 1967. Motivational and emotional controls of cognition. *Psychological Review* 74:29–39.

———. 1983. *Reason in human affairs*. Stanford, Calif.: Stanford University Press.

Sinatra, G. M., and P. R. Pintrich, eds. 2003. *Intentional conceptual change*. Mahwah, N.J.: Lawrence Erlbaum Associates.

Skyrms, B. 1996. *The evolution of the social contract*. Cambridge: Cambridge University Press.

Sloman, S. A. 1996. The empirical case for two systems of reasoning. *Psychological Bulletin* 119:3–22.

———. 1999. Rational versus arational models of thought. In *The nature of cognition*, ed. R. J. Sternberg, 557–85. Cambridge, Mass.: MIT Press.

———. 2002. Two systems of reasoning. In *Heuristics and biases: The psychology of intuitive judgment*, ed. T. Gilovich, D. Griffin, and D. Kahneman, 379–96. New York: Cambridge University Press.

Sloman, S. A., D. Over, and J. M. Stibel. 2003. Frequency illusions and other fallacies. *Organizational Behavior and Human Decision Processes* 91: 296–309.

Sloman, S. A., and L. J. Rips, eds. 1998. *Similarity and symbols in human thinking*. Cambridge, Mass.: MIT Press.

Slovic, P. 1995. The construction of preference. *American Psychologist* 50:364–71.

Slovic, P., M. L. Finucane, E. Peters, and D. G. MacGregor. 2002. The affect heuristic. In *Heuristics and biases: The psychology of intuitive judgment*, ed. T. Gilovich, D. Griffin, and D. Kahneman, 397–420. New York: Cambridge University Press.

Slovic, P., and A. Tversky. 1974. Who accepts Savage's axiom? *Behavioral Science* 19: 368–73.

Slugoski, B. R., and A. E. Wilson. 1998. Contribution of conversation skills to the production of judgmental errors. *European Journal of Social Psychology* 28:575–601.

Smith, E. A., M. Borgerhoff Mulder, and K. Hill. 2001. Controversies in the evolutionary social sciences: A guide for the perplexed. *Trends in Ecology and Evolution* 16:128–35.

Smith, E. E., A. L. Patalino, and J. Jonides. 1998. Alternative strategies of categorization. In *Similarity and symbols in human thinking*, ed. S. A. Sloman and L. J. Rips, 81–110. Cambridge, Mass.: MIT Press.

Smith, E. R., and J. DeCoster. 2000. Dual-process models in social and cognitive psychology: Conceptual integration and links to underlying memory systems. *Personality and Social Psychology Review* 4:108–31.

Sober, E., and D. S. Wilson. 1998. *Unto others: The evolution and psychology of unselfish behavior*. Cambridge, Mass.: Harvard University Press.

Spanos, N. P. 1996. *Multiple identities and false memories: A sociocognitive perspective*. Washington: American Psychological Association.

Speel, H. C. 1995. Memetics: On a conceptual framework for cultural evolution. Paper presented at the Conference on Einstein meets Magritte, Free University of Brussels, June 1995.

Sperber, D. 1985. Anthropology and psychology: Towards an epidemiology of representations. *Man* 20:73–89.

———. 1994. The modularity of thought and the epidemiology of representations. In *Mapping the mind: Domain specificity in cognition and culture*, ed. L. A. Hirschfeld and S. A. Gelman, 39–67. Cambridge: Cambridge University Press.

———. 1996. *Explaining culture: A naturalistic approach*. Oxford: Blackwell.

———. 2000a. An objection to the memetic approach to culture. In *Darwinizing culture: The status of memetics as a science*, ed. R. Aunger, 163–73. Oxford: Oxford University Press.

———, ed. 2000b. *Metarepresentations: A multidisciplinary perspective*. Oxford: Oxford University Press.

———. 2000c. Metarepresentations in evolutionary perspective. In *Metarepresentations: A multidisciplinary perspective*, ed. D. Sperber, 117–37. Oxford: Oxford University Press.

Sperber, D., F. Cara, and V. Girotto. 1995. Relevance theory explains the selection task. *Cognition* 57:31–95.

Stanovich, K. E. 1989. Implicit philosophies of mind: The dualism scale and its relationships with religiosity and belief in extrasensory perception. *Journal of Psychology* 123:5–23.

———. 1990. Concepts in developmental theories of reading skill: Cognitive resources, automaticity, and modularity. *Developmental Review* 10:72–100.

———. 1993. Dysrationalia: A new specific learning disability. *Journal of Learning Disabilities* 26:501–15.

———. 1994. Reconceptualizing intelligence: Dysrationalia as an intuition pump. *Educational Researcher* 23, no. 4: 11–22.

———. 1999. *Who is Rational? Studies of Individual Differences in Reasoning*. Mahwah, N.J.: Lawrence Erlbaum Associates.

———. 2001. Reductionism in the study of intelligence: Review of "Looking Down on Human Intelligence" by Ian Dreary. *Trends in Cognitive Sciences*, 5: 91–92.

———. 2003. The fundamental computational biases of human cognition: Heuristics that (sometimes) impair reasoning and decision making. In *The psychology of problem solving*, ed. J. E. Davidson and R. J. Sternberg, 291–342. New York: Cambridge University Press.

Stanovich, K. E., A. Cunningham, and R. F. West. 1981. A longitudinal study of the development of automatic recognition skills in first graders. *Journal of Reading Behavior* 13:57–74.

Stanovich, K. E., and R. F. West. 1983. On priming by a sentence context. *Journal of Experimental Psychology: General* 112:1–36.

———. 1997. Reasoning independently of prior belief and individual differences in actively open-minded thinking. *Journal of Educational Psychology* 89:342–57.

———. 1998a. Cognitive ability and variation in selection task performance. *Thinking and Reasoning* 4:193–230.

———. 1998b. Individual differences in framing and conjunction effects. *Thinking and Reasoning* 4:289–317.

———. 1998c. Individual differences in rational thought. *Journal of Experimental Psychology: General* 127:161–88.

———. 1998d. Who uses base rates and $P(D/{\sim}H)$? An analysis of individual differences. *Memory & Cognition* 28:161–79.

———. 1999. Discrepancies between normative and descriptive models of decision making and the understanding/acceptance principle. *Cognitive Psychology* 38:349–85.

———. 2000. Individual differences in reasoning: Implications for the rationality debate? *Behavioral and Brain Sciences* 23:645–726.

Starmer, C. 2000. Developments in non-expected utility theory: The hunt for a descriptive theory of choice under risk. *Journal of Economic Literature* 38:332–82.

Stein, E. 1996. *Without good reason: The rationality debate in philosophy and cognitive science.* Oxford: Oxford University Press.

Steiner, G. 1997. *Errata: An examined life.* London: Weidenfeld and Nicolson.

Stent, G. S. 1978. Introduction. In *Morality as a biological phenomenon,* ed. G. S. Stent, 1–18. Berkeley and Los Angeles: University of California Press.

Sterelny, K. 1990. *The representational theory of mind: An introduction.* Oxford: Blackwell.

———. 2001a. *Dawkins vs. Gould: Survival of the fittest.* Duxford, England: Icon Books.

———. 2001b. *The evolution of agency and other essays.* Cambridge: Cambridge University Press.

Sterelny, K., and P. E. Griffiths. 1999. *Sex and death: An introduction to philosophy of biology.* Chicago: University of Chicago Press.

Sternberg, R. J., ed. 1982. *Handbook of human intelligence.* Cambridge: Cambridge University Press.

———. 1988. Mental self-government: A theory of intellectual styles and their development. *Human Development* 31:197–224.

———. 1989. Domain-generality versus domain-specificity: The life and impending death of a false dichotomy. *Merrill-Palmer Quarterly* 35:115–30.

———. 1997a. The concept of intelligence and its role in lifelong learning and success. *American Psychologist* 52:1030–37.

———. 1997b. *Thinking styles.* Cambridge: Cambridge University Press.

———, ed. 1999. *The nature of cognition.* Cambridge, Mass.: MIT Press.

———, ed. 2000. *Handbook of intelligence.* Cambridge: Cambridge University Press.

———. 2001. Why schools should teach for wisdom: The balance theory of wisdom in educational settings. *Educational Psychologist* 36:227–45.

———, ed. 2002. *Why smart people can be so stupid.* New Haven, Conn.: Yale University Press.

Stich, S. P. 1983. *From folk psychology to cognitive science.* Cambridge: MIT Press.

———. 1990. *The fragmentation of reason.* Cambridge: MIT Press.

———. 1996. *Deconstruction of the mind.* New York: Oxford University Press.

Stolz, J. A., and J. H. Neely. 1995. When target degradation does and does not enhance semantic context effects in word recognition. *Journal of Experimental Psychology: Learning, Memory and Cognition* 21:596–611.

Stone, T., and A. W. Young. 1997. Delusions and brain injury: The philosophy and psychology of belief. *Mind and Language* 12:327–64.

Stone, V. E., L. Cosmides, J. Tooby, N. Kroll, and R. T. Knight. 2002. Selective impairment of reasoning about social exchange in a patient with bilateral limbic system damage. *Proceedings of the National Academy of Sciences* 99:11531–36.

Stout, M. 2000. *The feel-good curriculum: The dumbing down of America's kids in the name of self-esteem.* Cambridge, Mass.: Perseus Publishing.

Suddendorf, T., and A. Whiten. 2001. Mental evolution and development: Evidence for secondary representation in children, great apes, and other animals. *Psychological Bulletin* 127:629–50.

Sutherland, S. 1992. *Irrationality: The enemy within*. London: Constable.

Swartz, R. J., and D. N. Perkins. 1989. *Teaching thinking: Issues and approaches*. Pacific Grove, Calif.: Midwest Publications.

Swets, J. A., R. M. Dawes, and J. Monahan. 2000. Psychological science can improve diagnostic decisions. *Psychological Science in the Public Interest* 1:1–26.

Symons, D. 1992. On the use and misuse of Darwinism in the study of human behavior. In *The adapted mind*, ed. J. Barkow, L. Cosmides, and J. Tooby, 137–59. New York: Oxford University Press.

Szathmáry, E. 1999. The first replicators. In *Levels of selection in evolution*, ed. L. Keller, 31–52. Princeton, N.J.: Princeton University Press.

Taylor, C. 1989. *Sources of the self: The making of modern identity*. Cambridge, Mass.: Harvard University Press.

———. 1992. *The ethics of authenticity*. Cambridge: Cambridge University Press.

Tetlock, P. E., and B. A. Mellers. 2002. The great rationality debate. *Psychological Science* 13:94–99.

Thaler, R. H. 1980. Toward a positive theory of consumer choice. *Journal of Economic Behavior and Organization* 1:39–60.

———. 1987. The psychology of choice and the assumptions of economics. In *Laboratory experimentation in economics: Six points of view*, ed. A. E. Roth, 99–130. Cambridge: Cambridge University Press.

———. 1992. *The winner's curse: Paradoxes and anomalies of economic life*. New York: Free Press.

Thomas, M., and A. Karmiloff-Smith. 1998. Quo vadis modularity in the 1990s? *Learning and Individual Differences* 10:245–50.

Tishman, S., D. N. Perkins, and E. Jay. 1995. *The thinking classroom: Learning and teaching in a culture of thinking*. Needham, Mass.: Allyn and Bacon.

Todd, P. M., and G. Gigerenzer. 2000. Precis of *Simple heuristics that make us smart*. *Behavioral and Brain Sciences* 23:727–80.

Tomasello, M. 1998. Social cognition and the evolution of culture. In *Piaget, evolution, and development*, ed. J. Langer and M. Killen, 221–45. Mahwah, N.J.: Lawrence Erlbaum Associates.

———. 1999. *The cultural origins of human cognition*. Cambridge, Mass.: Harvard University Press.

Tooby, J., and Cosmides, L. 1990. On the universality of human nature and the uniqueness of the individual: The role of genetics and adaptation. *Journal of Personality* 58:17–67.

———. 1992. The psychological foundations of culture. In *The adapted mind*, ed. J. Barkow, L. Cosmides, and J. Tooby, 19–136. New York: Oxford University Press.

Toplak, M., and K. E. Stanovich. 2002. The domain specificity and generality of disjunctive reasoning: Searching for a generalizable critical thinking skill. *Journal of Educational Psychology* 94:197–209.

———. 2003. Associations between myside bias on an informal reasoning task and amount of post-secondary education. *Applied Cognitive Psychology* 17: 851–60.

Trivers, R. L. 1971. The evolution of reciprocal altruism. *Quarterly Review of Biology* 46:35–57.

———. 1974. Parent-offspring conflict. *American Zoologist* 14:249–64.

Tsimpli, I., and N. Smith. 1998. Modules and quasi-modules: Language and theory of mind in a polyglot savant. *Learning and Individual Differences* 10:193–216.

Turkle, S. 1984. *The second self: Computers and the human spirit.* New York: Simon and Schuster.

Turner, M. 2001. *Cognitive dimensions of social science.* Oxford: Oxford University Press.

Tversky, A. 1969. Intransitivity of preferences. *Psychological Review* 76:31–48.

———. 1975. A critique of expected utility theory: Descriptive and normative considerations. *Erkenntnis* 9:163–73.

———. 1996a. Contrasting rational and psychological principles of choice. In *Wise choices,* ed. R. Zeckhauser, R. Keeney, and J. Sebenius, 5–21. Cambridge, Mass.: Harvard Business School Press.

———. 1996b. Rational theory and constructive choice. In *The rational foundations of economic behaviour,* ed. K. J. Arrow, E. Colombatto, M. Perlman, and C. Schmidt, 185–97. London: Macmillan.

Tversky, A., and W. Edwards. 1966. Information versus reward in binary choice. *Journal of Experimental Psychology* 71:680–83.

Tversky, A., and D. Kahneman. 1973. Availability: A heuristic for judging frequency and probability. *Cognitive Psychology* 5:207–32.

———. 1974. Judgment under uncertainty: Heuristics and biases. *Science* 185:1124–31.

———. 1981. The framing of decisions and the psychology of choice. *Science* 211:453–58.

———. 1982. Evidential impact of base rates. In *Judgment under uncertainty: Heuristics and biases,* ed. D. Kahneman, P. Slovic, and A. Tversky, 153–60. Cambridge: Cambridge University Press.

———. 1983. Extensional versus intuitive reasoning: The conjunction fallacy in probability judgment. *Psychological Review* 90:293–315.

———. 1986. Rational choice and the framing of decisions. *Journal of Business* 59:251–78.

Tversky, A., S. Sattath, and P. Slovic. 1988. Contingent weighting in judgment and choice. *Psychological Review* 95:371–84.

Tversky, A., and E. Shafir. 1992. The disjunction effect in choice under uncertainty. *Psychological Science* 3:305–9.

Tversky, A., P. Slovic, and D. Kahneman. 1990. The causes of preference reversal. *American Economic Review* 80:204–17.

Tversky, A., and R. H. Thaler. 1990. Anomalies: Preference reversals. *Journal of Economic Perspectives* 4:201–11.

Uchitelle, L. 2002. Why it takes psychology to make people save. *New York Times,* January 13.

Uleman, J. S., and J. A. Bargh, eds. 1989. *Unintended thought.* New York: Guilford Press.

Velleman, J. D. 1992. What happens when somebody acts? *Mind* 101:461–81.

Vernon, P. A. 1991. The use of biological measures to estimate behavioral intelligence. *Educational Psychologist* 25:293–304.

———. 1993. *Biological approaches to the study of human intelligence.* Norwood, N.J.: Ablex.

von Neumann, J., and O. Morgenstern. 1944. *The theory of games and economic behavior.* Princeton, N. J.: Princeton University Press.

Vranas, P. B. M. 2000. Gigerenzer's normative critique of Kahneman and Tversky. *Cognition* 76:179–93.

Wason, P. C. 1966. Reasoning. In *New horizons in psychology,* ed. B. Foss, 135–51. Harmondsworth, England: Penguin.

———. 1968. Reasoning about a rule. *Quarterly Journal of Experimental Psychology* 20:273–81.

Wasserman, E. A., W. W. Dorner, and S. F. Kao. 1990. Contributions of specific cell information to judgments of interevent contingency. *Journal of Experimental Psychology: Learning, Memory, and Cognition* 16:509–21.

Watson, G. 1975. Free agency. *Journal of Philosophy* 72:205–20.

Weber, M. 1968. *Economy and society.* New York: Bedminster Press.

Wegner, D. M. 2002. *The illusion of conscious will.* Cambridge, Mass.: MIT Press.

Wegner, D. M., and T. Wheatley. 1999. Apparent mental causation: Sources of the experience of will. *American Psychologist* 54:480–92.

Weiskrantz, L. 1986. *Blindsight: A case study and implications.* Oxford: Oxford University Press.

———. 1995. Blindsight: Not an island unto itself. *Current Directions in Psychological Science* 4:146–51.

Welch, D. A. 2002. *Decisions, decisions: The art of effective decision making.* Amherst, N.Y.: Prometheus Books.

Wellman, H. M. 1990. *The child's theory of mind.* Cambridge, Mass.: MIT Press.

West, R. L. 1996. An application of prefrontal cortex function theory to cognitive aging. *Psychological Bulletin* 120:272–92.

West, R. F., and K. E. Stanovich. 2003. Is probability matching smart? Associations between probabilistic choices and cognitive ability. *Memory & Cognition* 31:243–51.

Wetherick, N. E. 1993. Human rationality. In *Rationality: Psychological and philosophical perspectives,* ed. K. Manktelow and D. Over, 83–109. London: Routledge.

———. 1995. Reasoning and rationality: A critique of some experimental paradigms. *Theory & Psychology* 5:429–48.

Whiten, A. 2001. Meta-representation and secondary representation. *Trends in Cognitive Sciences* 5:378.

Whiten, A., and R. W. Byrne, eds. 1997. *Machiavellian intelligence II: Extensions and evaluations.* Cambridge: Cambridge University Press.

Wilkes, K. V. 1988. *Real people: Personal identity without thought experiments.* Oxford: Oxford University Press.

Williams, G. C. 1957. Pleiotropy, natural selection, and the evolution of senescence. *Evolution* 11:398–411.

———. 1966. *Adaptation and natural selection.* Princeton, N.J.: Princeton University Press.

———. 1985. A defense of reductionism in evolutionary biology. *Oxford Surveys in Evolutionary Biology* 2:1–27.

———. 1988. Huxley's *Evolution and ethics* in sociobiological perspective. *Zygon* 23: 383–407.

———. 1992. *Natural selection: Domains, levels and challenges.* Oxford: Oxford University Press.

———. 1996. *Plan and purpose in nature.* London: Phoenix Paperbacks.

Williams, W., T. Blythe, N. White, J. Li, R. J. Sternberg, and H. Gardner. 1996. *Practical intelligence in school.* New York: HarperCollins.

Willingham, D. B. 1998. A neuropsychological theory of motor-skill learning. *Psychological Review* 105:558–84.

———. 1999. The neural basis of motor-skill learning. *Current Directions in Psychological Science* 8:178–82.

Wilson, D. S. 1994. Adaptive genetic variation and human evolutionary psychology. *Ethology and Sociobiology* 15:219–35.

———. 2002. *Darwin's cathedral.* Chicago: University of Chicago Press.

Wilson, E. O. 1978. *On human nature.* Cambridge, Mass.: Harvard University Press.

Wilson, M. 2002. Six views of embodied cognition. *Psychonomic Bulletin and Review* 9:625–36.

Wilson, M., and M. Daly. 1992. The man who mistook his wife for a chattel. In *The adapted mind,* ed. J. Barkow, L. Cosmides, and J. Tooby, 289–322. New York: Oxford University Press.

Wilson, R. A., and F. C. Keil, eds. 1999. *The MIT encyclopedia of the cognitive sciences.* Cambridge, Mass.: MIT Press.

Wilson, T. D. 2002. *Strangers to ourselves.* Cambridge, Mass.: Harvard University Press.

Wilson, T. D., and N. Brekke. 1994. Mental contamination and mental correction: Unwanted influences on judgments and evaluations. *Psychological Bulletin* 116:117–42.

Wilson, T. D., and J. W. Schooler. 1991. Thinking too much: Introspection can reduce the quality of preferences and decisions. *Journal of Personality and Social Psychology* 60:181–92.

Wolfe, A. 1989. *Whose keeper? Social science and moral obligation.* Berkeley and Los Angeles: University of California Press.

Wolford, G., M. B. Miller, and M. S. Gazzaniga. 2000. The left hemisphere's role in hypothesis formation. *Journal of Neuroscience* 20 (RC64): 1–4.

Wright, R. 1994. *The moral animal: Evolutionary psychology and everyday life.* New York: Vintage Books.

Yang, Y., and P. N. Johnson-Laird. 2000. Illusions in quantified reasoning: How to make the impossible seem possible, and vice versa. *Memory & Cognition* 28:452–65.

Yates, J. F., ed. 1992. *Risk-taking behavior.* Chichester, England: Wiley.

Young, A. W. 2000. Wondrous strange: The neuropsychology of abnormal beliefs. In *Pathologies of belief,* ed. M. Coltheart and M. Davies, 47–73. Oxford: Blackwell.

Zajonc, R. B. 2001. Mere exposure: A gateway to the subliminal. *Current Directions in Psychological Science* 10:224–28.

Zajonc, R. B., and H. Markus. 1982. Affective and cognitive factors in preferences. *Journal of Consumer Research* 9:123–31.

Zelazo, P. D., J. W. Astington, and D. R. Olson, eds. 1999. *Developing theories of intention.* Mahwah, N.J.: Lawrence Erlbaum Associates.

Zelazo, P. D., and D. Frye. 1998. Cognitive complexity and control: II. The development of executive function in childhood. *Current Directions in Psychological Science* 7:121–26.

Zimmerman, D. 1981. Hierarchical motivation and freedom of the will. *Pacific Philosophical Quarterly* 62:354–68.